S0-BWW-834

The Cell Surface: Mediator of Developmental Processes

THE CELL SURFACE: MEDIATOR OF DEVELOPMENTAL PROCESSES

The Thirty-Eighth Symposium of The Society for Developmental Biology

Vancouver, British Columbia, June 25-27, 1979

The Cell Surface: Mediator of Developmental Processes

Stephen Subtelny, *Editor*

Department of Biology
Rice University
Houston, Texas

Norman K. Wessells, *Co-Editor*

Department of Biological Sciences
Stanford University
Stanford, California

ACADEMIC PRESS 1980
A Subsidiary of Harcourt Brace Jovanovich, Publishers
New York London Toronto Sydney San Francisco

ACADEMIC PRESS, INC.
111 Fifth Avenue, New York, New York 10003

United Kingdom Edition published by
ACADEMIC PRESS, INC. (LONDON) LTD.
24/28 Oval Road, London NW1 7DX

LIBRARY OF CONGRESS CATALOG CARD NUMBER: 78-23508

ISBN 0-12-612984-3

PRINTED IN THE UNITED STATES OF AMERICA

80 81 82 83 9 8 7 6 5 4 3 2 1

Contents

I. The Cell Surface: Background and Perspectives

II. The Cell Surface and Early Development

List of Contributors

Boldface Denotes Chairpersons

W. Steven Adair, Department of Anatomy and Neurobiology, Washington University, St. Louis, Missouri

Samuel H. Barondes, Department of Psychiatry, University of California, La Jolla, California

Daniel Branton, The Biological Laboratories, Havard University, Cambridge Massachusetts

Leon Browder, Department of Biology, University of Calgary, Calgary, Alberta, Canada

Frank B. Dazzo, Department of Microbiology and Public Health, Michigan State University, East Lansing, Michigan

E.M. Eddy, Department of Biological Structure, University of Washington, Seattle, Washington

Richard P. Elinson, Department of Zoology, University of Toronto, Toronto M5S 1A1, Ontario, Canada

David Epel, Hopkins Marine Station, Department of Biological Sciences, Stanford University, Pacific Grove, California

Martha L. Farrance, Department of Biological Structure, University of Washington, Seattle, Washington

Cyril V. Finnegan, University of British Columbia, Vancouver, Canada

Charles A. Foerder, Department of Biological Structure, University of Washington, Seattle, Washington

Judah Folkman, Department of Surgery, Harvard Medical School, Boston, Massachusetts

Christopher A. Gabel, Department of Biological Structure, University of Washington, Seattle, Washington

John C. Gerhart, Department of Molecular Biology, University of California, Berkeley, California

Norton B. Gilula, Rockefeller University, New York, New York

Luis Glaser, Department of Biological Chemistry, Washington University School of Medicine, St. Louis, Missouri

Ursula W. Goodenough, Department of Biology, Washington University, St. Louis, Missouri

Laura B. Grabel, Department of Anatomy, University of California, San Francisco, California

Liang-Hsien E. Hahn, Laboratory of Molecular Biology, National Cancer Institute, Bethesda, Maryland

Koki Hara, Hubrecht Laboratory, Uppsalalaan, 8, Utrecht, Netherlands

Margaret H. Hardy-Fallding, Department of Biomedical Sciences, University of Guelph, Guelph, Ontario, Canada

John H. Hartwig, Department of Medicine, Harvard Medical School, Boston, Massachusetts

Gerald M. Kidder, Department of Zoology, University of Western Ontario, London, Ontario, Canada

Marc Kirschner, Department of Biochemistry and Biophysics, School of Medicine, University of California, Berkeley, California

S.J. Klebanoff, Department of Medicine, University of Washington, Seattle, Washington

Ilmo Leivo, Department of Pathology, University of Helsinki, Haartmaninkatu 3, Helsinki, Finland

H.F. Linskens, Department of Botany, University of Nijmegen, The Netherlands

Gail R. Martin, Department of Anatomy, University of California, San Francisco, California

Yoshio Masui, Department of Zoology, University of Toronto, Toronto, Ontario M5S 1A1, Canada

Peter G. Meyerhof, Department of Zoology, University of California, Berkeley, California

Margaret A. Miller, Department of Zoology, University of Toronto, Ontario M5S 1A1, Canada

Kenneth Olden, Laboratory of Molecular Biology, National Cancer Institute, Bethesda, Maryland

Raymond Reeves, Department of Zoology, University of British Columbia, Vancouver, Canada

Steven D. Rosen, Department of Anatomy, University of California, San Francisco, California

Robert W. Shackmann, Department of Biological Structure, University of Washington, Seattle, Washington

Bennett M. Shapiro, Department of Biological Structure, University of Washington, Seattle, Washington

Olle I. Stendahl, Department of Medicine, Harvard Medical School, Boston, Massachusetts

Thomas P. Stossel, Department of Medicine, Harvard Medical School, Boston, Massachusetts

Robert W. Tucker, Department of Medical Oncology and Cell Growth and Regulation, Harvard Medical School, Boston, Massachusetts

Geertje A. Ubbels, Hubrecht Laboratory, Uppsalalaan 8, Utrecht, Netherlands

Victor D. Vacquier, Scripps Institution of Oceanography, University of California, La Jolla, California

Antti Vaheri, Department of Virology, University of Helsinki, Haartmaninkatu 3, Helsinki Finland

Jorma Wartiovaara, Department of Medical Biology, University of Helsinki, Siltavuorenpenger 20A, Helsinki, Finland

Norman K. Wessells, Department of Biological Sciences, Stanford University, Stanford, California

Kenneth M. Yamada, Laboratory of Molecular Biology, National Cancer Institute, Bethesda, Maryland

Helen L. Yin, Department of Medicine, Harvard Medical School, Boston, Massachusetts

Preface

An historian of science recently remarked that during the past 70 years — the lifetime of the older members of the Society for Developmental Biology — more has been learned about the physical and biological world than in all the lifetimes and generations of humans that have gone before.

Of that 70 years, the 1970's must certainly be regarded as the single decade in which most has been learned about the eukaryotic cell. Our current picture of the structure of genes, of the cytoskeleton, and of the cell surface is remarkably advanced over that of just nine years ago. It has become increasingly evident that the cell surface plays a truly pivotal role in the life, development, and regulation of cells. On the one hand, the surface functions in the transmission of information from the environment to the cell — and here I mean not only molecular signals, but also mechanical forces stemming from adhesions and junctions that affect the cytoskeleton and so intracellular activities. The surface is also in a real sense an expression of the cell's genetic information and developmental state. Embryologists and developmental biologists must pay increasing heed to the cell surface and to its changing properties as a general explanation of development is constructed.

The 38th Symposium of the Society for Developmental Biology was broken into three parts — first, a heuristic session in which speakers summarized the current status of our knowledge about the cell surface — and here we include the plasma membrane, its associated cytoskeleton on one side, and the variety of surface-associated macromolecules on the other.

With this as background, the second portion of the Symposium concentrated upon the cell surface in early development — this topic was chosen because of the great advances made in the field and the increasing realization that the surface is a key to so many crucial events in early development of plants and animals. Unlike the usual pattern of recent Symposia, a larger number of speakers presented 30-minute talks — the purpose was to provide a wider spectrum of systems, techniques, and results, both at the Symposium and in this volume.

Finally, the third set of talks presented a variety of experimental systems in which the cell surface figures prominently in important developmental events. Though at first glance plant symbiosis, mammalian teratocarcinomas, adhesion and cell shape, and various extracellular macromolecules may appear to have only tenuous connections, the results of the speaker's experiments speak clearly to the generality that much of cell activity and development is a function of the surface and its associated molecules.

The 38th Symposium was held on the campus of the University of British Columbia in Vancouver, Canada in June 1979. Our primary hosts were Cyril V. Finnegan, longtime member and student of amphibian development, and Raymond Reeves, molecular developmentalist and past associate of Wilt and Gurdon. Ray, especially, devoted long hours to the detailed planning that made the stay at U.B.C. such a pleasure. Those attending the Symposium experienced three gorgeous days in the sun, with the mountains and sea spectacularly evident, and the stunning architecture of so many buildings on the campus an added delight. The banquet, held on the cliffs above the sea at the Anthropology Museum, will be an indelible memory for us all. All these things, in combination with the fine speakers whose reports comprise this volume, made this first visit of the Society to Canada a great success.

Norman K. Wessells

Acknowledgments

The 38th Symposium received very generous support from the National Science Foundation of the United States. The Society is appreciative of that support and of the helpful advice and guidance of the Program Officer for Developmental Biology at N.S.F., Dr. Mary E. Clutter. The University of British Columbia also contributed to the Symposium, and we express thanks to its Vice President, Michael Shaw. The Symposium would not have taken place had not Ms. Penelope A. Scandalios, our interim Business Manager, performed heroic service; Penny rendered order out of chaos, coordinated meetings and accommodations with skill, and lent warmth and a smile to the formalities of her office. The Society stands in her debt.

Young Investigator's Prize

For the first time, the Society for Developmental Biology awarded a "Young Investigator's Prize" for the outstanding poster display presented by an undergraduate student, graduate student, or postdoctoral fellow who were members of the Society. The Officers had great difficulty in choosing among the many excellent presentations, but in the end awarded two Honorable Mentions and the prize itself.

YOUNG INVESTIGATOR'S PRIZE 1979

A. Scott Goustin
Graduate Student
Department of Zoology
University of California, Berkeley
Sponsor: Fred H. Wilt

MULTIPLE LEVELS FOR THE CONTROL OF HISTONE SYNTHESIS DURING EARLY SEA URCHIN EMBRYOGENESIS

HONORABLE MENTION

Michael A. Harkey
Graduate Student
Department of Zoology
University of Washington
Sponsor: Arthur H. Whiteley

GENE EXPRESSION DURING DIFFERENTIATION OF SEA URCHIN PRIMARY MESENCHYME CELLS

Beatrice Holton
Graduate Student
Department of Biology
University of Oregon
Sponsor: James A. Weston

GLIAL CELL DIFFERENTIATION *IN VITRO*

Abstract of Young Investigator's Prize

MULTIPLE LEVELS FOR THE CONTROL OF HISTONE SYNTHESIS DURING EARLY SEA URCHIN EMBRYOGENESIS. *A. Scott Goustin,* Department of Zoology, University of California, Berkeley, California 94720.

Control over gene expression can theoretically be exercised at any of several levels between promoter "melting" and protein synthetic rate. This poster examines the expression of a defined set of genes [the repeated genes coding for the early (alpha-subtype) histones] in an early developmental system (the cleaving sea urchin embryo in its first day of development). I examined the expression of these genes under the guidance of a simple translational rate equation,

$$dP^*/dt = M \cdot f \cdot n \cdot \frac{A}{t},$$

where dP^*/dt = the rate of histone protein synthesis (in moles of amino acid incorporated into histone per second per embryo), M = the amount of histone mRNA (in moles of mRNA chains), f = the fraction of histone mRNA partitioning into polysomes (f is the polysomal partition coefficient, where $0<f<1$), n = the number of ribosomes per polysome, and A/t = the histone peptide elongation rate (in codons per sec, or amino acids per sec). In these experiments, I sought to learn how the early embryo solves this basic rate equation, in order to gain insight into which are the *pivotal* regulatory parameters. I have measured absolute rates of histone synthesis (dP^*/dt), histone polysome size (n), the partitioning of histone mRNAs into polysomes (f), and histone peptide elongation rates (A/t). I conclude that there are two temporal phases for the control of histone gene expression in cleaving sea urchin embryos, distinguished on the basis of their pivotal kinetic parameters: (1) Phase I (0-6 hr) during which dP/dt is proportional to f, and (2) Phase II (6-12 hr), during which dP/dt is proportional to M. (Supported by a Regents' Fellowship from the University of California, and NIH Grant GM 13882 to Dr. F.H. Wilt.)

I. The Cell Surface: Background and Perspectives

Molecular Interactions Governing Plasma Membrane Structure

Daniel Branton

Cell and Developmental Biology
The Biological Laboratories
Harvard University
Cambridge, Massachusetts 02138

I. INTRODUCTION

Development in muticellular eukaryotes must depend on mechanisms that extend beyond the usual notions inherent in our concepts of sequential gene activation. For example, development of an embryo requires that cells know where they are and where they should be. There must be mechanisms that regulate this social behavior of cells and more than intuition informs us that the cell membrane is involved both as the donor and the receptor of such social signals.

The important activities of the cell membrane have now been documented in terms of transport function, receptor function, and mechanical function. Study of receptor function and mechanical function have led to the realization that components of the cell surface are mobile in the plane of the membrane but that this lateral mobility is subject to regulation (Singer and Nicolson, 1972; Nicolson, 1976a). Such regulated, lateral mobility has been the basis for many hypotheses on the molecular mechanisms mediating cell recognition and growth control (Edelman, 1976; Nicolson, 1976a, b). In particular, it is often suggested that the interaction of extracellular ligands with their cell surface receptors alters the distribution of transmembrane elements that can bind to motility-related proteins such as actin or tubulin at the cytoplasmic surface of the

ISBN 0-12-612984-3

plasma membrane (Edelman, 1976; Bourguignon and Singer, 1977).

Although many observations give credence to such hypotheses, a direct chemical demonstration of the postulated binding between membrane and cytoplasmic components is wanting. If transmembrane elements can interact with components at the cytoplasmic surface of the membrane, one would like to know the precise nature of the binding sites, the affinities and specificities of the interaction, and the manner in which these affinities are regulated. To obtain such information from a complex eukaryotic cell is difficult when the interacting components cannot be examined under simple conditions. Unfortunately, the plasma membrane of many eukaryotic cells is hard to purify and few membrane preparations yield vesicles with a clearly defined orientation. Recent work with the human erythrocyte membrane illustrates some of the approaches one might use in studying protein interactions at the surface of more complex, nucleated cells.

The membrane of the human erythrocyte is eaily purified and vesicles can be prepared in either of two well-defined orientations: right-side-out and inside-out (Steck, 1974b). The proteins of the membrane have been extensively studied (Steck, 1974a), and at least 10 major membrane polypeptides can be enumerated. Electron microscopy and light microscopy using fluorescent labels demonstrate that the mobility of transmembrane elements is regulated by interaction with at least two proteins on the cytoplasmic surface of the cell membrane (Elgsaeter and Branton, 1974; Elgsaeter *et al.*, 1976; Fowler and Branton, 1977). These proteins include the two largest membrane polypeptides (band 1 and 2) which together make up the molecule called spectrin (Marchesi and Andrews, 1971) and a smaller polypeptide (band 5) which has been identified as erythrocyte actin (Sheetz *et al.*, 1976). Our studies of spectrin binding in erythrocyte ghosts provide an example of how interactions between the cell membrane and cytoplasmic proteins can be directly examined.

II. ASSAY FOR SPECTRIN BINDING PROTEINS

To measure spectrin binding, a binding assay had to be developed (Bennett and Branton, 1977). Because spectrin was known to be located on the cytoplasmic surface of the cell membrane, inside-out or inverted vesicles on which the cytoplasmic face and the putative binding sites would be easily accessible were used. Right-side-out vesicles served as controls. Radiolabelled spectrin was obtained by incubating intact cells in ^{32}P, releasing the spectrin at low ionic strength, and purifying the molecule on sucrose gradients. By challenging the inside-out vesicles with ^{32}P-labelled spectrin and then separating the vesicles with attached spectrin from the spectrin that remained unbound, spectrin binding could be easily measured under a variety of different conditions. It was found that the binding of [^{32}P]spectrin to inverted vesicles devoid of spectrin and actin was at least 10-fold greater than to right-side-out membranes, and exhibited different properties. Association with inside-out vesicles was slow, was decreased to the value for right-side-out vesicles at high pH, or after heating spectrin above 50° prior to assay, and was saturable with increasing levels of spectrin. Binding to right-side-out vesicles was rapid, unaffected by pH or by heating spectrin, and rose linearly with the concentration of spectrin. Scatchard plots of binding to inverted vesicles were linear at pH 7.6, with a K_D of 45 μg/ml, while at pH 6.6, plots were curvilinear and consistent with two types of interactions with a K_D of 4 and 19 μg/ml, respectively. The maximal binding capacity at both pH values was about 200 μg of spectrin/mg of membrane protein. Unlabeled spectrin competed for binding with 50% displacement at 27 μg/ml.

[^{32}P]Spectrin dissociated and associated with inverted vesicles with an identical dependence on ionic strength as observed for elution of native spectrin from ghosts. $MgCl_2$, $CaCl_2$ (1 to 4 mM) and EDTA (0.5 to 1 mM) had little effect on binding in the presence of 20 mM KCl, while at low ionic strength, $MgCl_2$ (1 mM) increased binding and inhibited dissociation to the same extent as 10 to 20 mM KCL.

Very mild treatment of the inside-out vesicles with low concentrations of trypsin or chymotrypsin destroyed their ability to bind spectrin. This suggested that the binding site was a protein and that the fragment of the protein which binds spectrin might be released from the membrane by incubation in protease. This suggestion was confirmed when the relased binding fragment was purified and shown to be a 72,000 dalton polypeptide capable of competing with inside-out vesicles for the binding of spectrin (Bennett, 1978).

Isolation of the 72,000 dalton binding fragment was the key step that made it possible to identify band 2.1 as the spectrin binding protein on the unproteolyzed erythrocyte membrane. Antibody to the 72,000 dalton fragment cross-reacted with the 210,000 dalton, band 2.1 protein on inside-out vesicles (Bennett and Stenbuck, 1979). Extensive comparison of tryptic and chymotryptic maps of the purified fragment with all the major membrane polypeptides confirmed that band 2.1 was the source of the 72,000 dalton fragment (Luna *et al.*, 1979; Yu and Goodman, 1979).

Band 2.1 has subsequently been purified in a water-soluble form and the binding between band 2.1 and spectrin has been verified by solution experiments using rate zonal sedimentation techniques (Tyler *et al.*, 1979). In the course of these experiments, it became clear that in addition to band 2.1, band 4.1 of the native membrane can also bind to spectrin. Although the role, if any, that band 4.1 plays in binding spectrin to the erythrocyte membrane remains to be assessed, it appears probable that band 4.1 is important in mediating the interaction between spectrin and other erythrocyte polypeptides.

Now that the protein which binds spectrin to the membrane has been purified, experiments to map the binding interactions have become possible. The spectrin heterodimer (band 1 + 2) can be visualized in the electron microscope by using low angle shadowing techniques (Shotton *et al.*, 1979). It presents itself as an elongate *ca.* 100 nm long molecule in which the two component subunits (presumably band 1 and band 2) are visualized as floppy intertwined chains that bind to each other at their termini. Tetramers containing 2 heterodimer molecules form by the end-to-end association of the subunits so as to produce a molecule approximately 200nm long. The tetramers show no measurable overlap between the strands of the two component heterodimers. Higher order oligomers are not seen, suggesting that tetramer formation is a head-to-head rather than a head-to-tail association of heterodimers. Band 2.1 can also be visualized by low angle shadowing. It is a much less extended molecule than spectrin and its longest axis is between 5 and 8 nm. Because the two are so different from each other, spectrin and band 2.1 are readily distinguishable when mixed together. Furthermore, when band 2.1 binds to spectrin, the site of interaction is readily perceived as an extra lump along the length of the spectrin molecule. Studies of spectrin which have bound band 2.1 show that the binding of band 2.1 occurs on one of the two strands that make up the heterodimer and that this binding occurs at a site approximately 20 nm from one end of the molecule (Tyler *et al.*, 1979). By examining band 2.1 bound to spectrin tetramers, it was possible to identify the end of the heterodimer closest to the band 2.1 binding site as the end that participates in tetramer formation.

III. CONCLUSIONS

Our experiments show that direct verification of the interaction between membranes and fiber-forming elements at its cytoplasmic surface is possible. In the case of the human erythrocyte, the work to date has focussed on band 2.1 and spectrin rather than the transmembrane proteins and motility-related proteins that are presumed to interact in many nucleated eukaryotes. Preliminary experiments suggest there are specific links between band 2.1 and band 3, the major transmembrane protein, and also links between spectrin and erythrocyte actin. Just as the availability of inside-out vesicles and purified components has made it possible to apply direct biochemical assays to measure interactions between spectrin and its membrane attachment site, so too should the availability of such preparations allow one to explore the full continuum of interactions and specificities extending between band 3 and actin. However, the approaches outlined here are not limited to the human erythrocyte. Purified membrane preparations with a defined orientation are now available from other eukaryotic cells and similar attempts to investigate the interactions linking shape-determining or motility-related proteins to the plasma membrane should be undertaken.

REFERENCES

Bennett, V. (1978). *J. Biol. Chem.* **253**, 2292-2299.

Bennett, V. and Branton, D. (1977). *Science* **195**, 302-304.

Bennett, V. and Stenbuck, P. J. (1979). *J. Biol. Chem.* **254**, 2533-2541.

Bourguignon, L. Y. and Singer, S. J. (1977). *Proc. Nat. Acad. Sci. U.S.A.* **74**, 5031-5035.

Edelman, G. (1976). *Science* **192**, 218-226.

Elgsaeter, A. and Branton, D. (1974). *J. Cell Biol.* **63**, 1018-1030.

Elgsaeter, A., Shotton, D. and Branton, D. (1976). *Biochim. Biophys. Acta* **426**, 101-122.

Fowler, V. and Branton, D. (1977). *Nature* **268**, 23-26.

Luna, E. J., Kidd, G. H. and Branton, D. (1979). *J. Biol. Chem.* **254**, 2526-2532.

Marchesi, V. T. and Andrews, E. P. (1971). *Science* **174**, 1247-1248.

Nicolson, G. L. (1976a). *Biochim. Biophys. Acta* **457**, 57-108.

Nicolson, G. L. (1976b). *Biochim. Boiphys. Acta* **458**, 1-72.

Sheetz, M. P., Painter, R. G. and Singer, S. J. (1976). *Biochem.* **15**, 4486-4492.

Shotton, D. M., Burke, B. and Branton, D. (1979). *J. Mol. Biol.*, in press.

Singer, S. J. (1971). *In* "Membrane Structure and Function " (L. F. Rothfeld, ed.), pp. . Academic Press, New York.

Singer, S. J. and Nicolson, G. L. (1972). *Science* **175**, 720-731.

Steck, T. L. (1974a). *J. Cell Biol.* **62**, 1-19.

Steck, T. L. (1974b). *Methods in Membrane Biology* **2**, 245-281.

Tyler, J. M., Hargreaves, W. R. and Branton, D. (1979). *Proc. Nat. Acad. Sci. U.S.A.*, in press.

Yu, J. and Goodman, S. (1979). *Proc. Nat. Acad. Sci. U.S.A.* **76**, 2340-2344.

The Motor of Ameboid Leukocytes

Thomas P. Stossel, John H. Hartwig
Helen L. Yin and Olle I.Stendahl

*Hematology-Oncology Unit, Massachusetts
General Hospital, Department of Medicine
Harvard Medical School
Boston, Massachusetts 02114*

I. INTRODUCTION

Ameboid leukocytes are the polymorphonuclear leukocytes, the lymphocytes and mononuclear phagocytes. They are united in lineage, arising from a common hematopoietic stem cell, and look similar when engaged in some of their motile functions. Important work done by the motor of ameboid leukocytes is to drive locomotion, endocytosis, (phagocytosis and pinocytosis), secretion and cell division. During cell division and locomotion, the motor acts to change the entire shape of the cell. In secretion, the motor alters the thickness of the cell cortex in places, allowing secretory granules to approach the plasma membrane with which they fuse. During spreading, locomotion, and endocytosis, the motor extends pseudopodia. In the capping response of lymphocytes and polymorphonuclear leukocytes, the motor redistributes surface

ISBN 0-12-612984-3

membrane receptors (Bessis, 1973). As its proper office, the motor activity of ameboid leukocytes contributes to immunity, the defense of the host against infection; when inappropriately activated, it inflicts inflammation. This review focuses on the motor of cells important for immunology, but the similarity between the cortical morphology of leukocytes and that of cells involved in embryogenesis is considerable, suggesting that the same principles of motor activity are applicable in developmental biology. For example, certain embryonic cells from peripheral blebs and pseudopodia that resemble counterpart structures on the surface of leukocytes (Holtfreter, 1947; Trinkaus, 1973).

II. ELEMENTS OF THE MOTOR

The elements of a motor are: 1) a force-generating system, 2) orientation of the force to confer directionality, 3) a mechanism for control. This discussion reviews studies aimed at understanding the elements of motor activity in ameboid leukocytes. The studies on which the conclusions are based were done with rabbit lung macrophages and human peripheral blood polymorphonuclear leukocytes. Because many similar findings are obtainable with cells throughout phylogeny (Stossel, 1978; Taylor and Condeelis, 1979), the extension of the conclusions to ameboid leukocytes in general and probably to many other cells as well, is reasonable. The approach has been to crush cells, to isolate the parts of the motors, and to put them back together again. Abetting this approach has been the fact that isolated motor proteins assemble into structures which may resemble their functional units in the cell.

III. MORPHOLOGY OF THE MOTOR

One can compare the ameboid leukocyte to a ball of yarn covered with a thin layer of jelly. The ball of yarn contains the nucleus, secretory granules and other organelles. The jelly layer is the motor. Morphologists, observing ameboid leukocytes with the light microscope, came to this conclusion over 40 years ago and speculated that assemblies of "large asymmetrical molecules" were constituents of the motor (Lewis, 1939). The microscopists perceived that this cortical jelly layer, called the hyaline ectoplasm, which comprises pseudopodia extended by the cells, had a different composition than the cell interior, because granules and other visible organelles were excluded from this region. They fancied that the hyaline cortex exerted tension on the center of the cell and controlled the movement of the cell's internal contents. Equivalent ideas in developmental biology arose from extensive studies defining the mechanical properties of the gelatinous cortex of sea urchin eggs (Hiramoto, 1970). Following development of the electron microscope, microscopists tended to ignore this region because it was morphologically "structureless," although a few investigators commented on the wispy thin filament meshworks that comprised the pseudopodia of ameboid leukocytes (Keyserlingk, 1968; Reaven and Axline, 1973). Even after the knowledge arose that cortical filaments were actin polymers (Ishikawa *et al.*, 1969), most morphologists overlooked these cortical meshworks and focussed on "stress fibers," thick bundles of actin fibers in fibroblasts and other cultured cells (Buckley and Porter, 1967). Such stress fibers are rarely visible in ameboid leukocytes. Although some worry that the cortical filament networks seen in electron micrographs may be artefacts resulting from oxidative effects of osmium used during preparation of cells for electron microscopy, (Maupin-Szamier and Pollard, 1978), sufficient evidence attests to the reality of the cortical meshwork in ameboid leukocytes: 1) The meshwork is perceivable by electron microscopy techniques that do not require osmication, such as high voltage electron microscopy (Wolosewick and Porter, 1976), or scanning electron microscopy of cell remnants attached to surfaces (Boyles and Bainton, 1979). 2) The cell periphery is optically isotropic — that is, it scatters polarized light equally in all directions, indicating that fibers therein do not have a parallel order. 3) The cortical meshwork persists in transmission electron micrographs of ameboid leukocytes when osmication is minimized or abolished (Boyles and Bainton, 1979). 4) As amplified below, the fiber meshwork is a reasonable entity for ameboid leukocytes to use as their motor.

IV. FORCE-GENERATION BY THE MOTOR

The muscle protein, actin, comprises about 10% of the total protein of ameboid leukocytes (Boxer and Stossel, 1976; Hartwig and Stossel, 1975). Actin subunits are globular monomers (G-actin) which can aggregate in a helical fashion to assemble long (0.1 - 10 μ) linear fibers (F-actin). Both states of actin, monomer and filament, exist in equilibrium with one another, the relative proportion of both states being governed by the ambient salts and by other proteins (Oosawa and Kasai, 1971). The distribution between the two states of actin in the cytoplasm of ameboid leukocytes is unknown, but many actin fibers, oriented in a random fashion, populate the cortex of these cells. Preliminary evidence indicates that a substantial fraction of the total actin, however, may by unpolymerized, suggesting a role for monomer-polymer equilibrium in the control of motor activity. (Blikstad *et al.*, 1978).

Another muscle protein, myosin, represents about 1% of the total protein of ameboid leukocytes (Boxer and Stossel, 1976; Hartwig and Stossel, 1975; Stossel and Pollard, 1973). The myosin molecules are large asymmetric proteins with globular heads and helical tails. The tails aggregate to form characteristic bipolar filaments in the test tube (Boxer and Stossel, 1976; Hartwig and Stossel, 1975; Stossel and Pollard, 1973; Tatsumi *et al.*, 1973; Takeuchi *et al.*, 1975) and possibly within the ameboid leukocytes as well (Albertini *et al.*, 1977; Hoffstein and Weissman, 1978). The bipolarity arises from the globular heads which stick out from the ends of the shaft of the aggregated tails. If one incubates a mixture of leukocyte myosin filaments and actin filaments in the presence of ATP, the mixture becomes turbid and forms a precipitate that compacts itself into a tight aggregate (Boxer and Stossel, 1976; Stossel and Hartwig, 1976; Tatsumi *et al.*, 1973).

This reaction, called "superprecipitation," is an analogue of the contractile event occurring in intact muscle (Szent-Györgyi, 1947). The movement seen in the superprecipitation occurs because the heads of the myosin molecules projecting from their filaments sequentially make and break linkages with adjacent actin monomers in the actin filaments. The myosin heads bind the actin monomers, bend toward the center of the myosin filament, let go, bind the next monomer, and repeat the cycle continuously. The structure of actin filaments is such that they restrict the binding of myosin molecules in a way so as to assure a unidirectional polarity of movement. Actin filaments bound at opposite ends of the bipolar myosin filaments can only move towards each other, that is, toward the center of the myosin filaments. The energy for the movement arising from the interaction between actin and myosin derives from breakdown of ATP to ADP and phosphate by the myosin molecules. This mechanism for force-generation by the muscle proteins, actin and myosin, is the widely accepted "sliding filament" model (Huxley and Hanson, 1954; Huxley and Niedergerke, 1954).

V. ORIENTATION OF THE MOTOR

In striated muscle, a particular orientation of the force arises from the structure of the sarcomere where a spatial separation exists between parallel arrays of actin and myosin filaments. Contraction occurs in the direction of the parallel fibers. However, the isotropic actin meshworks in the periphery of ameboid leukocytes resemble the morphology of superprecipitating mixtures of actin and myosin (D'Haese and Komnick, 1972) rather than the morphology of contracting muscle. Superprecipitation occurs only in a centripetal direction, because the force vectors are random and directed inward from the walls of the container of the solution. In the periphery of ameboid leukocytes, the consequence of such a contraction would be a net force exerting itself away from the plasma membrane toward the center of the cell. Unlike the actomyosin solution in a test tube, however, the actin filaments in the leukocytes seem to have some kind of firm association with the plasma membrane and therefore would attempt to draw it toward the center of the cell with the cortial actin. Presumably an equal and opposite hydrostatic force would oppose this inward movement because this system is closed. Therefore, no net movement could occur. The same argument is applicable to the dynamics in the plane of the membrane. If the force-generation by myosin throughout the cortex and the mass upon which the force acts (actin and membranes) are equal throughout the cell periphery, no net movement occurs, according to Newton's first law. But local alteration of either the force or the mass in this cortical "tug-of-war" would result in net acceleration and could produce net movement in some direction.

Knowledge about any changes in the power of the force-generating mechanism, the myosin-actin interaction in the cytoplasm of leukocytes is very incomplete at the present time. The contractile activity of mixtures of actin and myosin purified from leukocytes is very weak compared with the force-generating capacity of actin and myosin from skeletal muscle (Boxer and Stossel, 1976; Hartwig and Stossel, 1975; Stossel and Pollard, 1973; Tatsumi *et al.*, 1973). The poor activity of leukocyte proteins is ascribable to the purified myosins and not the actins which are functionally nearly equivalent to skeletal muscle actin. A crude protein fraction, designated "cofactor," from rabbit lung macrophages activates the contractile power of macrophage actin-myosin mixtures to a level comparable to that of the skeletal muscle proteins (Stossel and Hartwig, 1975). Since cofactor is not now available in a pure state, we do not know if it is a regulator or modulator of macrophage contractile force. Calcium does not influence its activity. The idea that kinase enzymes phosphorylate some of the light chains of smooth muscle and certain nonmuscle cell myosins and thereby activate the myosins' contractile activity is currently popular (Adelstein, 1978; Dabrowska and Hartshorne, 1978; Lebowitz and Cooke, 1978), although controversial (Hirata *et al.*, 1977; Mikawa *et al.*, 1977). The macrophage cofactor might possibly be such a kinase. Since kinases that phosphorylate proteins and phosphatases that remove the phosphates from proteins establish a reciprocal control mechanism in many biological systems, the activity of myosin in leukocytes is potentially controllable in this fashion. However, evidence for this idea is lacking.

A different concept is that the control of orientation arises from changes in the actin filaments mass. Gelation or crosslinking of actin filaments controls these changes. Crude soluble extracts of ameboid leukocytes solidify (gel) slowly (hours) in the cold or rapidly (minutes) when warmed to 25-27°C (Boxer and Stossel, 1976; Stossel and Hartwig, 1976). Different rearrangements of molecular structures can cause liquid macromolecular solutions to solidify, and all the rearrangements involve interlocking of the macromolecules into a network to the extent that they become relatively insoluble. The insoluble network structure must have sufficient rigidity to counteract the ambient forces to remain solid. The molecular rearrangement occurring in the cytoplasmic extracts of ameboid leukocytes is the crosslinking of long actin fibers to form such a network. Solidification of diverse cytoplasmic extracts was reported many years ago (Anderson, 1957; Hultin, 1950) but first ascribed to actin only in 1975 (Kane, 1975). Since filaments assembled from purified actin do not solidify in this way, some other factor is responsible for crosslinking the filaments in extracts of ameboid leukocytes. This other factor is a high molecular weight protein called actin-binding protein (ABP) which noncovalently binds actin filaments together (Hartwig and Stossel, 1975; Stossel and Hartwig, 1975; Boxer and Stossel, 1976).

The foregoing conclusion is based on two properties which are useful for characterizing gels: 1) critical conditions, and 2) mechanical properties (Flory, 1953). The gel state occurs when a critical number of molecules become linked into a continuous network. Obviously, long fibers have a greater chance of contacting one another than small round molecules. Therefore, a small concentration of long fibers, such as actin filaments, is required to create a gel. In fact, actin filaments are so huge that any actin concentration capable of forming filaments can gel. However, as stated above, the interaction between actin filaments are very weak, and only if the actin filament concentration is very high or if exogenous crosslinker molecules tie the filaments together does a relatively strong network form.

A critical concentration of exogenous crosslinking molecules exists, at which a continuous network of the polymers first occurs. A very small change in the crosslinker concentration near the critical point abruptly alters the number of chains involved in the network, accounting for very abrupt sol-gel transformations. Above the critical crosslinker concentration, the mechanical rigidity of the polymer solution rises abruptly as the lattice forms. The degree to which a crosslinker increases the rigidity of a polymer gel depends on the number of crosslinks introduced and on the uniformity of the distribution of crosslinkers within the network (Flory, 1953). The potency of a gelling agent depends on the capacity of as few crosslinking molecules to link as many fibers as possible. An inefficient agent may form many bridges between adjacent filaments. This redundant crosslinking can create a very stable and rigid aggregate of the involved fibers, but at the expense of failing to recruit more filaments into the network and to stabilize the network as a whole. Measurement of the critical crosslinker concentration is a good way, therefore, to compare the efficiency of different gelling agents. The critical concentration of ABP for crosslinking actin filaments into a continuous network or gel is much lower than that of other actin-crosslinking proteins (Brotschi, *et al.*, 1978) such as myosin or a protein from smooth muscle structurally related to ABP called filamin (Wang, 1977). Above the critical crosslinker concentration, small additional amounts of ABP increase the rigidity of actin much more than equivalent quantities of myosin or filamin (Brotschi, *et al.*, 1978). The simplest explanation for the relative potency of ABP as an actin crosslinking agent is that it preferentially crosslinks actin filaments at angles close to 90°. Such perpendicular binding would minimize the formation of redundant or incestuous crosslinks.

ABP comprises about 1% of the total protein of ameboid leukocytes (Boxer and Stossel, 1976; Hartwig and Stossel, 1975). The bulk of actin-crosslinking activity in macrophage extracts copurifies with ABP (Brotschi, *et al.*, 1978). Evidence for high (Bryan and Kane, 1978; Schollmeyer, *et al.*, 1978; Wallach *et al.*, 1978) and low molecular-weight (Maruta and Korn, 1977) actin-crosslinking proteins in various other cells exists but is incomplete. The slow development of definite information is due to the difficulty in purifying proteins present in low concentrations in these cells and to problems with proteolysis. Degradation of ABP in disrupted macrophages is only preventable with potent combinations of proteolysis inhibitors (Brotschi, *et al.*, 1978). If unimpeded, proteolysis can convert the actin-crosslinking activity in macrophage extracts from high to low molecular components (Brotschi, *et al.*, 1978). Proteolysis can confound the problem by causing protein solutions to gel by nonspecifically generating crosslinks between filaments, a mechanism analagous to the gelation of denatured globular proteins (Tombs, 1974). The gelation of cell extracts following trypsinization (Moore *et al.*, 1978) probably occurs by this process. The existence of ABP, a protein that is a highly potent crosslinker of actin strongly supports the idea that controlled lattice formation in cortical cytoplasm could provide for amplification and directionality of the force-generating mechanism. It could also account for local changes in the stability of the cortex underlying the lipid bilayer. If these notions are correct, then lattice state of actin must be under some kind of control.

VI. CONTROL OF THE MOTOR

A protein called gelsolin (Yin and Stossel, 1979) recently purified from rabbit lung macrophages dissolves gels of actin crosslinked by ABP or filamin when the calcium concentration exceeds 9×10^{-8} M. When the calcium concentration falls below this level, the sol gels again. The protein binds to actin filaments when the calcium concentration is above the threshold calcium concentration. The activity of gelsolin is maximal at a calcium concentration of 10^{-5}M. Our present understanding of how activated gelsolin works is that it reversibly breaks actin filaments into shorter pieces. The evidence for this conclusion is that gelsolin lowers the intrinsic viscosity of F-actin and decreases the contour length of actin filaments seen in the electron microscope.

The effect of gelsolin on actin is rapid and stoichiometric, one molecule of activated gelsolin causing one break in an actin filament. Since gelsolin appears to comprise less than 1% of the total macrophage protein and over 100 moles of F-actin exist per mole of gelsolin in macrophage extracts made 0.1 M in KCl, limited cleavage rather than large scale depolymerization of F-actin is its mechanism of action.

Since severing of actin filaments obligatorily increases their number, the ratio of crosslinker to actin filaments can fall abruptly below the critical value, and solvation of the gels then takes place. If the calcium concentration rises in one domain of the cortical actin network containing uniformly dispersed myosin filaments, the myosin filaments in that domain act on a smaller mass of actin and therefore draw it toward the domains in which the calcium concentration remains lower and in which the network is more intact. Hence, fluctuations in the calcium concentration could rapidly control directionality of movement. If the lattice beneath the plasma membrane has a supportive role in strengthening the lipid bilayer, discontinuity in the lattice strength could permit the hydrostatic pressure opposing the centripetal force to create bulges or blebs in the periphery. Since the swelling of gels is inversely proportional to the degree of crosslinking (Flory and Rehner, 1943), solvent might move from less to more crosslinked areas. Such movements could amplify the shape changes created by the alterations in the stability of the lattice and play some role in the translocation of organelles.

The hypothesis stated here predicts that a mass of cytoplasmic lattice will flow from high to low calcium concentrations. This prediction is experimentally verifiable in horizontal capillaries containing actin, myosin, ABP and gelsolin. If the calcium concentration rises on one side of the capillary, the actin gel flows to the other side (Stendahl, *et al.*, 1979).

Up to now, the discussion has dealt with movement of cortical *cytoplasm*. But the *membrane,* that is, the lipid bilayer and its associated proteins and carbohydrates, also must move. At present the nature of the connection between contractile proteins and the bilayer is obscure and difficult to analyze. Transmission electron micrographs of ameboid leukocytes show that numerous actin filaments do run right up to the bilayer (DePetris, 1978), and some even say that actin exists on the outer cell surface (Owen, *et al.*, 1978). If polymorphonuclear leukocytes or lymphocytes react with antibodies to cell surface antigens or with lectins, the cell surface antigens or lectin binding molecules aggregate into a polar "cap" (DePetris, 1978). The cap contains many actin filaments (DePetris, 1978; Albertini, *et al.*, 1977), and immunofluorescent staining indicates that high concentrations of actin, myosin, and other contractile proteins reside in them (Braun, *et al.*, 1978; Geiger and Singer, 1979; Gabbiani, et al., 1977; Bourguignon and Singer, 1977). After lysis of the cells with detergents, the aggregated cell surface antigens and a large fraction of the total cell actin are co-isolable, suggesting some kind of connection between the two (Flanagan and Koch, 1978). But since much cortical cytoplasm also goes into the caps and other protrusion, the quantity of actin and associated proteins actually attached to the bilayer remains unknown. Moreover, the molecular basis of the association between actin and cell surface actigens is unclear. Nevertheless, there is no question but that a bulk translocation of actin is able to move membrane by virtue of a linkage between the two structures.

VII. EVIDENCE FOR THE EXISTENCE OF THE MOTOR IN AMEBOID LEUKOCYTES

If the hyaline ectoplasm is the motor of ameboid leukocytes, ABP and myosin should reside there. Studies employing biochemical and immunocytochemical techniques do reveal an enrichment of ABP and myosin in the cell periphery of ameboid leukocytes and document the presence of these proteins in pseudopod extension of the cortical cytoplasm (Davies and Stossel, 1977; Hartwig, *et al.*, 1977; Boxer, *et al.*, 1976).

Cytochalasin B is a fungal metabolite that reversibly alters the morphology of ameboid leukocytes, concomitantly inhibiting movement ascribed to the cortical motor: locomotion, phagocytosis, and capping (Tannenbaum, 1978). It enhances secretion by exocytosis in the presence of secretagogues (Hoffstein and Weissman, 1978). Cytochalasin B has minor effects on various physical properties of actin filaments, but very effectively dissolves actin-ABP gels (Hartwig and Stossel, 1976). Like the inhibition of cell motility, the solvation of actin gels by cytochalasin B is reversible following removal of the drug. Cytochalasin B acts by severing actin filaments (Hartwig and Stossel, 1979), thereby having a similar effect on actin gels as those of calcium-activated gelsolin. However, as cytochalasin B unevenly penetrates cells, it enhances locally the tendency for contraction of the cortical actin lattice by myosin, thereby causing widespread focal superprecipitation rather than waves of directional contraction characteristic of the cell's own control mechanism. Therefore, the effects of cytochalasin B depend on active contraction which explains why the expression of the effects of cytochalasin B require that cellular energy metabolism be unimpaired (Tannenbaum, 1978).

Considering the central role of calcium in the hypothesized control of the motor of ameboid leukocytes, the effects of calcium on movement of ameboid leukocytes might seem interesting. Although the extracellular calcium concentration influences many actions of ameboid leukocytes (Braun, *et al.*, 1978; Gallin and Rosenthal, 1974; Goldstein, *et al.*, 1975; Stossel, 1973), these results are not readily interpretable in terms of *intracellular* actions of calcium. Ionophores, drugs that alter the permeability of cells to calcium ions, may change the cytoplasmic calcium concentration, and change the morphology of ameboid leukocytes (Hoffstein and Weissman, 1978; Braun, *et al.*, 1978). But these experiments, too, are not easily explained, because the ionophores transport ions other than calcium and also alter cytoplasmic pH.

VIII. SUMMARY

Our investigations lead us to the following conclusions. The motor of ameboid leukocytes is an actin lattice in the cell cortex (Fig. 1). The force-generating mechanism is a superprecipitation of actin and myosin filaments, a process requiring hydrolysis of ATP. This energy-dependent mechanism may be a major consumer of the ameboid leukocyte's metabolic activity and accounts for the susceptibility of motile functions of ameboid leukocytes to inhibition by metabolic poisons. Directionality and amplification of the force generated derives from controlled focal changes in the crosslinking of actin filaments. Gelsolin, a calcium-activated protein, controls the lattice rigidity of actin by severing actin filaments between points of crosslinking by actin-binding protein. The cytoplasm and the membrane then move from domains of less to those of more crosslinked filaments. Calcium is therefore the ignition or signal that regulates movement in the ameboid leukocyte. A major challenge is now to understand the "switch," the way in which the cell can regulate local calcium concentrations in response to specific stimuli.

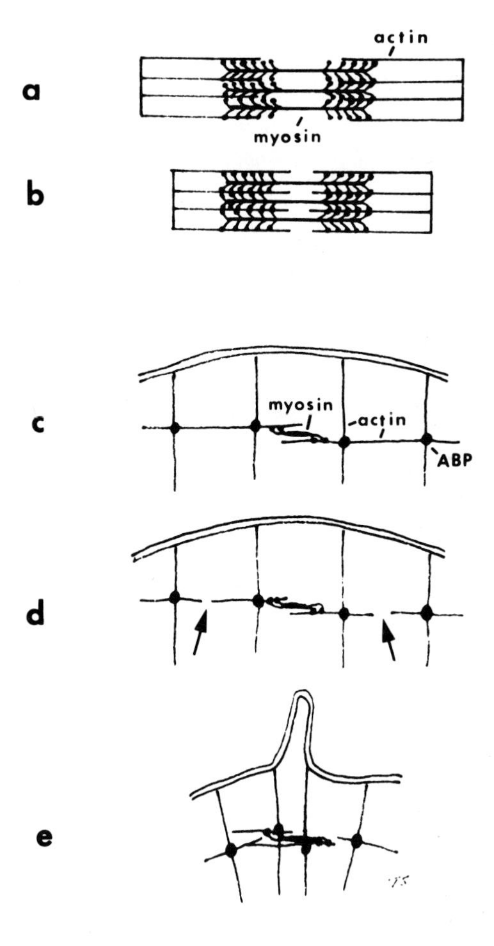

Fig. 1. Highly schematic rendition of motor units of striated muscle and ameboid leukocyte. a) Relaxed muscle. b) Contracted muscle. c) Cell periphery of ameboid leukocyte. Actin filaments, some attached to the membrane, are crosslinked by actin-binding protein (ABP). Myosin filaments generate isometric tension. d) Actin filaments are severed focally by calcium-activated gelsolin (arrows). e) Contraction occurs toward more highly crosslinked domains.

ACKNOWLEDGEMENTS

Supported by Grants from the U.S. Public Health Service (HL 19429), The Council for Tobacco Research, U.S.A., Inc., the Edwin S. Webster Foundation, United Ocean Services, and Edwin W. Hiam.

REFERENCES

Adelstein, R.S. (1978). *Trends Biochem. Sci.* **6**, 27-30.
Albertini, D.F., Berlin, R.D. and Oliver, J.M. (1977). *J. Cell Sci.* **26**, 57-75.
Anderson, N.G. (1957). *J. Cell Comp. Physiol.* **49**, 221-241.
Bessis, M. (1973). "Living Blood Cells and Their Ultrastructure." Springer Verlag, Berlin, Heidelberg, New York.
Blikstad, I., Markey, F., Carlsson, L., Persson, T. and Lindberg, U. (1978) *Cell,* **15**, 935-943.
Bourguignon, L.Y.W. and Singer, S.J. (1977). *Proc. Nat. Acad. Sci. U.S.A.* **74**, 5031-5035.
Boxer, L.A. and Stossel, T.P. (1976). *J. Clin. Invest.* **57**, 964-976.
Boxer, L.A., Floyd, A. and Richardson, S. (1976). *Nature* **263**, 259-261.
Boyles, J. and Bainton, D.F. (1979). *J. Cell Biol.* **82**, 347-368.
Braun, J., Fujiwara, K., Pollard, T.D. and Unanue, E.R. (1978). *J. Cell Biol.* **79**, 419-426.
Brotschi, E.A., Hartwig, J.H. and Stossel, T.P. (1978). *J. Biol. Chem.* **253**, 8988-8993.
Bryan, J. and Kane, R.E. (1978). *J. Mol. Biol.* **125**, 207-224.
Buckley, I.K. and Porter, K.R. (1967). *Protoplasma* **64**, 349-362.
Dabrowska, R. and Hartshorne, D.J. (1978). *Biochem. Biophys. Res. Commun.* **85**, 1352-1359.
Davies, W.A. and Stossel, T.P. (1977). *J. Cell Biol.* **75**, 941-955.
DePetris, S. (1978). *J. Cell Biol.* **79**, 235-251.
D'Haese, J. and Komnick, H. (1972). *Z. Zellforsch.* **134**, 411-426.
Flanagan, J. and Koch, G.D.E. (1978). *Nature* **273**, 278-281.
Flory, P.J. (1953). "Principles of Polymer Chemistry." Cornell University Press, New York.
Flory, P.J. and Rehner, J. Jr. (1943). *J. Chem. Phys.* **11**, 523-530.
Gabbiani, G., Chaponnier, C., Zumbe, A. and Vassalli, P. (1977). *Nature* **269**, 697-698.
Gallin, J.I. and Rosenthal, A.S. (1974). *J. Cell Biol.* **62**, 594-609.
Geiger, B. and Singer, S.J. (1979). *Cell* **16**, 213-222.
Goldstein, I.M., Hoffstein, S.T. and Weissman, G. (1975). *J. Immunol.* **115**, 665-670.
Hartwig, J.H. and Stossel, T.P. (1975). *J. Biol. Chem.* **250**, 5696-5704.
Hartwig, J.H. and Stossel, T.P. (1976). *J. Cell Biol.* **71**, 295-303.
Hartwig, J.H. and Stossel, T.P. (1979). *J. Mol. Biol.*, in press.
Hartwig, J.H., Davies, W.A. and Stossel, T.P. (1977). *J. Cell Biol.* **75**, 956-967.
Hiramoto, Y. (1970). *Biorrheology* **6**, 201-234.
Hirata M., Mikawa, T., Nonomura, Y. and Ebashi, S. (1977). *J. Biochem.* **82**, 1793-1796.
Hoffstein, S. and Weissmann, G. (1978). *J. Cell Biol.* **78**, 769-781.
Holtfreter, J. (1947). *J. Morphol.* **80**, 25-55.
Hultin, T. (1950). *Exp. Cell Res.* **11**, 272-283.
Huxley, A.F. and Niedergerke, R. (1954). *Nature* **173**, 971, 973.
Huxley, H.E. and Hanson, J. (1954). *Nature* **173**, 973-976.
Ishikawa, H., Bischoff, R. and Holtzer, H. (1969). *J. Cell Biol.* **43**, 312-328.
Kane, R.E. (1975). *J. Cell Biol.* **66**, 305-315.
Keyserlingk, D.G. (1968). *Exp. Cell Res.* **51**, 79-91.

Lebowitz, E.A. and Cooke, R. (1978). *J. Biol. Chem.* **253**, 5443-5447.
Lewis, W.H. (1939). *Arch Exp. Zellforsch.* **23**, 1-7.
Maruta, H. and Korn, E.D. (1977). *J. Biol. Chem.* **252**, 399-402.
Maupin-Szamier, P. and Pollard, T.D. (1978). *J. Cell Biol.* **77**, 837-852.
Mikawa, T., Nonumura, Y. and Ebashi, S. (1977). *J. Biochem.* **82**, 1789-1791.
Moore, P.B. and Carraway, K.L. (1978). *Biochem. Biophys. Res. Commun.* **80**, 560-567.
Oosawa, F. and Kasai, M. (1971). *In* "Subunits in Biological Systems" (S.N. Timasheff, G.D. Fasman, eds.), Part A, pp. 261-322. Marcel Dekker Inc., New York.
Owen, M.J., Auger, J., Barber, B.H., Edwards, A.J., Walsh, F.S. and Crumpton, M.J. (1978). *Proc. Nat. Acad. Sci. U.S.A.* **75**, 4484-4488.
Reaven, E.P. and Axline, S.G. (1973). *J. Cell Biol.* **59**, 12-27.
Schollmeyer, J.V., Rao, G.H.R. and White, J.G. (1978). *Am. J. Pathol.* **93**, 433-446.
Stendahl, O., Yin, H. and Stossel, T. (1979). *Clin Res.* **27**, 465a.
Stossel, T.P. (1973). *J. Cell Biol.* **58**, 346-356.
Stossel, T.P. (1978). *Annu. Rev. Med.* **29**, 427-457.
Stossel, T.P. and Pollard, T.D. (1973). *J. Biol. Chem.* **248**, 8288-8294.
Stossel, T.P. and Hartwig, J.H. (1975). *J. Biol. Chem.* **250**, 5706-5712.
Stossel, T.P. and Hartwig, J.H. (1976). *J. Cell Biol.* **68**, 602-619.
Szent-Gyorgyi, A. (1947). "Chemistry of Muscular Contraction." Academic Press, New York.
Takeuchi, K., Shibata, N. and Senda, N. (1975). *J. Biochem.* **78**: 93-103.
Tannenbaum, S.W. (1978). "Cytochalasins-Biochemical and Cell Biological Aspects." North Holland-Elsevier Co., Amsterdam.
Tatsumi, N., Shibata, N., Okamura, Y., Takeuchi, K. and Senda, N. (1973). *Biochim. Biophys. Acta.* **305**, 433-444.
Taylor, D.L. and Condeelis, J.S. (1979). *Int. Rev. Cytol.* **56**, 57-144.
Tombs, M.P. (1974). *Farad. Disc. Chem. Soc.* **57**, 158-163.
Trinkaus, J.P. (1973). *Develop. Biol.* **30**, 68-103.
Wallach, D., Davies, P.J.A. and Pastan, I. (1978). *J. Biol. Chem.* **253**, 3328-3335.
Wang, K. (1977). *Biochemistry* **16**, 1857-1865.
Wolosewick, J.J. and Porter, K.R. (1976). *Am. J. Anat.* **147**, 303-324.
Yin, H.L. and Stossel, T.P. (1979). *Nature,* **281**, 583-586.

Cell-to-Cell Communication and Development

Norton B. Gilula
The Rockefeller University
New York, New York 10021

I. INTRODUCTION TO JUNCTIONAL CELL-TO-CELL COMMUNICATION

Animal cells utilize a variety of mechanisms for interacting with and transmitting information to other cells. These mechanisms are important for regulating or synchronizing the activities of cells in multicellular systems. In most organisms, these mechanisms are initially operational at the time of fertilization, and they are retained throughout

ISBN 0-12-612984-3

the lifetime of the organism(Moscona, 1974; Poste and Nicolson, 1976). Some organisms, such as slime molds (Lerner and Bergsma, 1978) and sponges (Maslow, 1976) have highly specialized mechanisms for cellular interactions, and the most specific type of interaction mechanism is associated with the vertebrate immune system (Rosenthal, 1975; Lerner and Bergsma, 1978).

One of the most general types of cellular interaction is cell-to-cell communication (Sheridan, 1976; Bennett, 1977; Gilula, 1978; Loewenstein, 1979). This property appears to be expressed by almost all cells in both vertebrate and invertebrate organisms. Cell-to-cell communication can be defined as the transfer of low-molecular weight substances from cell-to-cell via a specialized low-resistance pathway. The low-resistance pathway appears to be a cell surface membrane specialization that is called the gap junction or nexus (Gilula, 1977). In essence, junctional pathways are established between cells to permit the exchange of small cytoplasmic molecules or to facilitate the propagation of an action potential from cell-to-cell (Bennett, 1977). An in-depth treatment of this subject can be found in several published volumes (Cox, 1974; De Mello, 1977; Feldman *et al.*, 1978).

II. BIOLOGY OF CELL-TO-CELL COMMUNICATION

A. *Ionic Coupling*

Communication between cells via a low-resistance pathway was first described physiologically as an electrical or electrotonic synapse between excitable cells in the invertebrate nervous system (Furshpan and Potter, 1959) and the mammalian myocardium (Woodbury and Crill, 1961; Weidmann, 1966). The cell-to-cell pathways were defined as "low resistance" because a current pulse could be transmitted from one cell to the next with very little voltage attenuation. In addition, it was observed that the presence of such pathways permitted the rapid propagation of electrical impulses between excitable cells. In the nervous system, this electrical synaptic property has obvious temporal advantages over chemical synapses for required rapid responses (Bennett, 1977), and in the myocardium, the electrotonic synapse provides the fundamental mechanism for synchronizing the cells into a functional syncytium (Weidman, 1952). Similar pathways were also discovered between non-excitable epithelial cells (Loewenstein *et al.*, 1965; Loewenstein, 1966; Furshpan and Potter, 1968; Johnson and Sheridan, 1971). In addition to the low-resistance spread of current, the pathways were also defined as permeable to injected low molecular weight dyes (Loewenstein, 1966; Sheridan, 1976) such as sodium fluorescein (M. Wt. 330). In the past 15 years, these low-resistance pathways have been described between cells from a wide range of animal organisms and higher plants, both *in vivo* and in culture (for reviews, see Bennett, 1977; Loewenstein, 1979). At present, it is possible to conclude that most animal cells can form low-resistance pathways, with the notable exception of circulating erythroid and lymphoid cells. In this regard it is interesting to note that there have been several reports of low-resistance pathways between lymphoid cells (Hülser and Peters, 1972; Oliveira-Castro and Dos Reis, 1977); however, these reports have remained ambiguous because of a failure to demostrate communication between these cells by other independent methods (Cox *et al.*, 1976).

On the basis of the physiological observations, it is possible to project the following properties for the low-resistance pathway. (1) The pathways are specialized regions of the cell surface membranes with resistance properties that are similar to cytoplasm (10-100 Ω cm^2) rather than the nonjunctional cell surface membrane (10^6-10^8 Ω cm^2). (2) The current that is passed from cell-to-cell is probably in the form of inorganic ions such as Na^+, K^+, Cl^-, etc. (3) The pathways must have finite dimensions or channels with appropriate polar or hydrophilic properties that would permit the movement of inorganic ions with these hydrated molecular dimensions. Channels of 12-15 A in diameter would be required for this purpose.

B. *Metabolic Coupling*

Communication has also been detected between cells in culture as the contact-dependent transfer of radiolabeled metabolites (Subak-Sharpe *et al.*, 1969; Cox *et al.*, 1970; Gilula *et al.*, 1972; Cox *et al.*, 1974; Pitts and Simms, 1977). This type of communication can be demonstrated by utilizing one cell type with a known metabolic capacity as a pre-radiolabeled donor and another cell type without the same metabolic capacity as a recipient. After co-culturing the donor and recipient, the communication of this metabolic capacity is revealed in autoradiographs as a contact-dependent transfer of radiolabeled molecules to the recipient. In the initial studies, recipient cells were deficient in inosinic pyrophosphorylase activity (IPP^- or $HGPRT^-$) and the donor cells were prelabeled with the exogenous purine ^3H-hypoxanthine. Thus, the donor cells were able to "rescue" the recipient cells in HAT medium by transferring an important metabolite related to purine metabolism. In subsequent studies data was obtained to indicate that a small molecule, presumably a nucleotide, was transferred and not a large polynucleotide or enzyme-related macromolecule (Pitts, 1977; Pitts and Simms, 1977). Recently, an approach has been developed using ^{3}H–uridine non-specifically to demonstrate contact-dependent nucleotide transfer or exchange between metabolically competent cells (Pitts and Simms, 1977).

C. *Ionic Coupling and Metabolic Coupling*

A close relationship between ionic and metabolic coupling has been established in two separate studies (Gilula *et al.*, 1972; Azarnia *et al.*, 1972). In co-culture combinations of communication-competent and communication-defective cells, both ionic and metabolic coupling were examined. In co-cultures comprised of two communication-competent cell types, both ionic and metabolic coupling were present. However, ionic and metabolic coupling were not detected between communication-competent and communication-defective cells. Therefore, it was possible to conclude that both contact-dependent phenomena can occur simultaneously, and both require the presence of a similar pathway. Although likely, it was impossible to conclude from those studies that both inorganic ions and metabolites are transmitted through the same pathway. The gap junction was implicated as the structural pathway for communication in the co-cultures since it was not detected between cells that failed to communicate, whereas it was readily detected between cells that were able to communicate (Gilula *et al.*, 1972).

III. JUNCTIONAL PERMEABILITY

The permeability properties of low-resistance junctions have been primarily analyzed by utilizing: (1) dye injections (visualization of the cell-to-cell transfer of fluorescence or color) (Loewenstein, 1966; Sheridan, 1976); (2) injections of radiolabeled molecules (visualization of cell-to-cell transfer in autoradiographs or direct chromatographic analysis in tissue slices) (Rieske *et al.*, 1975; Tsien and Weingart, 1976); (3) transfer of radiolabeled molecules between cells in co-cultures (visualization in autoradiographs) (Subak-Sharpe *et al.*, 1969; Pitts, 1977); and (4) injections of synthetic molecules conjugated to fluorescein or rhodamine (visualization of fluorescence movement) (Simpson *et al.*, 1977). On the basis of the information from these approaches, several general statements can be made. (1) Cell-to-cell transfer appears to be a relatively passive process; there has been no direct demonstration that energy is directly required for channel function. (2) Cell-to-cell transfer occurs at a slow rate, on the order of passive cytoplasmic diffusion. (3) Junctional permeability appears to be determined primarily on the basis of molecular size, and perhaps charge. Molecules of 1200 daltons and smaller can pass through the junctional channels, whereas larger molecules cannot be transferred from cell-to-cell (Simpson *et al.*, 1977). In essence, the junctional channels appear to function as a molecular sieve. These data suggest that a large number of cytoplasmic molecules, such as inorganic ions, amino acids, nucleotides, sugars, etc., can move from cell-to-cell; however; macromolecular species, such as proteins, nucleic acids, polysaccharides, etc., cannot move through these pathways.

Three major physiological parameters have been implicated in regulating junctional permeability: (1) intracellular calcium concentration (Rose and Loewenstein, 1975; Rose and Rick, 1978), (2) intracellular pH (Turin and Warner, 1977) and (3) voltage-dependence (Spray *et al.*, 1979). It appears that all three parameters can influence junctional permeability in specific biological systems, and none of these parameters, except for pH, appear to have widespread regulatory effects. Both $[Ca^{++}]$ and pH can cause a complete loss of permeability. However, the $[Ca^{++}]$ effect is usually irreversible whereas the pH effect is readily reversible (Rose and Loewenstein, 1975; Turin and Warner, 1977; Rose and Rick, 1978). The voltage-dependence is characterized by a reversible modulation in the permeability or junctional resistance, without a complete loss in function (Spray *et al.*, 1979).

IV. STRUCTURAL BASIS FOR CELL-TO-CELL COMMUNICATION

A specific cell membrane specialization, the gap junction, or nexus, has been strongly implicated as a structural pathway for communication on the basis of several experimental observations. (1) The gap junction is present at the site of electrical synaptic activity between neurons in the nervous system (Bennett, 1972; Sotelo, 1977). (2) The gap junction connects adjacent myocardial cells for the electrotonic propagation of impulses in the myocardium (Dewey and Barr, 1964; Dreifuss *et al.*, 1966; Barr *et al.*, 1968). This was elegantly demonstrated by utilizing selective cell contact dissociation procedures such as EDTA treatment or perfusion with hypertonic sucrose; for example, EDTA treatment disrupts myocardial desmosome contacts but not gap junctions (or electrotonic coupling) without affecting desmosomes (Dreifuss *et al.*, 1966). (3) The gap junction is consistently present *in vivo* and in culture between cells that are communicating (for reviews, see Sheridan, 1976;

Gilula, 1977; Loewenstein, 1979). (4) The gap junction is not expressed by a mouse cell culture line that is not able to communicate (Gilula *et al.*, 1972; Azarnia *et al.*, 1974). This mouse A9 cell line is currently used as the universal "communication-defective" fibroblast. These observations have provided a substantial basis for gap junctional involvement in communication; however, other junctional structures, such as tight junctions or septate junctions, may also serve as low-resistance pathways. It has not yet been possible to resolve this issue since these junctions invariably coexist with gap junctions.

The gap junction was initially resolved by Revel and Karnovsky in 1967 (for review of early history see McNutt and Weinstein, 1973). It is currently synonymous with the structure that was called the nexus by Dewey and Barr in 1962. In thin-section electron microscopy, the gap junction can be detected as a unique apposition between adjacent cells. At the site of contact, the junction can be resolved as a seven-layered (septilaminar) structure. The entire width of the septilaminar structure is 15-19 nm. The septilaminar image represents the parallel apposition of two 7.5 nm unit membranes that are separated by a 2-4 nm 'gap' or electron-lucent space. This thin-section appearance led to the use of the term 'gap junction' to describe this structure.

The precise clarification of the gap junctional structure in thin sections relied on the use of electron-opaque materials, or tracer substances, that are able to fill the extracellular space. Currently, colloidal lanthanum hydroxide, pyroantimonate, and ruthenium red can all be utilized for this 'tracing' or 'staining' purpose (Revel and Karnovsky, 1967; Payton *et al.*, 1969; McNutt and Weinstein, 1970; Martinez-Palomo, 1970; Friend and Gilula, 1972). The tracer substances are capable of penetrating a central region of the junction that corresponds to the location of the 'gap'. In oblique or *en face* views of tracer-impregnated gap junctions, it is possible to visualize a unique polygonal lattice of 7-8nm subunits. The tracer outlines the subunits, which have a 9-10 nm center-to-center spacing, as a result of penetrating the regions of the lattice that are continuous with the extracellular space (Revel and Karnovsky, 1967). A 1.5-2 nm electron-dense dot is frequently present in the central region of these subunits. A similar lattice was described by Robertson (1963) at the site of a electrotonic synapse in a study that preceded the use of the tracer approaches. When gap junctions have been examined in detergent-treated isolated membrane fractions with negative stain procedures, a similar polygonal lattice of subunits has been observed (Beneditti and Emmelot, 1965, 1968; Goodenough and Revel, 1970; Goodenough and Stoeckenius, 1972; Goodenough, 1974, 1976; Hertzberg and Gilula, 1979).

The freeze-fracture technique has been utilized to obtain important complementary information about the gap junctional structure. The freeze-fracture procedure provides detailed information about the internal content of the junctional membranes. In general, specialized membranes, such as those present at the sites of cell junctions, have significant internal membrane structural modifications (for review, see McNutt and Weinstein, 1973; Staehelin, 1974). The freeze-fractured gap junctional membranes contain two complementary membrane halves or fracture faces. The cytoplasmic or inner membrane half contains a polygonal lattice of homogeneous 7-8 nm intramembrane particles, and the outer membrane half contains a complementary arrangement of pits or depressions. In many instances, a 2-2.5 nm dot is detectable in the central region of these junctional particles. These fracture face characteristics and membrane particle dispositions have now been documented as a constant feature of most non-arthropod gap junctions that have been examined (Goodenough and Revel, 1970; Chalcroft and Bullivant, 1970; McNutt and Weinstein, 1970; Friend and Gilula, 1972; Staehelin, 1974). The junctional membrane lattices can exist in a variety of pleiomorphic forms, but the variations surround a single theme; a plaque-like or localized (focal) contact between interacting cells. Gap junctions are usually present as oval or circular plaques; however, a variety of forms, including linear strands (Raviola and Gilula, 1973) have been reported (Larsen, 1977).

The gap junctions are structurally resistant to treatments with proteases and other agents that are used to dissociate cells (Muir, 1967; Berry and Friend, 1969; Amsterdam and Jamieson, 1974). However, there has been one satisfactory procedure reported for 'splitting' or separating the gap junctional membranes in intact tissues. This procedure involves the perfusion of tissues with hypertonic sucrose solutions (Barr *et al.*, 1965, 1968; Dreifuss *et al.*, 1966; Goodenough and Gilula, 1974). In intact mouse liver, the junctional membranes are separated by this treatment somewhere in the central region of the extracellular 'gap' (Goodenough and Gilula, 1974). The separated junctional membranes still contain the characteristic particle lattices in freeze-fractured replicas, and the particles appear to be more tightly packed when the membranes are separated.

Gap junctions have been described in a variety of arthropod tissues with both thin-section and freeze-fracture techniques (Flower, 1972; Peracchia, 1973b; Johnson *et al.*, 1973; Satir and Gilula, 1973; Gilula, 1974a; Dallai, 1975). The structural features of the arthropod gap junctions are sufficiently different from non-arthropod gap junctions to be considered as a unique structural variation. In thin sections, the arthropod gap junctions are quite similar to non-arthropod gap junctions, although the intercellular 'gap' is slightly larger (about 3-4 nm) (Payton *et al.*, 1969; Hudspeth and Revel, 1971; Rose, 1971; Peracchia, 1973a). In freeze-fracture replicas, the junctional particles are associated with the outer membrane half (fracture face E); the membrane particles are large (11 nm or larger in diameter) and often heterogeneous in size; and they are frequently present as fused aggregates of two or more particles. In addition, the particles are not usually in a highly ordered polygonal array (Flower, 1972; Johnson *et al.*, 1973; Dallai, 1975; Epstein and Gilula, 1977).

V. STRUCTURE-FUNCTION RELATIONSHIPS IN CELL-TO-CELL COMMUNICATION

There have been several attempts to relate the structure of gap junctions directly to low resistance physiological properties (see Bennett 1973, 1977). In general it has been virtually impossible to relate specific resistances or permeability measurements to the number of presumptive channels that can be detected ultrastructurally. The difficulties have been related primarily to the problem of finding a well-defined system (preferably a two-cell system) where the coupling and other membrane properties are relatively stable, and where the number of gap junctional particles can be reliably quantitated.

Recently, there have been several studies that have attempted to correlate particle lattice spacings within the junctions to a physiological low or high-resistance state (Peracchia and Dulhunty, 1976; Peracchia, 1978; Peracchia and Peracchia, 1978; Raviola *et al.*, 1979). Although there has not yet been an adequate integration of structural and physiological sampling, the initial experiments have led to the hypothesis that the junctional particles are tightly packed in the high-resistance or uncoupled conditions, and the particles are loosely packed in the low-resistance or coupled condition.

The precise location of the low-resistance channel in the gap junctional lattice has never been demonstrated with either ultrastructural or molecular probes. However, on the basis of the ultrastructural appearance of the gap junctional lattice, it has been suggested that the channel is located in the central region of the junctional particles (Payton *et al.*, 1969; McNutt and Weinstein, 1970). More recently, continuous electron-dense 'channels' at this site have been observed in negative stain preparations of isolated gap junctions (Gilula, 1974b; Goodenough, 1976; Hertzberg and Gilula, 1979). The penetration of aqueous stain into this region of the junctional particles and the recent evidence obtained by X-ray crystallography (Makowski *et al.*, 1977) is certainly consistent with the localization of a polar 'channel' within the junctional lattice.

VI. BIOCHEMICAL AND BIOPHYSICAL CHARACTERIZATION OF GAP JUNCTIONS

Gap junctions have been isolated as enriched subcellular fractions primarily from rat and mouse liver (Benedetti and Emmelot, 1968; Goodenough and Stoeckenius, 1972; Evans and Gurd, 1972; Gilula, 1974b; Duguid and Revel, 1975; Goodenough, 1974, 1976; Culvenor and Evans, 1977; Erhart and Chauveau, 1977). The gap junctional fractions have been obtained by starting with a plasma membrane fraction and reducing the nonjunctional membrane elements by treatment with detergents, such as deoxycholate or sarkosyl. Unfortunately, the detergent-resistant fraction that contains the gap junctions, is heavily contaminated with several elements (such as collagen, amorphous protein, uricase crystals) that interfere with the final enrichment on sucrose density gradients. Therefore, in most cases enzymatic treatments have been used to reduce the contaminating material; collagenase and hyaluronidase (both containing some non-specific protease activity) have frequently been employed for this purpose. With one exception (Hertzberg and Gilula, 1979), all of the gap junctional fractions that have been isolated from liver have been exposed to enzymatic treatment, so the reported polypeptide contents may be significantly altered by this treatment (see Duguid and Revel, 1975, for a discussion of the proteolysis). Since no endogenous activity has been detected in the junctional fraction, ultrastructural analysis (with thin sections and negative staining) is the major criterion for purity.

In the isolated gap junction both protein and lipid are present, but no carbohydrate has been reported. In spite of the possible proteolysis, it appears that a 25-27,000 daltons polypeptide can be detected, together with other polypeptides, as a prominent component in both mouse and rat liver preparations. Using a nonenzymatic isolation procedure that has been developed recently, two prominent polypeptides, 47,000 and 27,000 daltons, have been found in rat liver gap junctions (Hertzberg and Gilula, 1979). The isolated gap junctions contain all the major membrane phospholipids, with the exception of sphingomyelin, and some neutral lipid (Goodenough and Stoeckenius, 1972; Hertzberg and Gilula, 1979).

The isolated mouse liver gap junctions have been analyzed by correlated electron microscopy and X-ray diffraction (Casper *et al.*, 1977;

Makowski *et al.*, 1977) and both meridional and equatorial reflections have been phased. From the meridional reflections, the authors interpret the electron density profile to indicate the presence of two lipid bilayers (42 A peak-to-peak) that are spanned by protein, with protein also present in the 'gap'. In addition, gap junctional subunits with a six-fold substructure have been deduced from the equatorial reflections.

VII. ASSEMBLY AND TURNOVER OF GAP JUNCTIONS

There have been several recent studies both *in vivo* and in culture that have described a sequence of structural events that are associated with the formation of gap junctions (Revel *et al.*, 1973; Johnson *et al.*, 1974; Decker and Friend, 1974; Benedetti *et al.*, 1974; Albertini and Anderson, 1974; Decker, 1976). The formation process in freeze-fracture replicas generally consists of the following stages: (1) the appearance of formation plaques; (2) the appearance of large 'precursor' particles with a reduction of the intercellular space; (3) the appearance of smaller 'junctional' particles in polygonal arrangements; and (4) the enlargement of junctions. There is no information available yet about the biochemical or structural relationship of the large precursor particles to the smaller gap junctional particles. Such junctional precursors, however, probably do exist in intact cells since it has been difficult, if not impossible, in most systems to inhibit the generation of communication between cells by inhibiting protein synthesis (Epstein *et al.*, 1977; Sheridan, 1978). There has been only one reported exception (Decker, 1976) and, in this system, the formation of gap junctions that is apparently hormonally induced is inhibited by inhibiting protein and RNA synthesis.

In several studies, it has been reported that the gap junctional size and frequency may be influenced by treatments with hormones (Merk *et al.*, 1972; Decker, 1976; Burghardt and Anderson, 1979), proteases (Orci *et al.*, 1973; Shimono and Clementi, 1977; Polak-Charcon *et al.*, 1978), vitamins (Elias and Friend, 1976) and hepatectomy (Yee and Revel, 1978). In epithelial systems, there is also an apparent structural relationship between gap junctional particles and tight junctional elements during the formation process; in some cases the two junctional elements arise in close association with each other (Yee, 1972; Decker and Friend, 1974; Albertini *et al.*, 1975; Montesano *et al.*, 1975; Yee and Revel, 1978; Polak-Charcon *et al.*, 1978). This raises the interesting possiblity that there may be a biochemical and/or functional relationship between gap and tight junctions.

Gap junctions must be regarded as unusually stable (or static) elements of cell surface plasma membranes. This statement can be made primarily on the basis of a study on the relative degradation of mouse liver surface membrane proteins (Gurd and Evans, 1973). This study indicated that the degradation of proteins in a subcellular fraction containing gap junctions occurred very slowly. Gap junctions have been frequently observed as internalized elements or 'annular gap junctions' in a number of different tissues (Merk *et al.*, 1973; Albertini *et al.*, 1975; Amsterdam *et al.*, 1976), and in some instances the internalized structures are incorporated into phagolysosomal vacuoles (Larson, 1977).

VIII. SPECIFICITY OF JUNCTIONAL COMMUNICATION

The specificity of communication between cells from different tissues and from different organisms has been examined in considerable detail in the past few years. Unlike the specificities associated with cellular interactions in the immune system, communication appears to occur between all communication-competent cells if: (1) the cells contain the same non-arthropod gap junctional structure; and (2) there are no physical constraints to permitting interactions between cells. This property of communication has been defined by using co-culture approaches with cells from a variety of different organisms. Thus far, all communication-competent cells from vertebrate organisms that have been examined can communicate readily with each other in co-culture conditions (Michalke and Loewenstein, 1971; Epstein and Gilula, 1977). The only reported exceptions may be due to cellular arrangements or clonal morphology in the culture (Fentiman *et al.*, 1976; Pitts and Burk, 1976). Arthropod cells have provided an interesting exception with respect to communication specificity (Epstein and Gilula, 1977). Cells from arthropod organisms will not communicate in culture with cells from non-arthropod organisms or with cells from arthropod organisms in different orders. These cells can only communicate with cells from organisms within the same phylogenetic order.

IX. CO-CULTURE APPROACH FOR STUDYING COMMUNICATION PROPERTIES

A model co-culture system was generated in order to study the potential role of cell communication in the transmission of hormonal stimulation in multicellular systems (Lawrence *et al.*, 1978). Two hormonally responsive cell phenotypes were selected for this purpose: the follicle stimulating hormone (FSH) responsive rat ovarian granulosa cell and the catecholamine responsive mouse neo-natal ventricular myocardial cell. In initial experiments, the two cell types were fully characterized for both hormonal responsiveness and communication competency. The rat ovarian granulosa cells were treated with a saturating dose of FSH and their response was measured as either the production of a secreted cell product, plasminogen activator, or a morphological change in cell shape (Lawrence *et al.*, 1979). Both FSH-stimulated responses are preceded by a dose-dependent increase in intracellular cyclic AMP; this effect is potentiated by cyclic nucleotide phosphodiesterase inhibitors and it is mimicked by dibutyryl cAMP. The mouse ventricular myocardial cells were stimulated with the catecholamine, nor-epinephrine, and the response was measured as an increase in beat frequency or as a change in the action potential properties. Upon treatment with nor-epinephrine, there is a significant increase in the amplitude and a decrease in the duration of the action potential. The catecholamine stimulation is accompanied by a rapid elevation of intracellular cyclic AMP. This stimulation can be potentiated by cyclic nucleotide phosphodiesterase inhibitors, such as theophylline and methylisobutylxanthine; and it can be mimicked by dibutyryl cyclic AMP. The communication properties of the two cell types were examined in both homologous and heterologous cultures. Both cell types could be characertized as communication-competent by ionic coupling, metabolic coupling, and the presence of gap junctional contacts. In fact, communication was as prominent between heterologous cells as it was between homologous cells. The transmission of hormonal communication between heterologous cells was examined initially by treating myocardial cells cultured alone or in contact with granulosa cells with FSH. The myocardial cells cultured alone had no detectable response to FSH, whereas the myocardial cells in contact with the granulosa cells had a marked increase in beat frequency together with the characteristic stimulated action potential properties. In a similar fashion, it was possible to stimulate the production of plasminogen activator by treating granulosa cells in contact with myocardial cells with nor-epinephrine. In both cases, the transmission of hormonal stimulation was directly dependent on direct contact between the heterologous cell type, and the observed effect does not appear to be related to either a hormone-induced change in electrical properties of the cell membrane or to a movement of hormone receptors between the heterologous cells. In this system both hormonal responses are cyclic AMP dependent; therefore, the cyclic nucleotide is a likely candidate for the mediator of hormonal stimulation.

X. COMMUNICATION DURING GROWTH AND DIFFERENTIATION

A. *Oogenesis*

The mammalian ovarian follicle represents a complete cycle of development and differentiation that involves gap junctional interactions between heterologous cell types. We utilized both electrophysiological and morphological techniques to establish that the granulosa cells of the cumulus oophorus communicate extensively with the oocyte at certain stages of follicular development (Gilula *et al.*, 1978). In addition, the extent of communication (ionic coupling) between the oocyte and cumulus cells was examined in preovulatory follicles during the ten-hour interval prior to ovulation. We observed that ionic coupling progressively decreased and was terminated by the time of ovulation. This temporal pattern of terminating communication between the cumulus and the oocyte may indicate that communication provides a mechanism for regulating the maturation of the oocyte during follicular development prior to ovulation.

B. *Embryogenesis*

Communication has been examined during mouse embryogenesis in both preimplantation and postimplantation stages (Lo and Gilula, 1978; Lo, 1979). In addition, the communication and differentiation properties of a mouse embryonal teratocarcinoma cell line, PCC4azal, were characterized for a potential *in vitro* reconstitution experiment (Mintz and Illmensee, 1975) in the future.

The onset of communication in the preimplantation mouse embryo was examined by monitoring ionic coupling, the transfer of injected fluorescein (330 daltons) and horseradish peroxidase (40,000 daltons) (Lo and Gilula, 1979a, 1979b). In the two-cell, four-cell, and pre-compaction eight-cell embryos, cytoplasmic bridges between sister blastomeres were responsible for ionic coupling and the transfer of injected fluorescein and horseradish peroxidase. In contrast, no communication was observed between blastomeres from different sister pairs. Junction mediated communication was unequivocally detected for the first time in the embryo at the early compact stages (late eight-cell embryo), and it was retained between all cells throughout compaction. At the blastocyst stage, trophoblast cells of the blastocyst were linked by junctional channels to other trophoblast cells, as well as to cells of the inner cell mass as indicated by the spread of injected fluorescein. In the postimplantation mouse embryos studies *in vitro,* both trophoblasts and cells of the inner cell mass (ICM) were ionically coupled and dye spread to one another. In older and more developed embryos, the spread of injected dye was more limited even though ionic coupling extended beyond these apparent boundaries. This partial segregation of cell-to-cell communication as indicated by the limited dye spread may parallel specific differentiation processes, in particular that of giant trophoblast, embryonic ectoderm and extraembryonic endoderm differentiation.

Culture conditions were defined for the spontaneous differentiation of PCC4azal embryonal carcinoma cells to endoderm-like cells and giant cells (Lo, 1979). Various enzymatic and antigenic markers, such as alkaline phosphatase, acid phosphatase, plasminogen activator, lactate dehydrogenase, stage specific embryonic antigen–1 (SSEA–1), and the histocompatibility antigen ($H–2^k$) were examined on the three different cell populations in this system. The ultrastructural morphology of the cells was also characterized and compared with the cells of the embryoid bodies. It was found that the ultrastructure of the PCC4azal cells is strikingly similar to the cells of the embryoid body. Finally, communication was expressed by all three PCC4azal cell types as indicated by ionic coupling, dye transfer, metabolic coupling, and the presence of gap junctions.

C. *Histogenesis*

Cell communication has been examined in chick embryonic myogenesis in culture and *in vivo* (Kalderon *et al.,* 1977). In a previous study (Rash and Fambrough, 1973), it was suggested that a gap junctional type of communication may play an important role in the fusion of myoblasts into multinucleated myotubes. We have recently found that prefusion myoblasts can interact via gap junctions, ionic coupling, and metabolic coupling. Communication was also examined in fusion-arrested cultures to determine its potential relationship to fusion competency. In cultures that were fusion-arrested by treatment with either 1.8mM EGTA, 3.3×10^{-6} M BUdR, or 1μg/ml cycloheximide, both gap junctions and ionic coupling were present. Therefore, it is possible to conclude that cell communication is not a sufficient property by itself, to generate fusion between myoblasts. Also, we have studied the membrane events involved in fusion in the same cultures with ultrastructural techniques (Kalderon and Gilula, 1979). The multinucleated muscle cells are generated by the fusion of two plasma membranes from adjacent cells, apparently by forming a single bilayer that is particle-free in freeze-fracture replicas. This single bilayer subsequently collapses, and cytoplasmic continuity is established between the cells. Gap junctional elements are not detectable participants in the membrane events that are associated with this process. Currently, we are studying the possibility that communication provides an important regulatory or recognition mechanism for the preparatory events associated with muscle fusion.

D. *Organogenesis*

The chick embryonic otocyst has been examined to determine the pattern of gap junctional communication that is associated with the differentiation of the epithelium (support and sensory cells) and its concomitant innervation (Ginzberg and Gilula, 1979). The differentiation of the epithelium is accompanied by alterations in the distribution of cell junctions. Prior to innervation of the epithelium, gap junctions existed between all cells. Upon differentiation, gap junctions are present only between support cells. Thus, some of the gap junctions that join homogeneous epithelial cells prior to innervation are removed as sensory cells differentiate, and a separate population of large gap junctions is formed between differentiating support cells. The morphological evidence suggests two possible mechanisms that may be responsible for the observed changes in gap junctional distribution: removal of gap junctions by internalization, and formation of gap junctions by aggregation of precursor particles. The temporal correlation between junctional modulation, cytological differentiation of sensory and support cells, and ingrowth of nerve fibers makes the latter event a likely developmental cue for differentiation of this epithelium.

XI. SUMMARY

In the past 10-15 years, there has been considerable progress in defining the property of cell-to-cell communication in many different tissues and organsims. Although most of the information has been descriptive, it has provided an overall framework for beginning to understand the potential regulatory role of communication in development and differentiation. Recent studies that have focused on the communication patterns during specified events in differentiation have provided access to understanding how communication might perform its function(s) and how it may be modulated. In this presentation, I have selected a few limited examples of this modulation that we have studied in our laboratory. Now it will be imperative to integrate genetic and immunological approaches, together with biochemistry and physiology, to determine the precise role(s) of communication in development.

ACKNOWLEDGEMENTS

This work has been supported by grant (HL 16507 and GM 24753) from the National Institute of Health. N.B.G. is the Recipient of a Research Career Development Award (HL 00110).

REFERENCES

Albertini, D.F. and Anderson, E. (1974). *J. Cell Biol.* **63**, 234-250.

Albertini, D.F., Fawcett, D.W. and Olds, P.J. (1975). *Tissue and Cell* **7**, 389-405.

Amsterdam, A. and Jamieson, J.D. (1974). *J. Cell Biol.* **63**, 1037-1056.

Amsterdam, A., Josephs, R., Lieberman, M.E. and Lindner, H.R. (1976). *J. Cell Sci.* **21**, 93-105.

Azarnia, R., Michalke, W. and Loewenstein, W.R. (1972). *J. Membr. Biol.* **10**, 247-258.

Azarnia, R., Larsen, W.J. and Loewenstein, W.R. (1974). *Proc. Nat. Acad. Sci. U.S.A.* **71**, 880-884.

Baldwin, K.M. (1979). *J. Cell Biol.* **82**, 66-75.

Barr, L., Dewey, M.M. and Berger, W. (1965). *J. Gen. Physiol.* **48**, 797-823.

Barr, L., Berger, W. and Dewey, M.M. (1968). *J. Gen. Physiol.* **51**, 347-368.

Benedetti, E.L. and Emmelot, P. (1965). *J. Cell Biol.* **26**, 299-305.

Benedetti, E.L. and Emmelot, P. (1968). *J. Cell Biol.* **38**, 15-24.

Benedetti, E.L., Dunia, I. and Bloemendal, H. (1974). *Proc. Nat. Acad. Sci. U.S.A.* **71**, 5073-5077.

Bennett, M.V.L. (1972). *In* "Structure and Function of Synapses" (G.D. Pappas and D.P. Purpura, eds.), pp. 221-256. Raven, New York.

Bennett, M.V.L. (1973). *Fed. Proc.* **32**, 65-75.

Bennett, M.V.L. (1977). *In* "Handbook of Physiology" (E.R. Kandel, ed.), I. The Nervous System, Sect. I: Cellular Biology of Neurons, pp. 357-416. Williams and Wilkins, Baltimore, Maryland.

Berry, M.N. and Friend, D.S. (1969). *J. Cell Biol.* **43**, 506-520.

Burghardt, R.C. and Anderson, E. (1979). *J. Cell Biol.* **81**, 104-114.

Caspar, D.L.D., Goodenough, D.A., Makowski, L. and Phillips, W.C. (1977). *J. Cell Biol.* **74**, 605-628.

Chalcroft, J.P. and Bullivant, S. (1970). *J. Cell Biol.* **47**, 49-60.

Cox, R.P. (1974). "Cell Communication." John Wiley and Sons, Inc., New York.

Cox, R.P., Kraus, M.R., Balis, M.E. and Dancis, J. (1970). *Proc. Nat. Acad. Sci. U.S.A.* **67**, 1573.

Cox, R.P., Kraus, M.R., Balis, M.E. and Dancis, J. (1974). *In* "Cell Communication" (R.P. Cox, ed.), pp. 67-96. John Wiley and Sons, Inc., New York.

Cox, R.P., Kraus, M.R., Balis, M.E. and Dancis, J. (1976). *Exp. Cell Res.* **101**, 411-414.
Culvenor, J.G. and Evans, W.H. (1977). *Biochem. J.* **168**, 475-481.
Dallai, R. (1975). *Submicr. Cytol.* **7**, 249-257.
Decker, R.S. (1976). *J. Cell. Biol.* **69**, 669-685.
Decker, R.S. and Friend, D.S. (1974) *J. Cell. Biol.* **62**, 32-47.
DeMello, W.C. (1977). "Intercellular Communication," Plenum, New York.
Dewey, M.M. and Barr, L. (1962). *Science* **137**, 670-672.
Dewey, M.M. and Barr, L. (1964). *J. Cell Biol.* **23**, 553.
Dreifuss, J.J., Girardier, L. and Forssman, W.G. (1966). *Pflugers Arch.* **292**, 13-33.
Duguid, J.R. and Revel, J.P. (1975). *Symp. Quant. Biol.* **XL**, 45-47.
Ehrhart, J.C. and Chaveau, J. (1977). *FEBS Lett.* **78**, 295-299.
Elias, P.M. and Friend, D.S. (1976). *J. Cell Biol.* **68**, 173-188.
Epstein, M.L. and Gilula, N.B. (1977). *J. Cell Biol.* **75**, 769-787.
Epstein, M.L., Sheridan, J.D., and Johnson, R.G. (1977). *Exp. Cell Res.* **104**, 25-30.
Evans, W.H. and Gurd, J.W. (1972) *Biochem. J.* **128**, 691-700.
Feldman, J., Gilula, N.B. and Pitts, J.D. (1978). "Intercellular Junctions and Synapses," Receptors and Recognition, Series B, Vol. II, Chapman and Hall, London.
Fentiman, I., Taylor-Papadimitriou, J. and Stoker, M. (1976). *Nature* **264**, 760-762.
Flower, N.E. (1972). *J. Cell Sci.* **10**, 683-691.
Friend, D.S. and Gilula, N.B. (1972). *J. Cell Biol.* **53**, 758-776.
Furshpan, E.J. and Potter, D.D. (1959). *J. Physiol.* **143**, 289-325.
Furshpan, E.J. and Potter, D.D. (1968). *Curr. Top. Develop. Biol.* **3**, 95-127.
Gilula, N. B. (1974a). *In* "Cell Communication" (R.P. Cox, ed.) pp. 1-29, John Wiley and Sons, New York.
Gilula, N.B. (1974b). *J. Cell Biol.* **63**, 111a.
Gilula, N.B. (1977). *In* "International Cell Biology (B.R. Brinkley and K.R. Porter, eds.), pp. 61-69, Rockefeller University Press, New York.
Gilula, N.B., Reeves, O.R. and Steinbach, A. (1972). *Nature* **235**, 262-265.
Gilula, N.B., Epstein, M.L. and Beers, W.H. (1978). *J. Cell Biol.* **78**, 58-75.
Ginzberg, R.D. and Gilula, N.B. (1979). *Develop. Biol.* **68**, 110-129.
Goodenough, D.A. (1974). *J. Cell Biol.* **61**, 557-563.
Goodenough, D.A. (1976). *J. Cell Biol.* **68**, 220-231.
Goodenough, D.A. and Revel, J.P. (1970). *J. Cell Biol.* **45**, 272-290.
Goodenough, D.A. and Stoeckenius, W. (1972). *J. Cell Biol.* **58**, 646-656.
Goodenough, D.A. and Gilula, N.B. (1974). *J. Cell Biol.* **61**, 575-590.
Gurd, J.W. and Evans, W.H. (1973). *Eur. J. Biochem.* **36**, 273-279.
Hertzberg, E.L. and Gilula, N.B. (1979). *J. Biol. Chem.* **254**, 2138-2147.
Hudspeth, A.J. and Revel, J.P. (1971). *J. Cell Biol.* **50**, 92-101.
Johnson, R.G. and Sheridan, J.D. (1971). *Science* **174**, 717-719.
Johnson, R.G., Herman, W.S. and Preus, D.M. (1973). *J. Ultrastruct. Res.* **43**, 298-312.
Johnson, R.G., Hammer, M., Sheridan, J., and Revel, J.P. (1974). *Proc. Nat. Acad. Sci. U.S.A.* **71**, 4536-4540.
Kalderon, N., Epstein, M.L. and Gilula, N.B. (1977), *J. Cell Biol.* **75**, 788-806.
Kalderon, N. and Gilula, N.B. (1979). *J. Cell Biol.* **81**, 411-425.
Larsen, W.J. (1977). *Tissue and Cell* **9**, 373-394.
Lawrence, T.S., Beers, W.H. and Gilula, N.B. (1978). *Nature* **272**, 501-506.
Lawrence, T.S., Ginzberg, R.D., Gilula, N.B. and Beers, W.H. (1979). *J. Cell Biol.* **80**, 21-36.
Lerner, R.A. and Bergsma, D. (eds.) (1978). "The Molecular Basis of Cell-Cell Interaction," Alan R. Liss, Inc., New York.
Lo, C.W. (1979). Ph.D. Thesis. The Rockefeller University.
Lo, C.W. and Gilula, N.B. (1978). *J. Cell Biol.* **79**, 18a.

Lo, C.W. and Gilula, N.B. (1979a). Cell., **18**, 399-409.
Lo, C.W. and Gilula, N.B. (1979b). Cell., **18**, 411-422.
Loewenstein, W.R. (1966). *Ann. N.Y. Acad. Sci.* **137**, 441-472.
Loewenstein, W.R. (1979). *Biochem. Biophys. Acta.* **560**, 1-65.
Loewenstein, W.R., Socolar, S.J., Higashino, S., Kanno, Y. and Davidson, N. (1965). *Science* **149**, 295-298.
Makowski, L., Caspar, L.D., Phillips, W.C. and Goodenough, D.A. (1977). *J. Cell Biol.* **74**, 629-645.
Martinez-Palomo, A. (1970). *Lab. Invest.* **22**, 605-614.
Maslow, D.E., (1976). *In* "The Cell Surface in Animal Embryogenesis and Development." Cell Surface Reviews, Vol. I, pp. 697-747 (G. Poste and G.L. Nicolson, eds.) North-Holland Publ. Co. New York.
McNutt, N.S. and Weinstein, R.S. (1970). *J. Cell Biol.* **47**, 666-687.
McNutt, N.S. and Weinstein, R.S. (1973). *Prog. in Biophys. and Mol. Biol.* **26**, 45-101.
Merk, F.B., Botticelli, C.R. and Albright, J.T. (1972). *Endocrinol.* **90**, 992-1007.
Merk, F.B., Albright, J.T., and Botticelli, C.R. (1973). *Anat. Rec.* **175**, 107-125.
Michalke, W., and Loewenstein, W.R. (1971). *Nature* **232**, 121-122.
Mintz, B. and Illmensee, U. (1975). *Proc. Nat. Acad. Sci. U.S.A.* **72**, 3585-3589.
Montesano, R., Friend, D.S., Perrelet, A. and Orci, L. (1975). *J. Cell Biol.* **67**, 310-319.
Moscona, A.A. (1974). "The Cell Surface in Development." J. Wiley and Sons, New York.
Muir, A.R. (1967). *J. Anat.* **101**, 239-262.
Oliveira-Castro, G.M. and Dos Reis, G.A. (1977). *In* "Intercellular Communication" (W.C. De Mello, ed.), pp. 201-230. Plenum, New York.
Orci, L., Amberdt, M., Henquin, J.C., Lambert, A.E., Unger, R.H. and Renold, A.E. (1973). *Science* **180**, 647.
Payton, B.W., Bennett, M.V.L. and Pappas, G.D. (1969). *Science* **166**, 1641-1643.
Peracchia, C. (1973a). *J. Cell Biol.* **57**, 54-65.
Peracchia, C. (1973b). *J. Cell Biol.* **57**, 66-76.
Peracchia, C. (1978). *Nature* **271**, 669-671.
Peracchia, C. and Dulhunty, A.F. (1976). *J. Cell Biol.* **70**, 419-439.
Peracchia, C. and Peracchia, L.L. (1978). *J. Cell Biol.* **79**, 217a.
Pitts, J.D. (1977). *In* "International Cell Biology" (1976-1977) (B.R. Brinkley, and K.R. Porter, eds.), pp. 43-49. The Rockefeller University Press, New York.
Pitts, J.D. and Burk, R.R. (1976). *Nature* **264**, 762-764.
Pitts, J.D. and Simms, J.W. (1977). *Exp. Cell Res.* **104**, 153-163.
Polak-Charcon, S., Shoham, J. and Ben-Shaul, Y. (1978). *Exp. Cell Res.* **116**, 1-13.
Poste, G., and Nicolson, G.L. (eds.) (1976). "The Cell Surface in Animal Embryogenesis and Development." Cell Surface Reviews, Vol. I. North-Holland Publ. Co., New York.
Rash, J.E. and Fambrough D. (1973). *Develop. Biol.* **30**, 166-186.
Raviola, E. and Gilula, N.B. (1973). *Proc. Nat. Acad. Sci. U.S.A.* **70**, 1677-1681.
Raviola, E., Goodenough, D.A. and Raviola, G. (1978). *J. Cell. Biol.* **79**, 229a.
Revel, J.P. and Karnovsky, M.J. (1967). *J. Cell Biol.* **33**, C7-C12.
Revel, J.P., Yip, P. and Chang, L.L. (1973). *Develop. Biol.* **35**, 302-317.
Rieske, E., Schubert, P. and Dreutzberg, G.W. (1975). *Brain Res.* **84**, 365-382.
Robertson, J.D. (1963). *J. Cell Biol.* **19**, 201-221.
Rose, B. (1971). *J. Membr. Biol.* **5**, 1-19.
Rose, B. and Loewenstein, W.R. (1975). *Nature* **254**, 250-252.
Rose, B. and Rick, R. (1978). *J. Membr. Biol.* **44**, 377-415.
Rosenthal, A.S. (ed.) (1975). "Immune Recognition" Academic Press, Inc., New York
Satir, P. and Gilula, N.B. (1973). *Ann. Rev. Entomol.* **18**, 143-166.

Sheridan, J.D. (1976). *In* "The Cell Surface in Animal Embryogenesis and Development." Cell Surface Reviews, Vol. I (G. Poste and G.L. Nicholson, eds.), pp. 409-443. North-Holland Publ. Co., New York.
Sheridan, J.D. (1978). *In* "Intercellular Junctions and Synapses" (J. Feldman, N.B. Gilula and J.D. Pitts, eds.) pp. 37-60. Chapman and Hall, London.
Shimono, M. and Clementi, F. (1977). *J. Ultrastruct. Res.* **59**, 101.
Simpson, I., Rose, B. and Loewenstein, W.R. (1977). *Science* **195**, 294-295.
Sotelo, C. (1977). *In* "International Cell Biology" (1976-1977) (B.R. Brinkley and K.R. Porter, eds.), pp. 83-92. The Rockefeller University Press, New York.
Spray, D.C., Harris, A.L. and Bennett, M.V.L. (1979). *Science* **204**, 432-433.
Staehelin, L.A. (1974). *Int. Rev. Cytol.* **39**, 191-283.
Subak-Sharpe, J.H., Burk, R.R. and Pitts, J.D. (1969). *J. Cell Sci.* **4**, 353-367.
Tsien, R.W., and Weingart, R. (1976). *J. Physiol.* London **260**, 117-141.
Turin, L. and Warner, A. (1977). *Nature* **270**, 56-57.
Weidmann, S. (1952). *J. Physiol.* **118**, 348-360.
Weidmann, S. (1966). *J. Physiol.* **187**, 323-342.
Woodbury, J.W. and Crill, W.E. (1961). *In* "Nervous Inhibition" (E. Florey, ed.), pp. 124-135. Pergamon Press, Oxford.
Yee, A.G. (1972). *J. Cell Biol.* **55**, 294a.
Yee, A.G. and Revel, J.P. (1978). *J. Cell Biol.* **78**, 554-564.

Cell Surface Protein and Cell Interactions

Kenneth M. Yamada, Kenneth Olden, and Liang-Hsien E. Hahn

Laboratory of Molecular Biology
National Cancer Institute
Bethesda, Maryland 20205

ISBN 0-12-612984-3

I. INTRODUCTION

To understand the role of the cell surface in cellular and developmental processes, it will be necessary to identify which of the many constituents of the cell surface are involved in each event. Leading candidates for these roles in vertebrate cells include a number of glycoproteins recently identified by cell surface labeling techniques (Fig. 1). Few of these proteins have been characterized in detail. The largest number of studies have concerned alterations in these glycoproteins after malignant

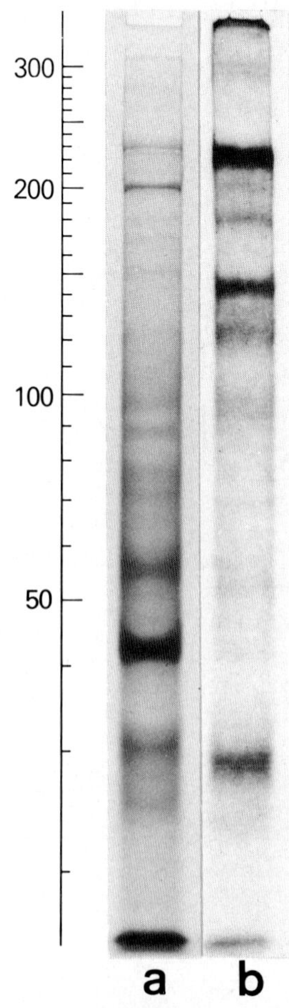

Fig. 1. Cell surface proteins of 3T3 mouse fibroblasts. Confluent cultures of 3T3 cells were labeled with ^{125}I catalyzed by lactoperoxidase, then homogenized in SDS and analyzed by SDS-polyacrylamide gel electrophoresis. a. Protein pattern by Coomassie blue staining. b. Autoradiogram of the same gel showing bands labeled by ^{125}I. Numbers at left indicate molecular weights (x 10^3).

transformation (reviewed by Robbins and Nicolson, 1975; Hynes, 1976; Yamada and Pouyssegur, 1979). In contrast, studies of alterations during embryonic development are only beginning.

A definitive understanding of how each of these glycoproteins acts in the many functions of the cell surface will probably require the laborious isolation and biochemical characterization of each protein. Several intensively studied vertebrate membrane proteins include histocompatibility antigens, transport molecules, and hormone receptors. In addition, the glycoprotein now known as fibronectin has also been intensively investigated recently. Interest in this glycoprotein arises from the finding that fibronectin is a major cell surface molecule of many cell types, and that its quantities are altered in several developmental events and in malignancy (for recent reviews, see Vaheri and Mosher, 1978; Yamada and Olden, 1978).

A. *Nomenclature*

A major source of confusion in the literature was recently resolved when most laboratories adopted the nomenclature in Table I. "Cellular fibronection" (or "cell surface fibronectin") has replaced other terms for the cell surface form of this protein. This form of fibronectin is synthesized in large amounts by cell types *in vitro* as diverse as fibroblasts, myoblasts, endothelial cells, and intestinal epithelial cells (see Yamada and Olden, 1978 for references).

"Plasma fibronectin" is used interchangeably with the older, somewhat misleading name "cold insoluble globulin." This plasma form of

TABLE I

Nomenclature of the Fibronectins

Cellular fibronectin
Large, external, transformation-sensitive (LETS) protein
Cell surface protein (CSP)
Galactoprotein a
Fibroblast surface antigen (SFA)
Zeta
Plasma fibronectin
Cold insoluble globulin
Cell attachment factor
Cell spreading factor
α_2 opsonic glycoprotein

fibronectin is similar, but clearly not identical, to cellular fibronectin in both structure and types of biological activity (for further discussion, see Yamada and Kennedy, 1979). Interestingly, a glycoprotein independently isolated by purifying for its activity in stimulating reticuloendothelial system phagocytosis has also been found to be this plasma form of fibronectin (Blumenstock *et al.*, 1978; Saba and Antikatzides, 1979).

B. *Properties of Cellular Fibronectin*

The general properties of cellular fibronectin are listed in Table II. It is a major cellular protein of cultured fibroblasts, in which it constitutes as much as 3% of total cellular protein of freshly explanted cells (Yamada and Weston, 1974). Quantities of cellular fibronectin produced by cells decrease after prolonged cultivation *in vitro* (Chen *et al.*, 1977; Mosher *et al.*, 1977) and can become particularly low during the establishment of many permanent cell lines (Yamada *et al.*, 1977b).

Fibronectin is an unusually large glycoprotein, with subunit polypeptides larger than 200,000 daltons. Although much interest in fibronectin originally derived from its reported loss after malignant transformation of fibroblasts, there are significant numbers of exceptions (Fig. 2; reviewed by Yamada and Olden, 1978; Kahn and Shin, 1979). Its role in malignancy, e.g., in the propensity of cells to metastasize, is still under investigation.

Cell surface fibronectin is characteristically highly sensitive to proteases, and is lost from living cells after treatment with low concentrations of proteases such as trypsin (Hynes, 1973; Yamada and Weston, 1974; Blumberg and Robbins, 1975). The use of proteases to dissect the structure and function of fibronectin will be discussed in detail in Section V.

Table II

General Properties of Cellular Fibronectin

Major cell protein
1-3% in fibroblasts *in vitro*
Subunit MW = 220 — 240,000
Glycoprotein
5% carbohydrate
Major concanavalin A receptor
Decreased after transformation
Protease-sensitive

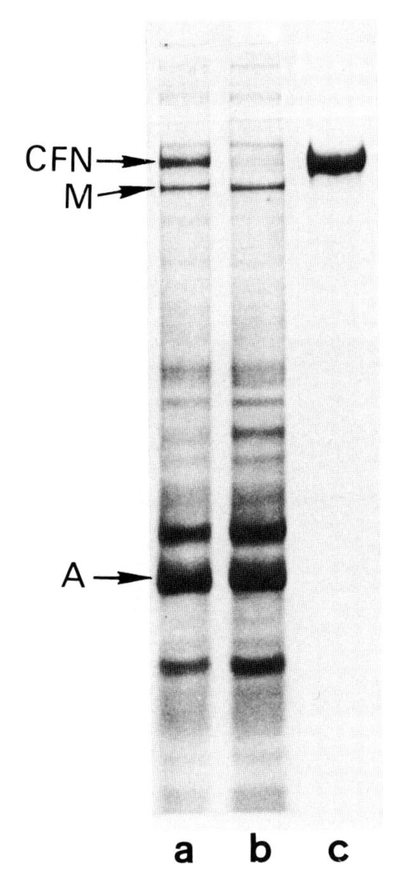

Fig. 2. Cellular fibronection on normal or transformed cells, and after isolation. a. Homogenate of chick embryo fibroblasts analyzed in an SDS gel stained for protein. b. Homogenate of a parallel culture of chick fibroblasts malignantly transformed by Rous sarcoma virus; note the decrease in fibronectin. c. Purified chick cellular fibronectin. CFN = cellular fibronectin; M = cellular (non-muscle) myosin; A = cellular actin.

II. FUNCTION: ADHESIVE GLYCOPROTEIN?

The initial purification of cellular fibronectin in milligram quantities (Yamada and Weston, 1974; Yamada *et al.*, 1975) initiated a series of studies of its possible function. These experiments are of four general types, attempting to determine whether or not it (a) possesses cell-cell adhesive properties in *in vitro* bioassays; (b) restores normal cell behavior when reconstituted onto cells missing the protein, i.e., onto transformed cells; (c) is inactivated by monospecific antibodies, thereby mimicking the behavior of cells that lose it after transformation; and (d) has a distinctive location on the cell surface that is consistent with some particular function.

A. *Adhesive Properties*

Hemagglutination has been utilized as a method to identify slime mold lectins thought to have a role in developmentally regulated cell-cell

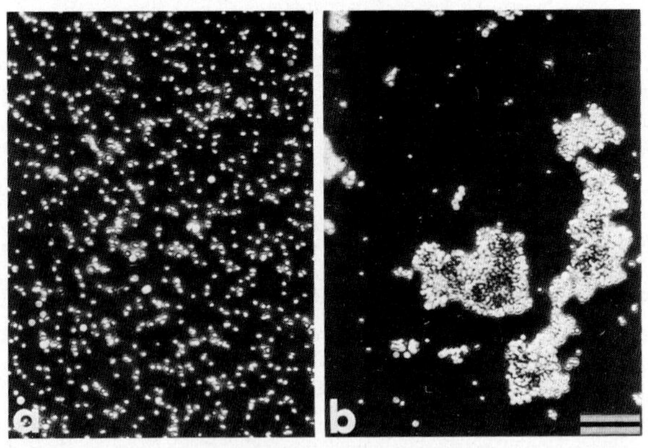

Fig. 3. Hemagglutination by cellular fibronectin. a. Formalin-fixed sheep erythrocytes incubated without fibronectin. b. Fixed erythrocytes plus 25 μg/ml fibronectin for 2 hours. Phase contrast microscopy; bar = 50 μ. From Yamada *et al.*, 1978.

adhesive interactions (Rosen *et al.*, 1974). Purified human and chick fibroblast cellular fibronectins will also hemagglutinate formalin-fixed sheep erythrocytes (Fig. 3; Yamada *et al.*, 1975). Fibronectin can half-maximally agglutinate these cells at concentrations as low as 1 μg/ml. Consistent with a role as an adhesive molecule, this agglutinating activity is destroyed by chelating agents and proteases (Yamada *et al.*, 1975).

In other assay systems, fibronectin also displays adhesion-stimulating activity, i.e., in a gyratory shaker assay and in the attachment of cells to cell monolayers (Yamada *et al.*, 1978; Roseman *et al.*, 1974). It is important to note that although fibronectin is also developmentally regulated in some tissues (Section III), there is no evidence that fibronectins are tissue-specific. This glycoprotein might therefore be considered as a general rather than a specific type of adhesive molecule, in contrast to other adhesive molecules assumed to be involved in events such as neural retinal development.

Interestingly, plasma fibronectins are virtually inactive in the hemagglutination bioassay (Yamada and Kennedy, 1979). However, both plasma and cellular fibronectins are highly active in cell-to-substratum adhesiveness. Certain cells, particularly permanent cell lines with low endogenous rates of cellular fibronectin synthesis, require fibronectin to attach to collagen or to spread on artificial tissue culture substrata (Klebe, 1974; Grinnell, 1978). Both forms of fibronectin are equally active in these cell-substratum adhesive assays (Pena and Hughes, 1978; Yamada and Kennedy, 1979).

B. *Reconstitution Experiments*

A more general approach to evaluating the function of a cell surface protein is to attempt to reconstitute it on cells with deficiencies in the molecule. Fibronectin can be reconstituted onto transformed cells lacking it, and it can also be incorporated into pre-existing fibrillar arrays of fibronectin on nontransformed cells (Yamada and Weston, 1975; Yamada *et al.*, 1976a; Ali *et al.*, 1977; Schlessinger *et al.*, 1977).

In these experiments, purified fibronectin is added to routine culture media containing 5–10% serum. The added fibronectin spontaneously reattaches to cells and to substrata from the media. Approximately 15% of the added fibronectin was found to reattach to the cell surface (Yamada *et al.*, 1976a), and its distribution is similar or identical to that found on normal fibroblasts (Ali *et al.*, 1977; Yamada, 1978).

Reconstitution of fibronectin onto transformed cells partially restores a number of aspects of the normal fibroblastic phenotype to transformed cells. It does not, however, revert either the abnormal growth rates or the elevated rates of nutrient transport of several transformed cell lines (Yamada *et al.*, 1976a; Yamada and Pastan, 1976; Ali *et al.*, 1977; Chen *et al.*, 1978).

The effects of fibronectin reconstitution onto transformed cells fall into three general classes, consisting of altered (1) cell social interactions, (2) morphology and cytoskeleton, and (3) cell motility and locomotion.

1. *Cell social interactions.* One of the simplest and more reliable indicators of neoplastic transformation of cultures of fibroblasts by tumor viruses is the loss of the normal appearance of cultures as determined by phase contrast microscopy. As normal fibroblasts proliferate and form a monolayer on the dish, they begin to align in characteristic parallel patterns (Fig. 4a). Transformed fibroblasts lose this ability to respond to neighboring cells, and instead they tend to overlap or underlap one another in random patterns (Fig. 4b; Abercrombie, 1970; Bell, 1977). In a

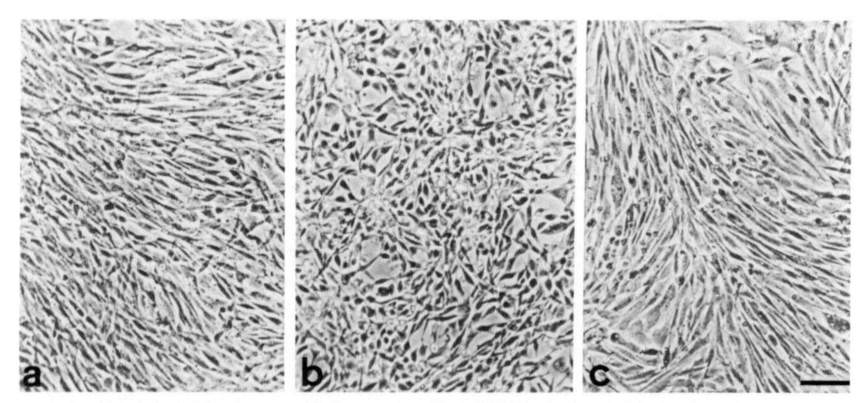

Fig. 4. Effects of fibronectin reconstitution on cell social interactions of transformed cells. a. Normal, untransformed chick fibroblasts. b. Chick fibroblasts transformed by Rous sarcoma virus. c. Parallel transformed cell culture treated with 25 μg/ml purified cellular fibronectin for 48 hours. Phase contrast; bar = 100 μ. From Yamada *et al.*, 1978.

number of cell types, fibronectin treatment can restore this aspect of normal cell social interaction, and the cells once again align in parallel arrays when they become confluent (Fig. 4c). This property is restored to approximately half of the 15 transformed cell types tested (Yamada *et al.*, 1976a; Ali *et al.*, 1977; Yamada *et al.*, 1978).

Since the random orientation and "overlapping" of transformed cells have actually been shown to result from "underlapping" of migrating cells under cells that have poor adhesiveness with which they collide, the simplest explanation of fibronectin's effects is that it restores a more normal adhesiveness to cells. Perhaps the more normal cell-substratum adhesiveness would eliminate cell-free spaces under transformed cells and thereby prevent underlapping of migrating cells that approach these areas (e.g., see scanning micrographs in Yamada *et al.*, 1976b). Additionally, increased cell-cell adhesiveness might stimulate alignment of cells in parallel arrays.

2. *Cell morphology and cytoskeleton.* Fibronectin reconstitution onto individual tumor cells *in vitro* also at least partially restores a normal cell morphology to all 15 transformed fibroblastic cell lines studied (Yamada *et al.*, 1976a, 1978; Ali *et al.*, 1977). These morphologic changes include restoring a flatter shape to abnormally rounded transformed cells, as well as a more fibroblastic appearance characterized by a roughly triangular outline and longer cell processes, and a decrease in the abnormal numbers of cell surface microvilli, blebs, and ruffles often found on transformed cells (Fig. 5). These results suggest that fibronectin plays a role in cellular and cell surface morphology as well as in affecting cell social interactions.

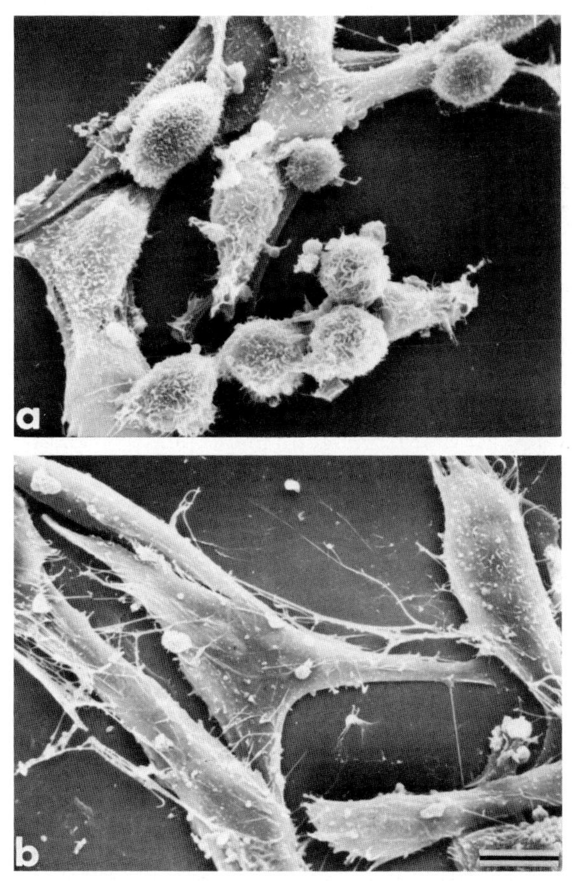

Fig. 5. Effects of fibronectin reconstitution on the morphology of transformed cells. a. Mouse SV1 tumor cells *in vitro*, control culture. b. SV1 cells treated with 25 μg/ml purified cellular fibronectin for 48 hours. Bar = 10 μ. From Yamada *et al.*, 1978.

The organization of certain intracellular structures, particularly microfilament bundles, is also restored towards normal in these reconstitution experiments. Untreated, rounded transformed cells are frequently defective in the capacity to form bundles of 4–6 nm diameter microfilaments; these bundles normally include actin, myosin, and filamin (actin-binding protein). The reorganization of bundles in fibronectin-treated cells appears to be associated with general increases in cell-substratum adhesiveness and a more flattened cell shape (Willingham *et al.*, 1977; Ali *et al.*, 1977).

An intriguing possibility is that fibronectin is a central component of cell adhesive plaques. Fibronectin and actin filaments are found to be roughly parallel or even coaxial in these attachment regions (see Singer, 1979 for a discussion of evidence), and fibronectin may therefore help to anchor microfilament bundle insertion regions of the plasma membrane to the substratum. Another related, perhaps more likely, possibility is that fibronectin increases close contacts in general (rather than solely acting on focal contacts that serve as microfilament anchoring points) to lead to a general increase in cell-substratum adhesiveness that would then secondarily facilitate the formation of the focal contacts.

3. *Cell motility.* Fibronectin treatment of certain transformed or normal fibroblastic cells can also increase cell motility, perhaps by augmenting traction to the substratum or by accentuating cell polarity (Pouyssegur *et al.*, 1977; Ali and Hynes, 1978; Yamada *et al.*, 1978). These effects on cell locomotion suggest that fibronectin might help to regulate cell migration *in vivo* (see Section III, B.)

C. *Inactivation Studies*

Further insight into the possible role of fibronectin in cell behavior can be obtained with inactivation studies using monospecific antibody. Treatment of normal cultured chick embryo fibroblasts with affinity-purified anti-fibronectin antibody results in rapid, noncytotoxic alterations in cell appearance (Yamada, 1978). Cells become rounded or spindle-shaped within 20 min, and morphologically begin to closely resemble transformed chick cells that are deficient in fibronectin (Fig. 6). Similar effects are produced by treatment of these cells with monovalent Fab fragments of anti-fibronectin, but not by various control antibodies. Removal of antibody-containing medium permits a restoration of the normal fibroblastic cell shape within 2–3 hours (Yamada, 1978). These results are complementary to those in which fibronectin is reconstituted on cells lacking it, further supporting its role in cell shape.

As discussed in detail later (Section III, A), treatment of L_6 myoblasts possessing cell surface fibronectin with affinity-purified anti-fibronectin produces results opposite to those obtained after the addition of exogenous fibronectin, i.e., myoblast fusion is stimulated by antibody treatment (Podleski *et al.*, 1979).

D. *Immunofluorescence Localization*

Antibodies to fibronectin have been employed by many laboratories to determine the localization of fibronectin *in vivo* and *in vitro. In vivo,*

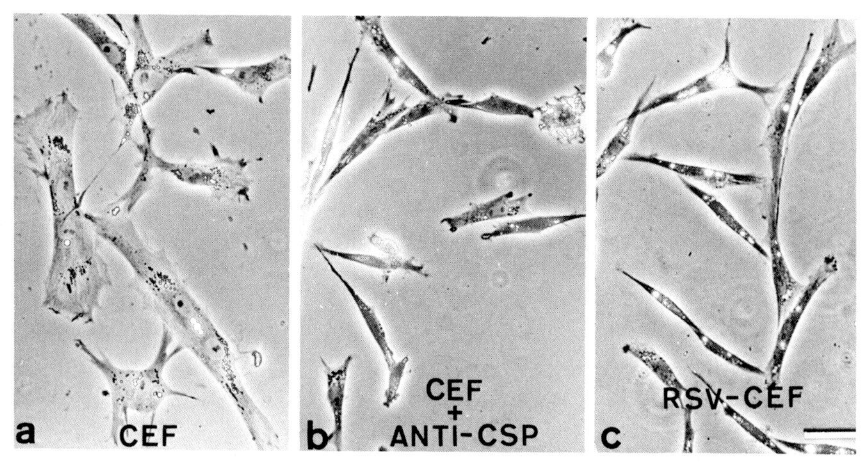

Fig. 6. Fibronectin antibody-inhibition experiment. a. Normal chick embryo fibroblasts (CEF). b. Parallel culture of normal fibroblasts treated with 50 μg/ml of affinity-purified anti-fibronectin (anti-CSP) for 1 hour. c. Chick fibroblasts transformed by Rous sarcoma virus (RSV). Phase contrast; bar = 50 μ.

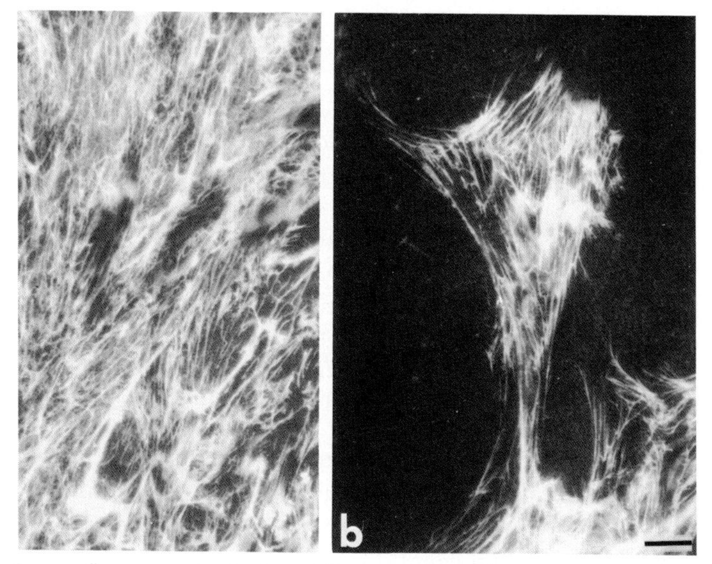

Fig. 7. Immunofluorescence localization of fibronectin in confluent or sparse cultures of chick fibroblasts. a. Densely confluent chick fibroblasts were examined by indirect immunofluorescence, using affinity-purified anti-fibronectin followed by fluorescein-labeled rabbit anti-goat IgG. b. Indirect immunofluorescence localization of fibronectin on an individual cell migrating away from a group of other cells. Epifluorescence microscopy; bar = 50 μ.

fibronectin is a predominant constituent of mesenchymal tissues and of basement membranes (Linder *et al.*, 1975; Stenman and Vaheri, 1978; Bray, 1978). A similar fibrillar localization in extracellular matrices is also characteristic of heavily confluent cultured cells (Fig. 7a). However, on single cells, it is concentrated strikingly in regions of cell-cell contact and in fibrils between cells and substrata (Fig. 7b). This localization has been interpreted as consistent with a role of fibronectin in cell-cell and cell-substratum adhesive interactions (Chen *et al.*, 1976; Mautner and Hynes, 1977; Hedman *et al.*, 1978; Yamada, 1978).

Fibronectin is also found in an intracellular, perinuclear location that represents a nonfibrillar precursor form of fibronectin. In pulse-chase studies using protein synthesis inhibitors, this mateial could be followed onto the cell surface within 2–3 hours (Yamada, 1978). Recent electron microscopic studies reveal that this intracellular precursor material is almost exclusively present in rough endoplasmic reticulum and Golgi apparatus (Yamada *et al.*, 1980). Fibronectin therefore appears to be synthesized and temporarily stored in rough endoplasmic reticulum in a nonfibrillar form, then is transferred to the cell surface to organize into its characteristic fibrillar patterns.

III. ROLES OF FIBRONECTIN IN DEVELOPMENT

Quantities of fibronectin are known to change in a number of developmental events (Wartiovaara, this volume; Linter *et al.*, 1975). Two particularly striking changes occur during differentiation of cartilage and muscle. Cell surface fibronectin is present in substantial amounts on mesenchymal cells and myoblasts, and is lost from differentiating chondrocytes and myotubes *in vitro* as *in vivo* (Lewis *et al.*, 1978; Chen, 1977; Furcht *et al.*, 1978). As chondrocytes dedifferentiate under certain conditions of tissue culture, they display increasing amounts of fibronectin (Dessau *et al.*, 1978). Likewise, inhibiting chondrogenesis of limb bud mesenchyme cells with vitamin A results in the retention of fibronectin (Hassell *et al.*, 1978). The effects of this drug on mature chondrocytes was recently found to be due primarily to an increased capacity of the cell surface to bind fibronectin, suggesting that changes in fibronectin receptors may play a significant role in developmental changes in fibronectin (Hassell *et al.*, 1979).

What role is played by such developmentally regulated alterations in fibronectin? Current hypotheses include functions such as (a) acting as an instructive or permissive regulatory molecule, (b) stimulating particular cell migratory events, or (c) playing a more passive role as a structural or scaffolding molecule.

A. *Regulatory Molecule?*

Direct tests of fibronectin's role in differentiation have been performed recently in two systems in which development is accompanied by the loss of fibronectin: chondrogenesis and myogenesis. Chick sternal chondrocytes were treated with purified chick fibroblast cellular fibronectin *in vitro,* and were shown to bind the molecule, and to begin to show an altered morphology within approximately two days of treatment. By four days, such treated chondrocytes were very similar in appearance to undifferentiated mesenchymal cells, with a flattened cell shape and elongated cell processes (Pennypacker *et al.,* 1979). The most important results concerned biosynthetic events. The synthesis of ^{35}S-labeled sulfated proteoglycans was found to be inhibited by 70%, and there was apparent cessation of the synthesis of a chondrocyte-specific 180,000 dalton cell surface protein. In addition, fibronectin treatment appeared to decrease the synthesis of chondrocyte-specific Type II collagen and to revert synthesis to Type I collagen (Pennypacker *et al.,* 1979; West *et al.,* 1979).

In addition, endogenously synthesized fibronectin also reappeared on such treated chondrocytes (Pennypacker *et al.,* 1979). These results suggest that the loss of fibronectin from chondrocytes during development may play a regulatory role in permitting the expression of differentiated function; artificially restoring fibronectin to the cell surface would then block the expression of the chondrocytic phenotype.

Extensive analyses of the process of myoblast fusion suggest a similar

Table III

Fibronectin Inhibition of L_6 Myoblast Fusion

Treatment	No. nuclei in myotubes/field	No. myotubes/ field
None	49 ± 6	4.7 ± 0.2
Fibronectin (20 μg/ml)	3 ± 1	0.8 ± 0.2
None	57 ± 4	9.3 ± 0.8
Fibronectin (40 μg/ml)	9 ± 3	1.4 ± 0.2
Control buffer	57 ± 5	9.7 ± 0.4

Data from Podleski *et al.,* 1979.

role for fibronectin in this developmental event. Fibronectin treatment of L_6 myoblasts inhibits their fusion (Table III; Podleski *et al.*, 1979). This inhibition is found after treatment of either mass cultures or of individual myoblast clones. The effects of fibronectin treatment can be circumvented by the concomitant addition of monospecific, affinity-purified anti-fibronectin, or by depleting the amounts of added fibronectin by prior affinity chromatography of solutions on anti-fibronectin columns.

Cultures in which fusion is inhibited by fibronectin display extensive arrays of parallel cells similar in appearance to those of densely confluent fibroblasts, or of transformed cells onto which fibronectin has been reconstituted. The number of cells (determined by counting of nuclei) is increased by 3-fold in fibronectin-treated cultures. This increase might be due to continued cell division by the nonfusing myoblasts, or alternatively, to action of fibronectin as a mitogen.

Interestingly, anti-fibronectin treatment alone appeared to *stimulate* myoblast fusion, suggesting that the antibody might be inactivating endogenous fibronectin. Gentle trypsinization of myoblasts to remove fibronectin enzymatically also stimulated fusion if performed on day 4 of the standard culturing protocol, which is the day on which cell surface fibronectin reaches a peak. In contrast, trypsinization earlier appeared to inhibit myogenesis (Podleski *et al.*, 1979), which is consistent with a possible requirement for the presence of fibronectin in early myoblast cultures.

The inhibition of chondrocyte and myoblast differentiation by fibronectin again suggests that the reason for the developmentally regulated decreases in cell surface fibronectin on these cells may be to permit subsequent differentiation. Fibronectin might act directly as a regulator of biosynthetic processes, or, perhaps more likely, indirectly by promoting cell spreading and migratory activity that might be incompatible with the more sedentary activities of chondrocyte synthesis of matrices of proteoglycans and Type II collagen, or of myoblast differentiation.

B. *Cell Migration*

Fibronectin can stimulate the motility of a variety of cells *in vitro* (see Section II, B, 3). Migration of certain cells *in vivo* during embryonic development might also involve this glycoprotein. An intriguing, but highly speculative role for fibronectin in development might be to help to direct cell migration by its regulated deposition along specific pathways of cell migration. For example, neural crest cells have been reported to have little or no endogenous fibronectin (Loring *et al.*, 1977); they might therefore be unusually sensitive to external sources of fibronectin, and they might therefore be stimulated to migrate into cell-free regions or along basal lamina that are enriched in this molecule.

C. *Extracellular Structural Molecule*

A general role for fibronectin in development may also be as a noncollagenous structural protein present in extracellular matrices and in a variety of basement membranes. This function appears likely, since substantial amounts of fibronectin are present in these regions *in vivo*, and since it binds to collagen Types I, II, III, and IV, as well as to hyaluronic acid and to other glycosaminoglycans with high affinity (see Section V, C). Organizing these extracellular materials may help to maintain and to regulate normal tissue architecture (e.g., see Banerjee *et al.*, 1977).

IV. MECHANISM OF ACTION

Table IV

Summary of Fibronectin's Biological Activities

1. Cell-cell aggregation (agglutination)
   A. Formalinized sheep erythrocytes
   B. Live cells (chick embryo and BHK)
2. Cell-substratum adhesion
   A. Adhesion to collagen
   B. Adhesion to plastic and glass
   C. Cell spreading
3. Reversion of transformed cells
   A. Alignment and overlapping
   B. Fibroblastic morphology
   C. Microvilli, blebs, and ruffles
   D. Microfilament bundles
   E. No restoration of growth control or decrease in nutrient transport
4. Cell motility
5. Inhibition of myoblast fusion and chondrogenesis
6. Non-antibody opsonic activity in reticuloendothelial system
7. Direct binding to collagen, heparin, and fibrin

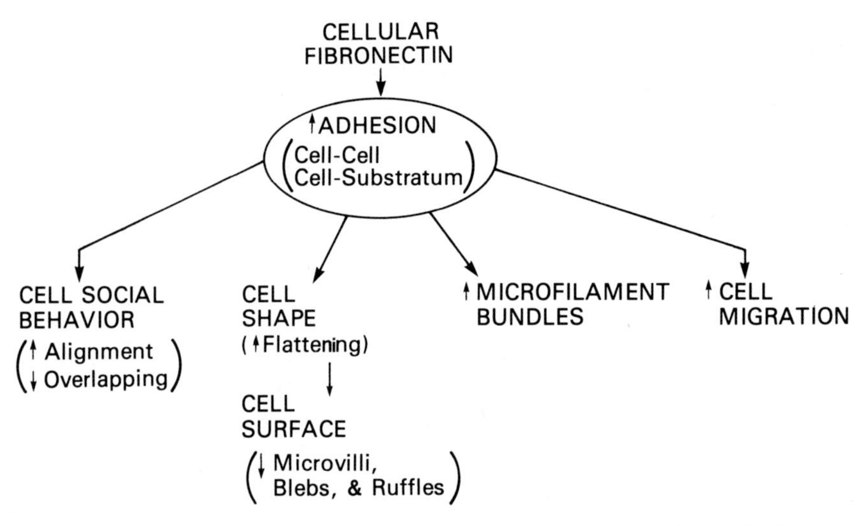

Fig. 8. Model for the role of fibronectin in cell behavior. Fibronectin-mediated adhesiveness is postulated to play a central role in the events indicated. See text for discussion.

A. *Biological Mechanisms*

The biological activities of fibronectin are summarized in Table IV. How does fibronectin mediate these pleiotropic effects on cell behavior? The simplest hypothesis is that it acts as a generally adhesive or sticky protein that increases cell-cell and cell-substratum adhesion and interactions. A proposed scheme for its biological mechanism of action is presented in Fig. 8. This suggested causal relationship is also based on several previous studies that have implicated cellular adhesion in helping to regulate cell shape, cell surface architecture, and cell motility (for recent discussions see Yamada and Olden, 1978; Letourneau *et al.*, 1979).

B. *Standard Assays*

Investigations into how this protein acts at the molecular level have been initiated recently. These biochemical studies of fibronectin's mechanism of action have depended upon a standard set of four *in vitro* assays of biological activity that are reviewed schematically in Fig. 9. It has been found to be necessary to use this entire battery of tests, since certain types or subfragments of fibronectin have important differences in biological specific activity depending upon the assay performed.

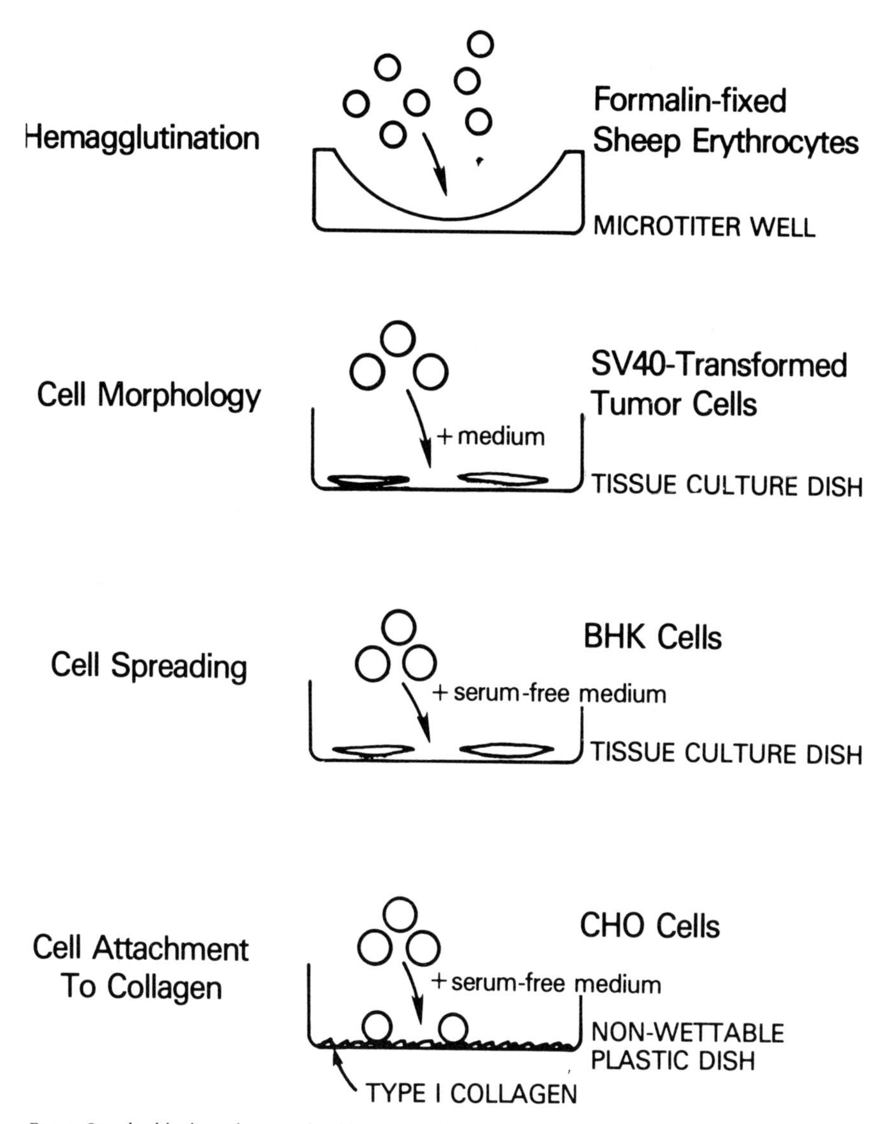

Fig. 9. Standard biological assays for fibronectin. See text for description.

Hemagglutination is quantitated in a routine microtiter assay of hemagglutinin activity that is used commonly in clinical laboratories. A fibronectin preparation to be tested is serially diluted, then the concentration resulting in half-maximal agglutination is determined

(Yamada *et al.*, 1975). This assay provides a quantitative measure of one aspect of cell-cell adhesiveness.

Fibronectin effects on the morphology of transformed fibroblasts in reconstitution experiments can be quantitated by determining the lowest concentration of fibronectin that results in detectable changes in morphology and alignment of the SV1 tumor cell (Yamada *et al.*, 1976b) in "blind" evaluations of sets of 3–4 random phase contrast photomicrographs. This assay appears to measure fibronectin effects on cell-cell interactions, plus effects on cell morphology.

The other two assays are conducted in the absence of serum (since serum contains the plasma form of fibronectin), and measure aspects of cell-to-substratum adhesiveness. The assay for cell spreading requires fibronectin, or other molecules that bind to cell surfaces and plastic, to mediate BHK cell spreading on plastic tissue culture dishes (Grinnel *et al.*, 1977). The percentage of cells that spread on the dishes and are surrounded by a zone of lamellar cytoplasm is determined by phase contrast microscopy.

The assay for cell attachment to collagen was introduced by Klebe (1974) and modified by Kleinman *et al.* (1979). We employ a CHO cell strain that has been selected for less background variability, which increases the sensitivity of the assay (Yamada *et al.*, 1979) These cells have an absolute requirement for fibronectin to attach to Type I collagen spread on petri dishes, and the assay consists of determining the percentage of cells that attach in various concentrations of fibronectin using a Coulter counter.

C. *Structure of Fibronectin*

A crucial first step in understanding the mechanism of action of a protein is to determine its structure. The overall structure of fibronectin is shown in Fig. 10. Cellular fibronectin exists as disulfide-bonded dimers and multimers of subunits of 220–240,000 daltons (e.g., see McConnell *et al.*, 1978). These subunits are highly unfolded and asymmetric (Yamada *et al.*, 1977a; Alexander *et al.*, 1978). The inter-subunit disulfide bonds are in an unusual location: they are confined to an end of the molecule (Chen *et al.*, 1977; Jilek and Hormann, 1978; Iwanaga *et al.*, 1978; Yamada *et al.*, 1978; Hynes *et al.*, 1978). This placement of disulfide bonds serves to further increase the length of the molecule. It is possible that this length may be necessary for fibronectin to bridge the distances between cells and other cells, or between cells and a substratum.

Another important feature of the molecule is that its carbohydrate is

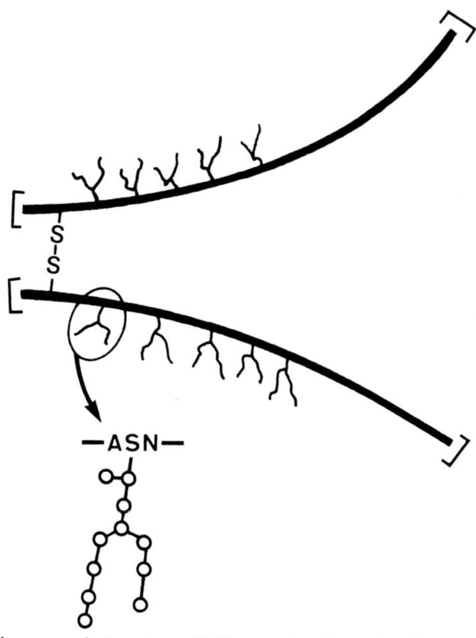

Fig. 10. Model of the general structure of fibronectin. Note the elongated subunits linked by asymmetrically placed interchain disulfide bonds. The oligosaccharide is of the "complex" class containing mannose, N-acetyl glucosamine and other sugars linked by N-glycosidic bonds to asparagine residues of the protein.

essentially of only one class — the "complex" mannose-containing oligosaccharide linked to asparagine residues (Yamada *et al.*, 1977a; Olden *et al.*, 1978; Carter and Hakomori, 1979). This uniformity of carbohydrate structure permits the experimantal inhibition of fibronectin glycosylation by the use of the new glycosylation inhibitor tunicamycin, which is highly specific for only this class of oligosaccharide.

1. *Role of fibronectin's carbohydrates.* Inhibition of fibronectin glycosylation by tunicamycin surprisingly did not inhibit its synthesis or secretion, but did lead to 2–3-fold increased rates of protein turnover (Olden *et al.*, 1978).

Recent experiments have evaluated the remaining possibility that the carbohydrate on this major cell surface glycoprotein is necessary for its biological function (Olden *et al.*, 1979). Glycosylated and nonglycosylated fibronectins were isolated from control and tunicamycin-treated cultures, respectively, by affinity chromatography on anti-fibronectin columns. The nonglycosylated fibronectin was shown to be at least 98% deficient in carbohydrate by chemical analyses for amino sugars and by corresponding decreases in the incorporation of the isotopically labeled

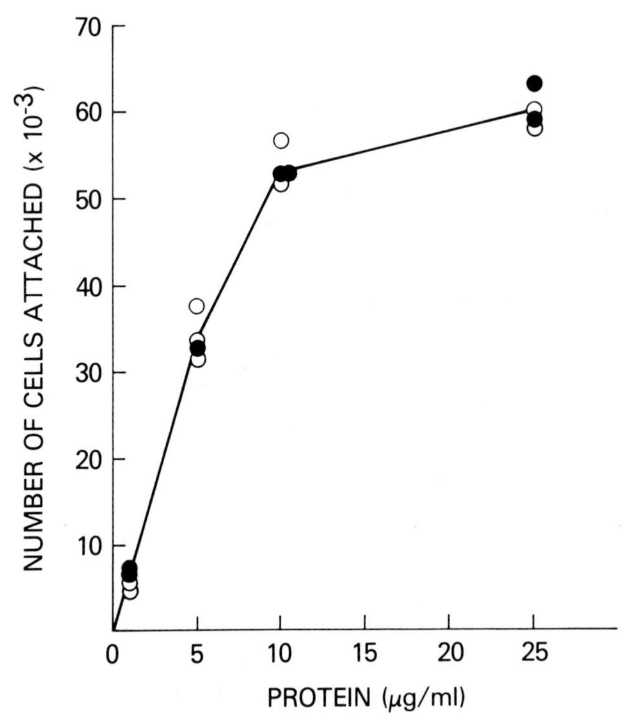

Fig. 11. Effects of glycosylated versus nonglycosylated fibronectins on the attachment of CHO cells to Type I collagen.

sugars mannose and glucosamine.

In all four assays of biological activity, nonglycosylated and native fibronectin have identical specific activities (Olden *et al.*, 1979). For example, Fig. 11 demonstrates that equal numbers of cells attach to collagen at each concentration of the native or the nonglycosylated molecule. The carbohydrate moiety therefore appears to play no detectable role in these biological activities of fibronectin.

However, one major difference does exist. Nonglycosylated fibronectin is significantly more sensitive than native fibronectin in susceptibility to proteolytic attack. In experiments with pronase, a broad-spectrum protease, the rate of release of acid-soluble amino acids and small peptides during digestion is about twice as rapid from the nonglycosylated molecule, indicating a greater sensitivity to hydrolysis (Olden *et al.*, 1979). This difference appears to account for the abnormal rates of protein turnover reported previously.

We therefore conclude that the carbohydrate moiety of fibronectin appears to play no role in its synthesis, secretion, or known biological activities. It instead serves to protect the molecule by unknown mechanisms against proteolytic attack and abnormal rates of turnover. Although this finding is instructive in identifying a relatively new function for carbohydrates on glycoproteins, it also indicates that it is necessary to turn to investigating the polypeptide portion of the molecule to understand its mechanism of action.

V. ACTIVE SITES OF FIBRONECTIN

If fibronectin functions by means of adhesiveness, it might act via specific binding sites such as those listed in Table V. For example, the polypeptide should contain one or more cell surface binding sites to mediate agglutination of cells and attachment of cells to substrates. A collagen-binding site should also be present to mediate cell attachment to collagen. In addition, one or more glycosaminoglycan binding sites should be present for binding to heparin (Stathakis and Mosesson, 1977) and possibly as a cellular receptor for hyaluronic acid or proteoglycans, whose attachment to cells can occur via glycoproteins that are protease-sensitive (Vogel and Kelly, 1977).

A. *Domain Hypothesis*

Recent studies have suggested that although cellular fibronectin is asymmetric and contains unfolded polypeptide regions, it also appears to contain domains of polypeptide structure by fluorescence and circular dicroism criteria (Alexander *et al.*, 1978; Colonna *et al.*, 1978). Our hypothesis was that these domains might represent fibronectin's specific binding sites (Fig. 12).

Table V

Possible Active Sites of Fibronectin

1. Cell surface interaction
2. Collagen binding
3. Glycosaminoglycan binding
4. Fibrin binding
5. Self-association

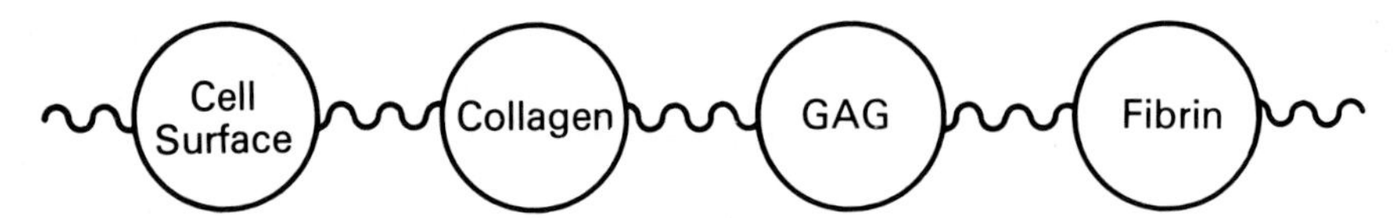

Fig. 12. Possible organization of fibronectin into structured polypeptide domains containing specific binding sites that are separated by regions of unfolded polypeptide. Such domains might include binding sites for the cell surface, for collagen, for glycosaminoglycans (GAG), or for fibrin.

Previous studies of molecules with multiple structural domains, such as fibrinogen, had shown that the unfolded parts of the molecule were much more protease-sensitive than the more globular, structured regions. Such a molecule can be treated with proteases to produce separate polypeptide fragments representing these protease-resistant domains. Our hope was that proteolysis of fibronectin would permit the isolation of a series of fragments containing various specific binding sites.

This approach was applied to fibronectin as indicated in Fig. 13. This

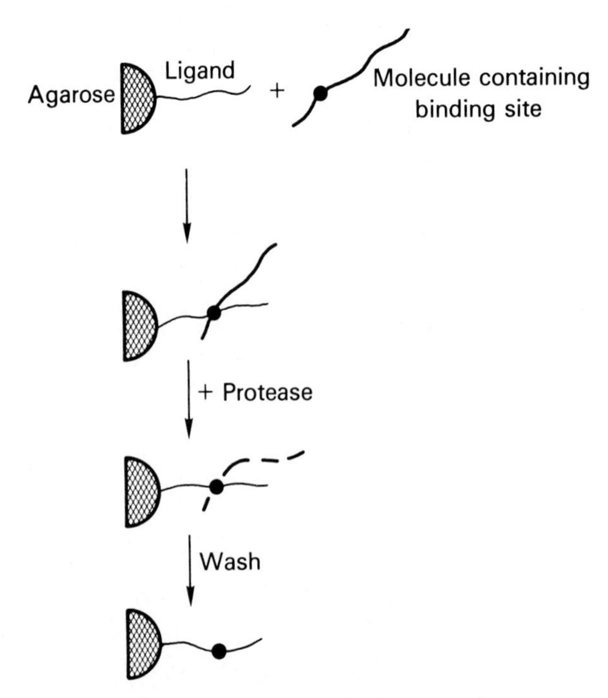

Fig. 13. General approach for the isolation of protease-resistant domains of a protein containing active binding sites. Half-circles represent an agarose bead to which a ligand is coupled via cyanogen bromide. See text for discussion.

general method may prove valuable for dissecting other cell surface components that bind to protease-resistant macromolecules. The molecule being bound, called the ligand, is coupled covalently to agarose beads via cyanogen bromide. The binding molecule, in this case fibronectin, is permitted to bind to the ligand. The gel column or suspensin of beads is then incubated with 2–10 μg/ml of various proteases.

Ideally, the protease will digest away all intervening, unfolded polypeptide regions, leaving only the protease-resistant fragment that contains the active site still bound to the column. Protease-resistant domains devoid of the active site should be recovered in the column

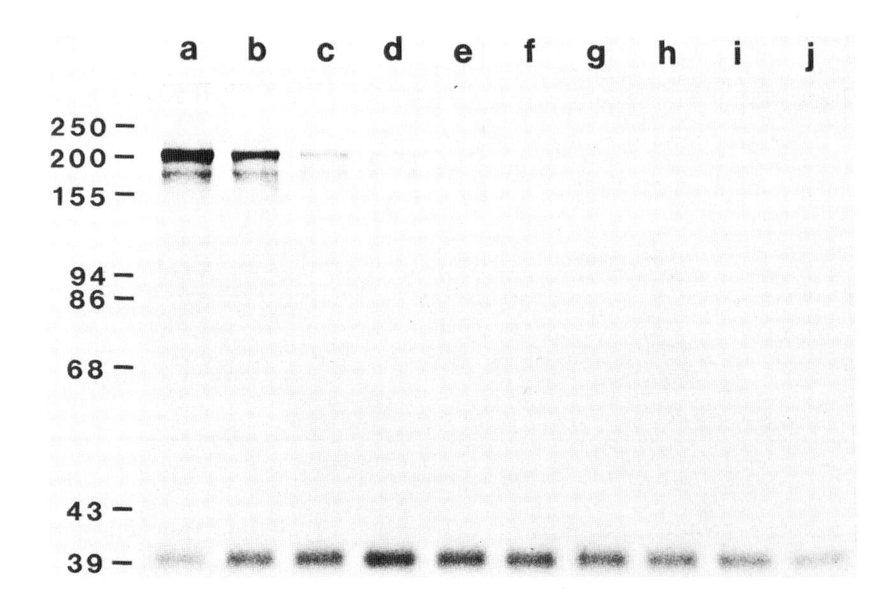

Fig. 14. Generation of collagen-binding fragments of fibronectin by proteolytic cleavage. ^{14}C-labeled fibronectin was permitted to bind to collagen-agarose, then subjected to proteolytic cleavage with chymotrypsin for varying periods of time. The complexes were washed extensively and the material remaining bound was eluted with boiling SDS. This autoradiogram was prepared after SDS gel electrophoresis. Lane a. 10 min of chymotrypsin digestion; b. 25 min; c. 50 min; d. 80 min; e. 2 hr; f. 3 hr; g. 5 hr; h. 7.5 hr; i. 10 hr; j. 20 hr. Numbers at left indicate molecular weight markers (x 10^3 daltons). From Hahn and Yamada, 1979a.

washes. The active fragment remaining bound on the agarose beads can then be eluted with SDS and analyzed, or released with 8 M urea for further purification.

B. *Collagen- and Cell-binding Sites*

An analysis of the results of such a digestion when the ligand is collagen is shown in Fig. 14. At early time points, e.g., 10 minutes of digestion with chymotrypsin, the major fragment remaining bound has a size of 205,000 daltons (vs. 220,000 for the original subunit and $\geq$ 440,000 for the native molecule). This fragment is further degraded to a relatively stable fragment of 40K daltons; after isolation, this 40K fragment rebinds to collagen with 70% efficiency (Hahn and Yamada, 1979a). The fragments that pass through the column with washing contain a number of polypeptides, including a major species of 160K.

The 205K, 160K, and 40K fragments were isolated by salt fractionation and gel chromatography, and were then characterized biologically. The results are summarized in Table VI (from Hahn and Yamada, 1979b).

Table VI

Biological Activities of Fibronectin Fragments

	Quantities required for half-maximal activity (μg/ml)			
Assay	Intact molecule	205K	160K	40K
Cell attachment to collagen	0.1	0.25	15	>20
Cell spreading	1*	1*	1	>40
Hemagglutination	3.9	34	185	>285

*70% of cells spread at 1 μg/ml.
Data in this table are from Hahn and Yamada, 1979b.

Only the 205K fragment is readily capable of mediating cell attachment to collagen, suggesting that it contains both cell- and collagen-binding sites. Although the 160K fragment is a poor mediator of the attachment of cells to collagen, both it and the 205K fragment are highly active in cell spreading, and both still contain residual hemagglutinating activity (even the 160K fragment is more active than native plasma fibronectin; Table VI).

In competitive inhibition experiments with these fragments, the 160K fragment inhibits cell attachment to collagen mediated by intact fibronectin if it is pre-incubated with cells, but not if pre-incubated with the collagenous substratum (Hahn and Yamada, 1979b). These results indicate that the 160K fragment contains binding site(s) for the cell surface.

In contrast, the 40K fragment has no biological activity by itself, even though it can bind directly to collagen. However, it can inhibit fibronectin-mediated attachment of cells to collagen, particularly if it is pre-incubated with the collagenous substratum before the fibronectin and cells are added. This fragment therefore appears to contain only a collagen-binding site. The competitive inhibition experiment strongly suggests that this site alone is of critical importance for fibronectin-mediated attachment to collagen.

A tentative mapping of these sites is indicated in Fig. 15, indicating separate locations of collagen- and cell-binding sites. Biochemical results consistent with ours have also been obtained by Ruoslahti and Hayman (1979) and Balian *et al.*, (1979).

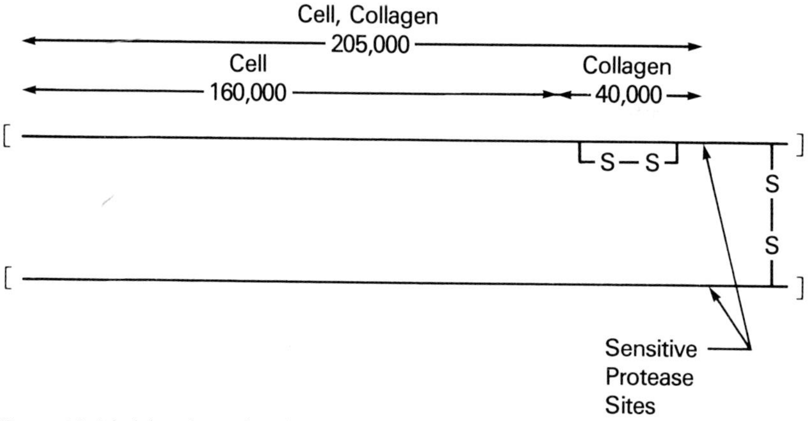

Fig. 15. Model of the relationship of cell-binding and collagen-binding fragments of fibronectin. Note that chymotryptic cleavage separates these sites, and abolishes the capacity to attach cells to collagen. From Hahn and Yamada, 1979b.

C. *Glycosaminoglycan-binding Sites*

We next attempted to identify glycosaminoglycan binding sites on fibronectin. This possibility was important because (1) it is conceivable that fibronectin may be a receptor for glycosaminoglycans thought to be involved in various developmental events (Toole *et al.*, 1972); (2) plasma fibronectin is known to bind to several glycosaminoglycans, particularly heparin; and (3) hyaluronic acid and proteoglycans are constituents of extracellular matrices and basement membranes; since these regions also often contain fibronectin, the interactions of these macromolecules might be important in determining their organization.

To analyze glycosaminoglycan binding to fibronectin, we adopted a filter-binding assay often used in studying hormone receptors and membrane transport. This procedure is of general usefulness if the ligand is soluble and the binding molecule is not. Purified cellular fibronectin is incubated with various purified ^{3}H-labeled glycosaminoglycans under conditions of physiological salt and pH (Fig. 16). The cellular fibronectin becomes insoluble rapidly, and is retained on the Millipore filters. In contrast, the glycosaminoglycans pass through the filter unless they are bound by the fibronectin. The amount bound can be determined by scintillation counting of the radioactivity remaining bound to the filters.

GLYCOSAMINOGLYCAN BINDING ASSAY

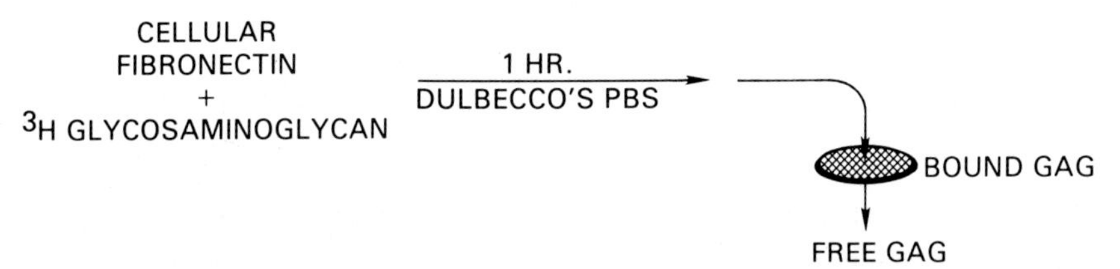

Fig. 16. Millipore filter binding assay for binding of labeled glycosaminoglycans to fibronectin. Purified cellular fibronectin and isolated glycosaminoglycans are incubated with Dulbecco's phosphate-buffered saline for 1 hour in test tubes at 23°C. The mixtures are filtered through 0.45 μ nitrocellulose (Millipore) filters, then washed and counted in a scintillation counter to determine radioactivity bound to fibronectin.

Using this assay, we found that of a series of ^{3}H-glucosamine-labeled glycosaminoglycans isolated from chick embryo fibroblasts, only hyaluronic acid bound efficiently (Fig. 17). Heparan sulfate was bound in small but reproducible amounts; the significance of this binding is not yet clear. In addition, ^{3}H-heparin from hog intestinal mucosa also bound efficiently, as expected from previous studies (Stathakis and Mosesson, 1977). The binding of hyaluronic acid and heparin was shown to be at clearly separate sites by competition experiments with unlabeled material (Yamada *et al.*, 1980).

This type of assay also permits determination of binding affinities. We found by the use of Scatchard analysis that the binding affinity (K_D) of

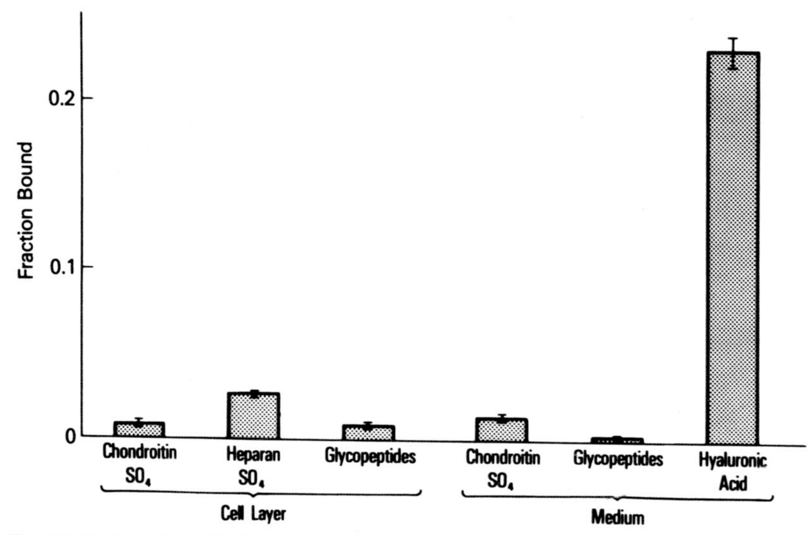

Fig. 17. Binding of specific glycosaminoglycans to fibronectin. Chick fibroblasts were labeled for 24 hours with ^{3}H-glucosamine, then glycosaminoglycans were isolated from the cell layer or the culture medium by Dr. Robert Pratt (National Institute of Dental Research) using DEAE column chromatography following pronase digestion. Ordinate indicates proportion of total input radioactivity that remained bound to fibronectin after exhaustive washing. Bars indicate standard error. From Yamada *et al.*, 1980.

heparin and hyaluronic acid was approximately 10^{-7} M (Yamada *et al.*, 1980). This relatively high affinity supports the notion that these binding interactions are of physiological significance.

This tight binding of fibronectin to glycosaminoglycans has permitted the isolation of fragments that contain a heparin-binding site. For these experiments, we utilized a heparin-agarose column (ligand in Fig. 13 = heparin) and digested with chymotrypsin or pronase. The results are opposite to those obtained with collagen-agarose columns (Fig. 18). The fragments that do not bind to the column include the 40K collagen-binding fragment and a 90K pronase-generated fragment (Fig. 18a, b).

The major chymotryptic fragment that remains bound to the heparin column is the 160K cell-binding fragment (Fig. 18d) that did not bind to collagen columns. This fragment, or the entire molecule, can be further digested by the broad-spectrum protease pronase to yield a 50K heparin-binding fragment (Fig. 18c). This fragment remains bound to the heparin column after extensive washing, and therefore contains the heparin-binding domain.

Our final map for the various identified proteolytic fragments of fibronectin is presented in Fig. 19. The assignment of the order of the

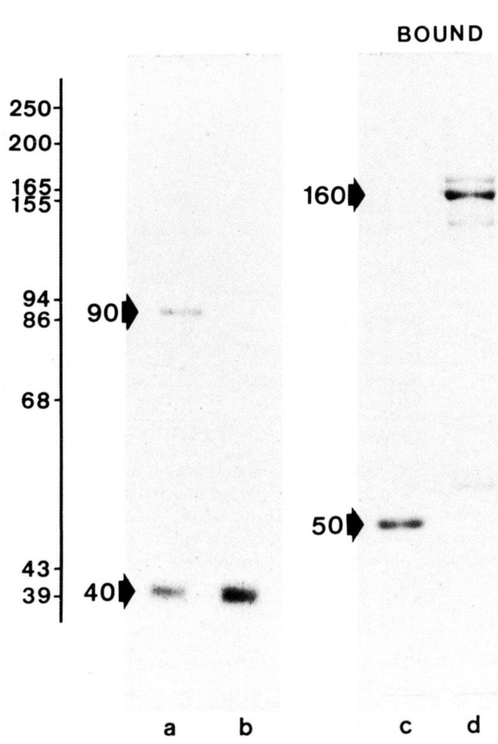

Fig. 18. Isolation of heparin-binding fragments of fibronectin. a,b) Material passing through heparin-agarose columns after proteolytic digestion. c,d) Material remaining bound on the same columns after digestion and extensive washing. a,c) Digested by 6 μg/ml pronase for 1 hour at 23°C. b,d) Digested by 6 μg/ml chymotrypsin for 1 hour. Figures at left indicate molecular weight markers (x 10^3 daltons).

160K and 40K fragments should be considered as tentative until the identification of amino and carboxy termini is definitive.

D. *Cellular Receptor for Fibronectin*

It is of obvious importance to determine the nature of the binding site for fibronectin itself on the cell surface. Recent experiments by Kleinman *et al.*, (1979) have suggested that a ganglioside-like molecule might be the cell surface receptor for plasma fibronectin. These experiments showed a highly specific inhibition of cell attachment to collagen by certain gangliosides. In collaborative experiments with Kleinman, we tested

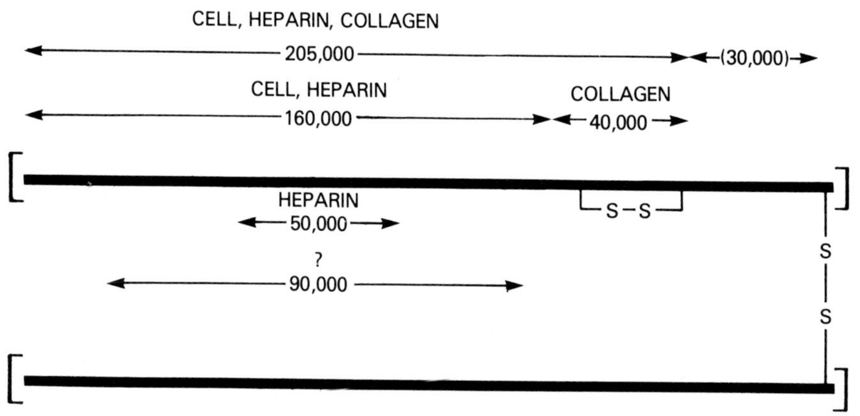

Fig. 19. Tentative map of various binding-site domains of fibronectin.

mixed brain gangliosides as competitive inhibitors of cellular fibronectin-mediated cell spreading and hemagglutination to evaluate the generality of this effect. The addition of gangliosides to a standard cell spreading assay employing 5 μg/ml cellular fibronectin resulted in a dose-dependent inhibition of cell spreading; spreading was maximally inhibited at 0.6 mM gangliosides (Fig. 20).

Gangliosides were also a particularly potent inhibitor of hemagglutination (Table VII). The inhibition was half-maximal at as low as 6 μM; this concentration should be compared to that of previously identified sugar inhibitors, whice were generally active only at 30–40 mM (Yamada *et al.*, 1975). The inhibition of hemagglutination by gangliosides suggests that the inhibition of adhesion by these lipids is not merely a result of effects on some critical cell metabolic event or on cytoskeletal organization, but is instead a more direct competitive inhibition at a cellular receptor site. These results therefore suggest that a major receptor for fibronectin on cells is a ganglioside or a ganglioside-like molecule.

VI. FUNCTIONAL MODEL FOR FIBRONECTIN

A schematic model for the functional organization of fibronectin active sites is presented in Fig. 21. This adhesive glycoprotein appears to consist

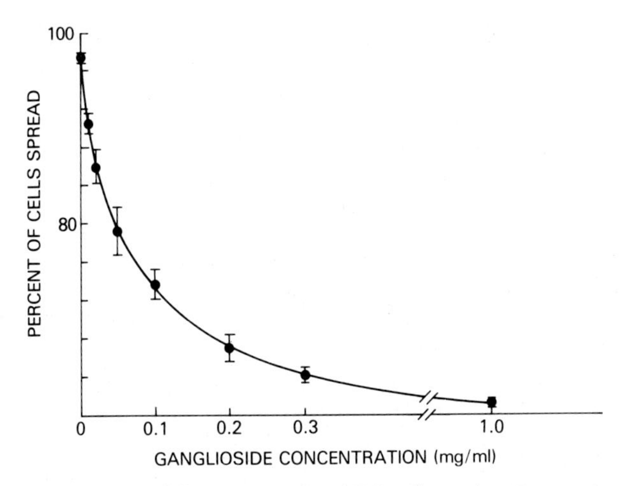

Fig. 20. Ganglioside inhibition of fibronectin-mediated BHK cell spreading. Trypsin-dissociated BHK cells were plated into 35 mm tissue culture dishes in the presence of 5 μg/ml purified cellular fibronectin plus increasing concentrations of mixed brain gangliosides (Sigma Type II). The proportion of fully spread cells was determined by phase contrast microscopy of cultures fixed with 2% glutaraldehyde. Bars indicate SEM; 1000 cells were scored for each point.

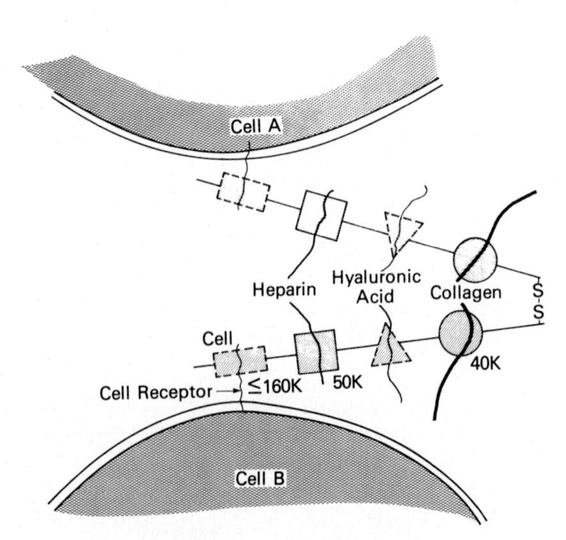

Fig. 21. Model for the structure and mechanism of action of fibronectin. Fibronectin is postulated to consist of structural domains mediating specific binding interactions. Combinations of pairs of sites would account for the binding or adhesive activities of fibronectin. For example, cell adhesive events would be mediated by binding of a fibronectin cell-binding site to the cellular receptor (e.g., a ganglioside) and binding of another site to another cell or to an extracellular molecule. Numbers indicate known apparent molecular weights of protease-resistant domains containing the indicated binding sites.

Table VII

Sensitivity of Standard Fibronectin Bioassays to Inhibition by Gangliosides

Assay	Ganglioside required for half-maximal inhibition
Cell attachment to collagen[a]	2-4 x 10^{-4}M
Hemagglutination[b]	0.6-2 x 10^{-5}M
Cell spreading[c]	2-4 x 10^{-5}M

[a] With 0.25% serum-coated dishes. From Kleinmann *et al.*, 1979.
[b] With 10 μg/ml purified cellular fibronectin; unpublished results.
[c] With 5 μg/ml cellular fibronectin.

of specific, protease-resistant domains of structure, each of which contains separate binding sites for the cell surface (within the 160K fragment), hyaluronic acid (not yet isolated), heparin (50K fragment), and collagen (40K fragment).

This organization into a series of specific binding sites could explain how fibronectin acts in various adhesive events. Fibronectin would be able to link cells to other cells or to various macromolecules; conversely, it might provide a means for cells to bind certain soluble macromolecules. This multiplicity of binding sites could thereby account for fibronectin's effects on a wide range of cellular events.

Several predictions arise from this model. For example, if one cell- and one collagen-binding site are located on each subunit, even the monomer of fibronectin should readily mediate the attachment of cells to collagen. This interpretation appears to be correct, since this attachment activity is retained in the monomeric 205K fragment at approximately half of the original activity of intact fibronectin.

On the other hand, cell-cell interaction should be defective if the fibronectin is monomeric. The 205K fragment is approximately 10-fold less active than the intact molecule and the 160K fragment is 50-fold less active in mediating hemagglutination. The retention of residual activity suggests that besides retaining cell-binding capacity, these fragments may also retain the capacity to self-associate into multivalent, noncovalent aggregates that can permit weak cell agglutination.

Fibronectin on the cell surface may help to bind heparin, which may be important in the non-antibody opsonic activity of the reticuloendothelial system that has been attributed to plasma fibronectin, and which is enhanced by heparin (Saba and Antikatzides, 1979). *In vivo*, another important consequence of these multiple binding sites on fibronectin may be to permit secreted firbonectin to cross-link a variety of extracellular molecules. Forming fibronectin-collagen-proteoglycan networks in cell-free extracellular spaces or in basement membranes may therefore be another important role for this molecule.

VII. CONCLUSIONS

The fibronectins are multi-subunit adhesive glycoproteins on cell surfaces and in blood. The plasma form of fibronectin appears to be deficient in certain cell-cell adhesive properties, but is otherwise structurally and functionally very similar to cellular fibronectin. Cellular fibronectin appears to mediate or to stimulate a wide variety of cell-cell and cell-substratum adhesive interactions. Cellular fibronectin also appears to be involved in the cell social interactions that maintain normal cell alignment and minimal overlapping, which are frequently lost after transformation.

Developmentally regulated decreases in this molecule have also been recently implicated in normal myoblast and chondrocyte differentiation. The results suggest that fibronectin may have direct roles in development, perhaps as a permissive type of regulatory molecule.

Finally, the mechanism of action of this protein has been examined in detail. The carbohydrate portion is not required for its known major biological activities, although it does stabilize the protein against proteolytic attack and abnormal turnover rates. The polypeptide backbone of fibronectin is organized into separate, protease-resistant domains that contain specific binding sites for the cell surface, hyaluronic acid, heparin, and collagen.

The cellular receptor for fibronectin's surface binding site may involve a ganglioside, since gangliosides are effective competitive inhibitors of fibronectin action in several assays. The identification and initial successes in isolating several of fibronectin's active site regions should facilitate a molecular understanding of the pleiotropic effects of this major cell surface glycoprotein on cell behavior.

ACKNOWLEDGEMENTS

We thank Drs. Robert Pratt, Hynda Kleinman, Thomas Podleski, and many other colleagues for fruitfull collaborations and Mrs. Dorothy Kennedy for invaluable technical assistance.

REFERENCES

Abercrombie, M. (1970). *In Vitro* **6,** 128-142.

Alexander, S.S., Jr., Colonna, G., Yamada, K.M., Pastan, I., and Edelhoch, H. (1978). *J. Biol. Chem.* **253,** 5820-5824.

Ali, I.U., and Hynes, R.O. (1978). *Cell* **14,** 439-446.

Ali, I.U., Mautner, V.M., Lanza, R., Hynes, R.O. (1977). *Cell* **11,** 115-126.

Balian, G., Click, E.M., Crouch, E., Davidson, J., and Bornstein, P. (1979). *J. Biol. Chem.* **254,** 1429-1432.

Banerjee, S.D., Cohn, R.H., and Bernfield, M.R. (1977). *J. Cell Biol.* **73,** 445-463.

Bell, P.B. (1977). *J. Cell Biol.* **74,** 963-982.

Blumenstock, F.A., Saba, T.M., Weber, P., and Laffin, R. (1978). *J. Biol. Chem.* **253,** 4287-4291.

Blumberg, P.M., and Robbins, P.W. (1975). *Cell* **6,** 137-147.

Bray, B.A. (1978). *J. Clin. Invest* **62,** 745-752.

Carter, W.G., and Hakomori, S. (1979). *Biochemistry* **18,** 730-738.

Chen, L.B. (1977). *Cell* **10,** 393-400.

Chen, L.B., Gallimore, P.H., and McDougall, J.K. (1976). *Proc. Nat. Acad. Sci. U.S.A.* **73,** 3570-3574.

Chen, A.B., Amrani, D.L., and Mosesson, M.W. (1977). *Biochim. Biophys. Acta* **493,** 310-321.

Chen, L.B., Moser, F.G., Chen, A.B., and Mosesson, M.W. (1977). *Exp. Cell Res.* **108,** 375-383.

Chen, L.B., Murray, A., Segal, R.A., Bushnell, A., and Walsh, M.L. (1978). *Cell* **14,** 377-391.

Colonna, G., Alexander, S.S., Jr., Yamada, K.M., Pastan, I., and Edelhoch, H. (1978). *J. Biol. Chem.* **253,** 7787-7790.

Dessau, W., Sasse, J., Timpl, R., Jilek, F., and von der Mark, K. (1978). *J. Cell Biol.* **79,** 342-355.

Furcht, L.T., Mosher, D.F., and Wendelschafer-Crabb, G. (1978). *Cell* **13,** 263-271.

Grinnell, F. (1978). *Int. Rev. Cytol.* **53,** 65-144.

Grinnell, F., Hays, D.G., and Minter, D. (1977). *Exp. Cell Res.* **110,** 175-190.

Hahn, L.-H. E., and Yamada, K.M. (1979a). *Proc. Nat. Acad. Sci. U.S.A.* **76,** 1160-1163.

Hahn, L.-H.E., and Yamada, K.M. (1979b). *Cell,* **18,** 1043-1051.

Hassell, J.R., Pennypacker, J.P., Yamada, K.M., and Pratt, R.M. (1978). *Ann. N.Y. Acad. Sci.* **312,** 406-409.

Hassell, J.R., Pennypacker, J.P., Kleinman, H.K., Pratt, R.M., and Yamada, K.M. (1979). *Cell,* **17,** 821-826.

Hedman, K., Vaheri, A., and Wartiovaara, J. (1978). *J. Cell Biol.* **76,** 748-760.

Hynes, R.O. (1973). *Proc. Nat. Acad. Sci. U.S.A.* **70,** 3170-3174.

Hynes, R.O. (1976) *Biochim. Biophys. Acta* **458,** 73-107.

Hynes, R.O., Ali, I.U., Destree, A.T., Mautner, V., Perkins M.E., Senger, D.R., Wagner, D.O., and Smith, K.K. (1978). *Ann. N.Y. Acad. Sci.* **312,** 317-342.

Iwanaga, S., Suzuki, K., and Hashimoto, S. (1978). *Ann. N.Y. Acad. Sci.* **312,** 56-73.

Jilek, F., and Hormann, H. (1978). *Hoppe-Seyler's Z. Physiol. Chem.* **359,** 247-250.

Kahn, P., and Shin, S. (1979). *J. Cell Biol.* **82,** 1-16.

Klebe, R.J. (1974). *Nature* **250,** 248-251.

Kleinman, H.K., Martin, G.R., and Fishman, P.H. (1979). *Proc. Nat. Acad. Sci. U.S.A.,* **76,** 3367-3371.

Kleinman, H.K., McGoodwin, E.B., Rennard, S.I., and Martin, G.R. (1979). *Anal. Biochem.* **94,** 308-312.

Letourneau, P.C., Ray, P.N., and Bernfield, M.R. (1979). *In* "Biological Regulation and Control" (R. Goldberger, ed.). Plenum Press, New York, in press.

Lewis, C.A., Pratt, R.M., Pennypacker, J.P., and Hassell, J.R. (1978). *Develop. Biol.* **64,** 31-47.

Linder, E., Vaheri, A., Ruoslahti, E., and Wartiovaara, J. (1975). *J. Exp. Med.* **142,** 41-49.
Loring, R., Erickson, C., and Weston, J. (1977). *J. Cell Biol.* **75,** 71a (abstract).
Mautner, V., and Hynes, R.O. (1977). *J. Cell Biol.* **75,** 743-768.
McConnell, M.R., Blumberg, P.M., and Rossow, P.W. (1978). *J. Biol. Chem.* **253,** 7522-7530.
Mosher, D.F., Saksela, O., Keski-Oja, J., and Vaheri, A. (1977). *J. Supramolec. Struct.* **6,** 551-664.
Olden, K., Pratt, R.M., and Yamada, K.M. (1978). *Cell* **13,** 461-473.
Olden, K., Pratt, R.M., and Yamada, K.M. (1979). *Proc. Nat. Acad. Sci. U.S.A.* **76,** 3343-3347.
Pena, S.D.J., and Hughes, R.C. (1978). *Cell Biol. Internat. Rep.* **2,** 339-344.
Pennypacker, J.P., Hassell, J.R., Yamada, K.M., and Pratt, R.M. (1979). *Exp. Cell Res.* **121.,** 411-415.
Podleski, T.R., Greenberg, I., Schlessinger, J., and Yamada, K.M. (1979). *Exp. Cell Res.* **122**, 317-326.
Pouyssegur, J., Willingham, M., and Pastan, I. (1977). *Proc. Nat. Acad. Sci. U.S.A.* **74,** 243-247.
Robbins, J.C., and Nicolson, G.L. (1975). *In* "Cancer: A Comprehensive Treatise" (F.F. Becker, ed.), Vol. IV, pp. 3-54. Plenum, New York.
Roseman, S., Rottmann, W., Walther, B., Ohman, R., and Umbreit, J. (1974). *Methods Enzymol.* **32B**, 597-611.
Rosen, S.D., Simpson, D.L., Rose, J.E., and Barondes, S.H. (1974). *Nature* **252,** 128, 149-150.
Ruoslahti, E., and Hayman, E.G. (1979). *FEBS Lett.* **97,** 221-224.
Saba, T.M., and Antikatzides, T.G. (1979). *Ann Surg.* **189,** 426-432.
Schlessinger, J., Barak, L.S., Hammes, G.G., Yamada, K.M., Pastan, I., Webb, W.W., and Elson, E.L. (1977). *Proc. Nat. Acad. Sci. U.S.A.* **74,** 2909-2913.
Singer, I.I. (1979). *Cell* **16,** 675-685.
Stathakis, N.E., and Mosesson, M.W. (1977). *J. Clin. Invest.* **60,** 855-865.
Stenman, S., and Vaheri, A. (1978). *J. Exp. Med.* **147,** 1054-1064.
Toole, B.P., Jackson, G., and Gross, J. (1972). *Proc. Nat. Acad. Sci. U.S.A.* **69,** 1384-1386.
Vaheri, A., and Mosher, D.F. (1978). *Biochim. Biophys. Acta* **516,** 1-25.
Vogel, K.G., and Kelley, R.O. (1977). *J. Cell Physiol.* **92,** 469-480.
West, C.M., Lanza, R., Rosenbloom, J., Lowe, M., Holtzer, H., and Avdalovic, N. (1979). *Cell* **17**, 491-501.
Willingham, M.C., Yamada, K.M., Yamada, S.S., Pouyssegur, J., and Pastan, I. (1977). *Cell* **10,** 375-380.
Yamada, K.M. (1978). *J. Cell Biol.* **78,** 520-541.
Yamada, K.M., and Weston, J.A. (1974). *Proc. Nat. Acad. Sci. U.S.A.* **71,** 3492-3496.
Yamada, K.M., and Weston, J.A. (1975). *Cell* **5,** 75-81.
Yamada, K.M., and Pastan, I. (1976). *J. Cell Physiol.* **89,** 827-830.
Yamada, K.M., and Olden, K. (1978). *Nature* **275,** 179-184.
Yamada, K.M., and Kennedy, D.W. (1979). *J. Cell Biol.* **80,** 492-498.
Yamada, K.M., and Pouyssegur, J. (1979). *Biochimie* **60,** 1221-1233.
Yamada, K.M., Yamada, S.S., and Pastan, I. (1975). *Proc. Nat. Acad. Sci. U.S.A.* **72,** 3158-3162.
Yamada, K.M., Yamada, S.S., and Pastan, I. (1976a). *Proc. Nat. Acad. Sci. U.S.A.* **73,** 1217-1221.
Yamada, K.M., Ohanian, S.H., and Pastan, I. (1976b). *Cell* **9,** 241-245.
Yamada, K.M., Schlesinger, D.H., Kennedy, D.W., and Pastan, I. (1977a). *Biochemistry* **16,** 5552-5559.
Yamada, K.M., Yamada, S.S., and Pastan, I. (1977b). *J. Cell Biol.* **74,** 649-654.
Yamada, K.M., Olden, K., and Pastan, I. (1978). *Ann. N.Y. Acad. Sci.* **312,** 256-277.
Yamada, K.M., Hahn, L.E., and Olden, K. (1979). *Prog. Clin. Biol. Res.*, in press.

Yamada, S.S., Yamada, K.M., and Willingham, M.C. (1980a), submitted.
Yamada, K.M., Kennedy, D.W., Kimata, K., Pennypacker, J., and Pratt, R.M. (1980b)., submitted.

From Cell Adhesion to Growth Control

Luis Glaser

Department of Biological Chemistry
Division of Biology and Biomedical Sciences
Washington University School of Medicine
St. Louis, Missouri 63110

I. CHEMICAL BASES FOR CELL ADHESION

In a number of systems, it is possible to show under laboratory conditions that isolated single cells adhere to each other with specificity; that is that some combinations of cells can adhere to each other, while others cannot.

Such adhesion can be due to the presence of specific molecules in the plasma membrane of these cells, or it could be due to the extracellular matrix surrounding these cells. Adhesive phenomena related to the extracellular matrix have been discussed in this volume by K. Yamada and we will confine the present review to adhesive events in which plasma membrane components are involved in "direct" cell to cell adhesion.

In considering the nature of the molecules on the cell surface which are candidates for adhesive molecules, one can consider proteins, carbohydrates (on glycolipids and glycoproteins) and the head group of phospholipids. There is not enough variability in phospholipid head groups to account for specific cell adhesion, and carbohydrate-carbohydrate interactions of the requisite specificity are not known. We

ISBN 0-12-612984-3

therefore are left with two possible candidates for specific cell adhesion, either carbohydrate-protein interaction or protein-protein interaction.

It is therefore the task of investigators in this field to elucidate the nature of the molecules involved in cell adhesion. It is not immediately obvious that the same molecules will be involved in cell adhesion in different systems, or that different cells use the same molecular arrangement to establish cell adhesion. For example, current evidence strongly suggests that carbohydrate is an important component in cell adhesion in slime molds (see chapter by S. Barondes), sponges (Burger and Jumblatt, 1977; Muller *et al.*, 1979) and probably the adhesion of dissociated liver cells to each other (Obrink, 1977; Schnaar *et al.*, 1978). There is evidence that adhesion of retinal cells to halves of the optic tectum is in part mediated by a protein carbohydrate interaction (Marchase, 1977), but the possible involvement of carbohydrates in other systems is less obvious.

Cell adhesion requires, as a minimal component, a protein that can bind to a complementary ligand on an adjacent cell, in the same way as an enzyme binds to a substrate or an antibody to an antigen. It is not clear that this simple model actually applies to any system that has been investigated so far. A specific version of such a model utilizes cell surface glycosyl transferases that can bind to glycosyl acceptors on adjacent cells (Schur and Roth, 1975), while specific cell surface glycosyl transferases have been demonstrated (see for example Porzig, 1978), their role in cell-to-cell adhesion remains speculative.

There is evidence that cell adhesion may be a multi-stage process and that multiple ligands on the cell surface must interact to form a stable cell-to-cell adhesion (Umbreit and Roseman, 1975; Moya *et al.*, 1979). In addition, any given cell may contain several different adhesive molecules, each of which would allow this cell to interact specifically with a different cell type under appropriate circumstances (Santala *et al.*, 1977).

If cell-to-cell adhesion is a multi-stage process, it is not clear at which step specificity is manifest. The simplest model is one in which cells bind to each other by specific ligands, followed by the formation of multiple cell-to-cell attachment which need not be specific, although they can involve the same ligands as are involved in the initial adhesive event (Fig. 1). Some of the existing evidence can be interpreted to suggest that initial ligand binding results in the induction or exposure of additional sites on the cell surface (Santala *et al.*, 1977; Weigel *et al.*, 1978). More complex models can, however, be generated, including those in which reversal of cell-to-cell adhesion plays an important role in determining cellular adhesive specificity, i.e. models in which certain adhesive events are only temporary while others become more permanent.

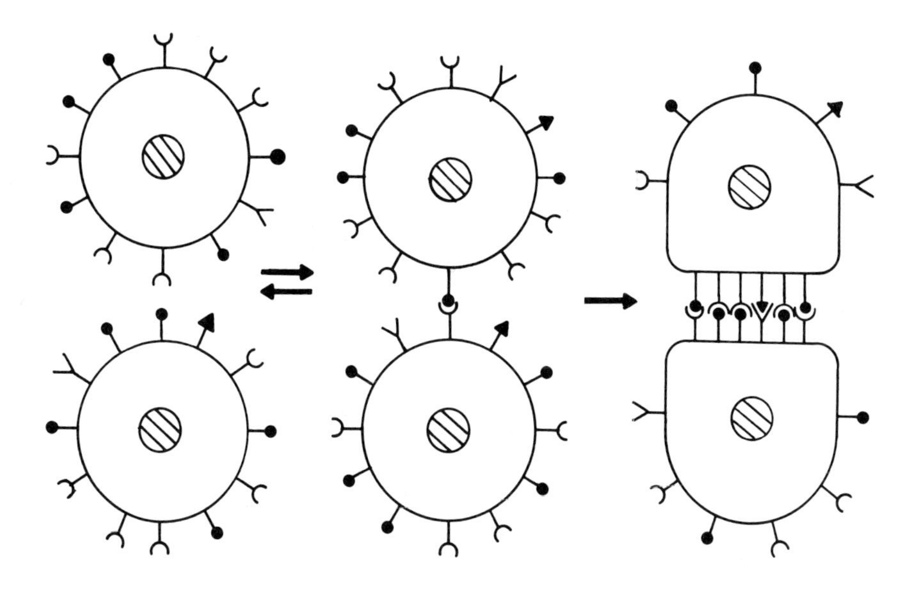

Fig. 1. Model of cell to cell adhesion. This model is to be considered only as a diagrammatic illustration of the numerous steps in cell adhesion. For the cells shown, the ligands Y ↑ are presumed to represent complementary molecules on the cell surface which are involved in cell adhesion. Adhesion is pictured as involving an initial reversible step, followed by an apparently irreversible step (under assay conditions), in which multiple complementary ligands bind to each other, and other alterations in the cell surface may take place, diagrammatically illustrated by the change in cell shape. The cell surface may contain additional complementary molecules on the cell surface, diagrammatically illustrated by Y ↑. These are presumed to be ligands of relatively low affinity or concentration so that by themselves they cannot mediate cell-to-cell adhesion. After cell to cell adhesion, these ligands occur next to each other at a high local concentration and will bind. Such ligands could be responsible for changes in cell function following cell-to-cell adhesion without themselves being the major components involved in cell-to-cell binding.

In discussing cellular adhesion in these terms, the architecture of the cell surface tends to be ignored. Since a cell is not a smooth sphere from which various ligands protrude, changes in cell surface architecture and in the arrangements of the molecules involved in cell adhesion must by necessity change the adhesive patterns of cells. Finally, it should be clear, as will be described below, that the usual methods of measuring cell adhesion fundamentally measure cell-to-cell adhesion against a shear force, and on a time scale that may be very short compared to physiological events. *In vivo* the time scale on which cells can interact may be long and no substantial shear force may exist.

II. SYSTEMS FOR THE STUDY OF CELL ADHESION

Cell-to-cell adhesion is usually determined by measuring the rate of adhesion of labelled cells to immobilized cells or very large aggregates of cells. The measurement is the rate of adhesion against a known shear force, and the absolute values of the rate will vary depending on the geometry of the vessel, the temperature, the concentration of single cells, etc. Therefore, only comparative values, rather than absolute values, can be obtained by these methods. A summary of some of the methods used is shown in Fig. 2.

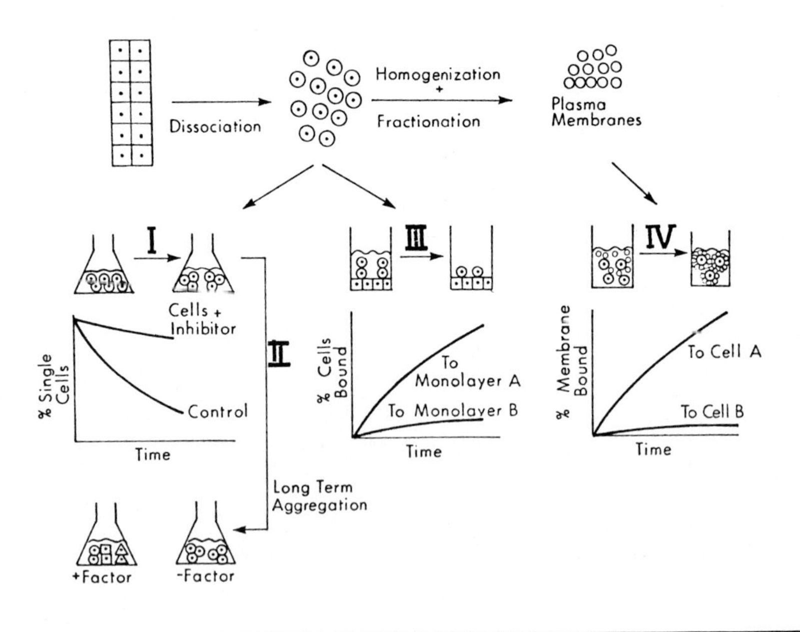

Fig. 2. Systems for the study of cell adhesion. This figure provides a graphic summary of systems used to study cell adhesion. I. Measures the disappearance of single cells. Specificity in this case can only be obtained for inhibitors. II. Measures the formation of large aggregates in response to soluble factors obtained from cells and measures the tissue specificity of these factors. III. Measures the binding of radioactively labeled single cells to a monolayer of cells, and specificity can be determined both for cells in the monolayer and for single cells in suspension. IV. Measures the ability of plasma membranes to adhere to cells. Idealized data for two cell types (A and B) are illustrated by the graph in each case.

This type of methodology is convenient, and yields quantitative information. However, the interpretation of these results needs to be carried out with a great deal of caution. The measurement that is made is a rate, but the conclusion that is drawn is related to the number and types of sites present on the cell surface. Rarely has it been possible to obtain equilibrium information in these systems (Beug and Gerisch, 1972; Steinberg, 1978). The availability of adhesive sites on the surface of the cell, as well as their arrangement relative to other cell surface components, may influence this type of assay.

With this type of methodology a great deal of information has been obtained regarding specific cell adhesion, starting with pioneering work by Roth and Weston (1967; see also Roth, 1968) almost 12 years ago. Several comprehensive reviews have appeared on this subject (Frazier and Glaser, 1979; Gottlieb and Glaser, 1980) and the results can be summarized briefly as follows:

Specific cell adhesion can be demonstrated in a variety of systems, including single cells obtained from specific tissues in embryos, cultured cells, dissociated cells from sponges, aggregating cells obtained from the cellular slime mold, etc.

In the embryonal system, homotypic association is usually preferred when the distinction to be made is between grossly unrelated areas of the embryo, such as liver and retina (McGuire and Burdick, 1976), but within a complex and closely related area like the nervous system, heterotypic adhesion is often preferred in initial cell adhesion experiments (see for example Gottlieb and Arington, 1979). Cell-to-cell adhesion is not uniform within any given region within the embryos and striking differences in the adhesive properties of cells within any given region of the nervous system can be demonstrated, most notably the presence of adhesive gradients, such as occur in the neural retina illustrated in Fig. 3 (Barbera *et al.*, 1973; Gottlieb *et al.*, 1976, and Marchase, 1977). While the conceptual framework that generated these experiments are attempts to study retina-tectal connectivity, the relation of the observed gradients to retina-tectal connectivity has not been established.

In the study of cell adhesion by single cells obtained by dissociation of a preformed tissue, there is concern that the initial dissociation has altered the cell surface. Although a recovery period is often included in these experiments (see for example, Barbera *et al.*, 1973; Marchase, 1977), it is not certain that the cell surface regenerated *in vitro* is identical to that present in the original embryo.

Adhesion measured *in vitro* usually takes place in the absence of a defined extracellular matrix, which may well have important effects on cellular adhesion and recognition (see for example the recent results on neuromuscular junction by Sanes *et al.*, 1978). Finally, the reversal of adhesion and the ability of a cell to compare similar but not identical

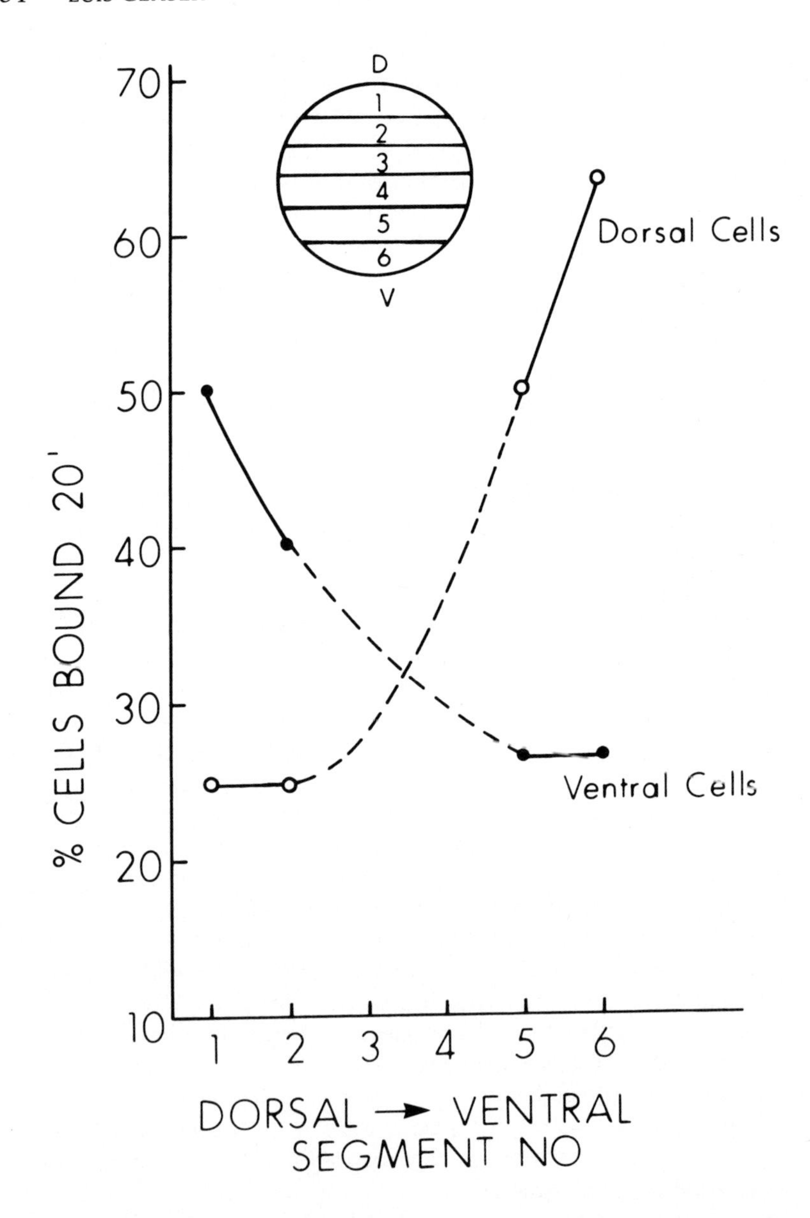

Fig. 3. Adhesive gradient in the chick neural retina. The figure summarizes the observations of Gottlieb *et al.* (1976) and illustrates the rate of adhesion of neural retinal cells prepared from the extreme dorsal part (region 1) of the chick neuroretina and the extreme ventral portion (region 6) to monolayers prepared from the region indicated schematically in the insert. D = dorsal; V = ventral. Ventral cells bind fastest to monolayers prepared from the extreme dorsal region of the retina and the converse is true for cells from the extreme dorsal region of the retina.

surfaces and make a preferential choice is absent from most *in vitro* adhesion studies, at least in a form that is easily amenable to study. Such adhesion and selection must occur in long-term aggregate cultures, where cells differentiate and segregate in various patterns within the aggregate resembling the original tissue (Moscona, 1965; Hausman and Moscona, 1975). The formation of large aggregates can be stimulated in a highly specific manner by aggregation-promoting factors, some of which have been obtained in pure form (Hausman and Moscona, 1975, 1976, 1979). The mode of action of these factors is not understood in detail, and will be discussed in section III.

III. APPROACHES TO THE MOLECULAR BASIS OF CELL ADHESION

The approaches to the molecular basis of cell adhesion fall into several categories. The first is the characterization of these molecules by attempts to interfere with adhesion by selective modification of cell surface components. Thus, for example, if treatment of cells with proteolytic enzymes interferes with cell adhesion, one concludes that a protein or proteins are involved in cell adhesion. Since proteins are probably involved in all specific cell adhesion phenomena, this particular experiment is by itself not very informative. Useful information in several systems has been obtained by the use of specific glycosidases (Marchase, 1977; Muller *et al.*, 1979). When glycosidase treatment of whole cells interferes with cell adhesion, this evidence suggests that certain carbohydrates may participate in cell adhesion. Since many molecules usually contain any given monosaccharide, this approach provides useful but possibly limited information on the specific molecules involved in cell adhesion.

Antibodies can be used to try to isolate molecules involved in cell-to-cell adhesion. In principle, monovalent antibody fragments directed against surface components involved in cell adhesion should block adhesion, if they bind to these molecules in a manner that interferes with cell adhesion (Muller and Gerisch, 1978; Rutishauser *et al.*, 1978a, b; McClay, *et al.*, 1977). If suitable methods are available for absorbing out other cell surface antibodies, or if such antibodies have been prepared as monoclonal antibodies, then it should be possible to purify the adhesive molecules using these antibodies as affinity ligands. Monoclonal antibodies have so far not been prepared against such ligands. Antibodies have been prepared that interfere with cell adhesion in slime molds (Muller and Gerisch, 1978) and in the nervous system (Rutishauser *et al.*, 1978), and these have been used to purify the cell surface antigens with which these antibodies react. In both cases, there is considerable evidence of specificity, in that binding of equivalent quantities of other antibodies to the cell surface does not inhibit cell aggregation. Since cell adhesion is a complex multi-step phenomenon, the fact that an antibody inhibits cell adhesion does not by itself define the step in the adhesion sequence in which the antigen participates. The limitations of this approach have been discussed in detail in a recent review (Gottlieb and Glaser, 1980).

An alternative approach to the identification of molecules involved in cell adhesion is the isolation of molecules which promote cell adhesion under defined conditions. Such molecules were originally isolated from sponges, and have since been characterized in neural retina and other parts of the nervous system, as well as liver. In the case of sponges, the molecules appear to act as bridges between cells (Burger and Jumblatt, 1977; Muller *et al.*, 1979). In the nervous system, the assay used is based on the formation of large aggregates in 24 hours rotation assays, where, in the absence of factors, relatively smaller aggregates are formed. A 50,000 molecular weight glycoprotein has been isolated from retina that promotes large aggregate formation. This molecule has been shown to be located on the surface of retinal cells (Hausman and Moscona, 1975, 1976, 1979). Similar molecules, which have not as yet been purified, have been obtained from other regions of the embryonal nervous system.

The precise function of these molecules is not clear, and the possibility that they represent very specific trophic factors must be considered, since a factor with similar activity for dissociated liver cells has been identified as taurine (Sankaran *et al.*, 1977).

If one assumes that cell surface carbohydrates are involved in cell adhesion, then it would seem reasonable to attempt to isolate developmentally regulated lectins, i.e. multivalent carbohydrate binding proteins. This is an approach that has been pursued successfully in the study of slime mold aggregation (see chapter in this volume by S. Barondes). Developmentally regulated lectins have been identified in embryonal systems, but their function is as yet not known. Similarly the combined use of polyacrylamide gel electrophoresis with specific staining by plant lectins has revealed the presence of a number of developmentally regulated glycoproteins in retina but their relation to cell adhesion remains speculative (Mintz and Glaser, 1978).

A very general approach to the isolation of adhesive molecules on the cell surface assumes that plasma membranes, prepared by gentle procedures, retain *(at least in part)* the adhesive molecules present on the cell surface and responsible for specific cell adhesion. If this is correct, then membranes should be able to bind to cells with specificity. These membranes would be a convenient starting material for the fractionation of the membrane components involved in cell adhesion.

That plasma membranes can bind to cells with specificity has been demonstrated with dissociated cells from the embryonal nervous system, with dissociated liver cells, and with cultured cells (Fig. 4). The purification of the proteins involved in this binding phenomenon has until recently only had modest success. The problem is complex because the proteins in the membranes responsible for adhesive specificity are likely to be hydrophobic and their extraction from the membranes will require detergents which have to be removed before these proteins can be presented to live cells. In addition, plasma membrane vesicles are almost certainly multivalent; therefore, they can bind to cells with high affinity, even if the intrinsic binding constant for each ligand molecule is relatively weak. Under these circumstances, to demonstrate binding of the solubilized molecules to cells may be very difficult.

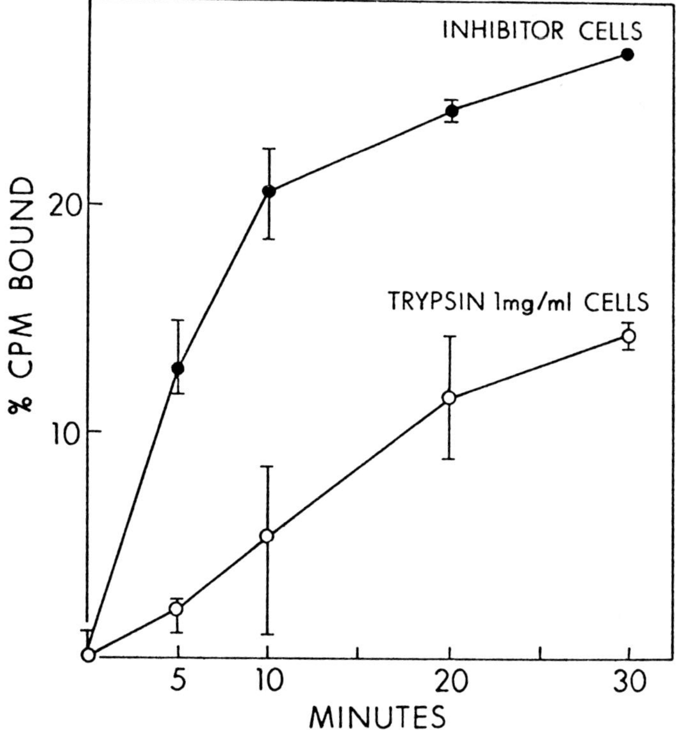

Fig. 4. Binding of plasma membranes to cultured cells. The figure illustrates the rate of adhesion of plasma membranes prepared from B103 cells (an established rat neuronal cell line, Schubert *et al.* 1974) to homologous cells. Membranes were prepared from cells grown in the presence of [^{3}H]-leucine. •, binding to cells previously incubated with trypsin and trypsin inhibitor. O, binding to cells previously incubated with trypsin. The figure illustrates the sensitivity of the binding site on the cells to trypsin, and its apparently rapid recovery rate. Reproduced with permission from Santala *et al.* (1977).

One of the adhesive components present in plasma membranes obtained from chick neuronal retina has been identified as ligatin, a small protein which can be obtained in pure form in relatively large quantities. Its precise function remains to be determined, but it blocks cell-to-cell adhesion in retina in a number of assay systems (Jakoi and Marchase, 1979).

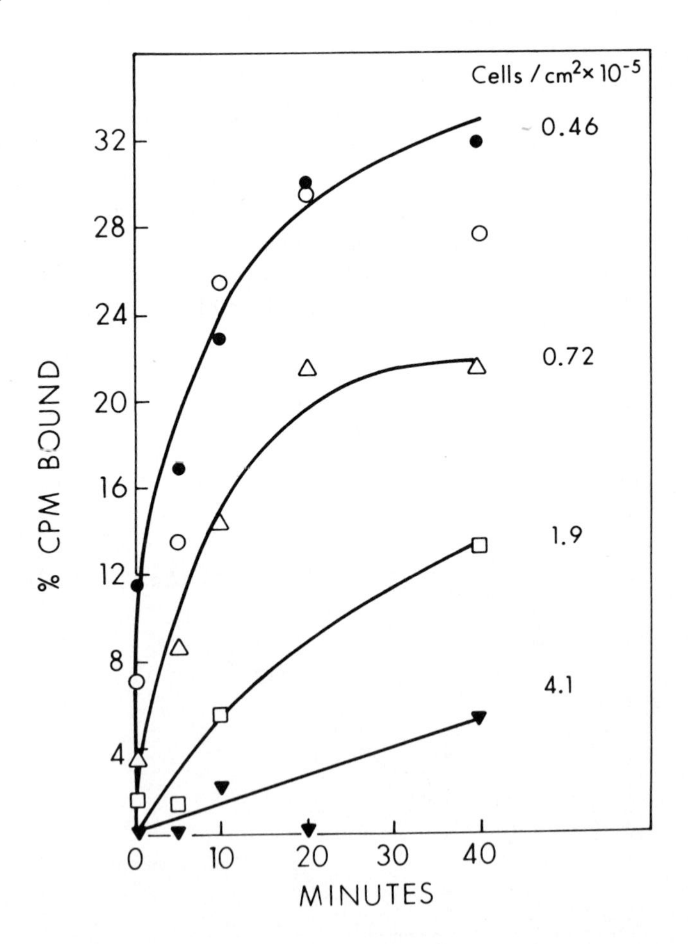

Fig. 5. Changes in adhesive properties of cells with culture conditions. The figure illustrates the binding of plasma membranes from C-6 cells, a rat astrocytoma to homologous cells, obtained from cultures grown to various densities. For details see Santala and Glaser (1977). As cell density increases, cells lose their ability to bind membranes, although membranes prepared from these cells can still bind to cells grown to low density. Neuronal cells in culture have a series of components designated A, B, a, b, that are responsible for their ability to bind plasma membranes (Santala *et al.*, 1977). A is complementary to a and B is complementary to b. The component lost from C-6 cells at high density is B, while the complementary component present in plasma membranes from C-6 cells is b.

Finally, specific cell adhesion can be studied by examining the functional consequences of cell adhesion. In situations where the binding of cells to each other results in a change in a cellular function, this change in function can be used to monitor the presence of such surface ligands on isolated plasma membranes when these are added to cells, and can subsequently be used to monitor the purification of the relevant surface components. An example of this approach will be discussed in detail below.

It should be realized that cell-to-cell adhesion is not a static phenomenon. Not only do cells "de-adhere" but also they change their apparent adhesive specificity during development (Merrell *et al.*, 1976; Rutishauser *et al.*, 1976), or under the influence of trophic factors or culture conditions in the laboratory. A simple example of this type of behavior is illustrated in Fig. 5.

IV. GROWTH CONTROL BY CELL-TO-CELL INTERACTION

Much of the work summarized in the previous section, both our own and that of other laboratories, has recently been reviewed in detail Frazier and Glaser, 1979; Gottlieb and Glaser, 1980), and we will concentrate in the remainder of this chapter on a more detailed discussion of one particular approach to the study of cell adhesion.

In a very superficial way, we may consider the interaction of two cells as a phenomenon similar to the binding of a hormone to a cell surface receptor. This implies that, as a consequence of the adhesion between two cells, a transmembrane event is triggered in one or both of the adhering cells that results in an alteration of cellular function. It is not necessary to assume that initial cell adhesion (specific cell adhesion) is determined by the same ligands that are involved in this functional interaction, although obviously this would represent the simplest model for this system (see Fig. 1). It follows from such a model that it should be possible to use this change in cell function to measure the presence of these factors in isolated plasma membranes and subsequently in purified components isolated from such membranes. This approach has been initiated in several systems and will be illustrated by a brief discussion of two systems from our laboratory, a) contact inhibition of growth in 3T3 cells and b) mitogenicity of the axonal cell surface for Schwann cells.

Swiss 3T3 cells grow on a plastic surface to a defined cell density, which is determined by a number of variables, including the concentration of mitogenic factors in the medium, but also cell-to-cell contact (Todaro *et al.*, 1963; Dulbecco, 1970; Holley, 1975; Whittenberger and Glaser, 1978).

Plasma membranes from 3T3 cells, when added to sparse 3T3 cells, inhibit growth in a time and concentration-dependent manner which resembles that observed at high cell density in a number of ways, as illustrated in Table I (Whittenberger and Glaser, 1977; Whittenberger *et al.*, 1979; Lieberman *et al.*, 1979). There are also some notable differences. Most striking is that, at high cell density but not in cells inhibited by membranes, the rate of uptake of glucose and phosphate is inhibited. We conclude from this discrepancy that additional factors are involved in the shut-down of these two uptake systems, which are therefore not obligatorily linked to contact inhibition of growth, a conclusion that has also been reached by others based on very different evidence (Barsh and Cunningham, 1978). It is important to emphasize that inhibition of cell growth by membranes is not due to the removal of mitogenic factors from growth medium. Plasma membranes have no effect on the rate of growth of SV-40 transformed 3T3 cells which do not show contact inhibition of growth.

TABLE I

Comparison of the Inhibition of Growth of 3T3 Cells by High Cell Density and Plasma Membranes

Function	Cell Growth Arrested by Plasma Membrane	Cell Growth Arrested by High Cell Density
Cells arrested early in G_1 portion of cell cycle	+	+
Maximum fraction of cells arrested per cell cycle	50%	Possibly 50%*
Reversibility:		
By trypsinization and replating	+	+
By increase in the concentration of defined Mitogens such as PDGF or by addition of serum	+	+
Decrease in rate of uptake of:		
α-Aminoisobutyric acid	+	+
Uridine	+	+
Glucose	—	+
Phosphate	—	+

*This is based on our interpretation of the data of Martz and Steinberg (1972).

Membranes labelled with ^{125}I can bind to cells under the same conditions as they inhibit cell division (M. Lieberman and T. Woolsey, unpublished observations). This binding does not occur when the membranes have been heat-inactivated so that they lose biological activity. Binding is also prevented by high concentrations of serum, and appears to be prevented by increasing concentrations of more defined mitogens, specifically platelet derived growth factor (Ross and Vogel, 1978; Antoniades *et al.*, 1979).

Contact inhibition of growth has been subject to a great deal of debate. The original experiments of Dulbecco (1970) showed quite clearly that medium depletion could not account for these observations. More recently it has been suggested that growth at confluency becomes diffusion limited (Stoker and Piggott, 1974; Whittenberger and Glaser, 1978) or anchorage limited (Folkman and Moscona, 1978; O'Neill *et al.*, 1979); that is that the limit of cell growth is the requirement for cells like 3T3 cells to be attached to the substratum for growth. The experiments with membranes present additional information relevant to the controversy in that they provide a direct test for the idea of contact inhibition of growth under conditions where neither the area of substratum available to the cells or diffusion of high molecular weight medium compounds are likely to be limiting. While membranes binding to cells may generate a diffusion barrier, this barrier must not be effective against small molecules such as glucose or phosphate. The solubilization of the inhibitory activity from membranes with the detergent octylglucoside, to a form which is non-sedimentable at 160,000 x g for 1 hour (Whittenberger, *et al.*, 1978), provides additional information in this regard, since this more disperse form of the membrane-derived inhibitor when bound to cells is far less likely to provide a diffusion barrier for molecules to reach the cell surface (Fig. 6).

The kinetics of inhibition are striking in that in each cell cycle only 50% of the cells can be inhibited by membranes and, by inference, by cell-to-cell contact. This implies that early in the G_1 portion of the cell cycle the sensitivity of cells to contact is set such that 50% of the cells can be inhibited by membranes or contact with other cells, while the other 50% will be resistant until they again reach the early portion of G_1, when the sensitivity will be reset.

The observation that mitogenic factors decrease the binding of membranes to cells and, by inference, the binding of cells to each other, suggests that one mechanism by which mitogens overcome contact inhibition of growth is by changes in cell surface recognition molecules. As an oversimplified picture, one may consider that cells like 3T3 cells receive a variety of signals from the environment, including mitogenic signals and inhibitory signals for growth, including those derived from cell-to-cell contact. The sum of those signals ultimately determines at some point early in G_1 whether the cell will proceed through the cell cycle

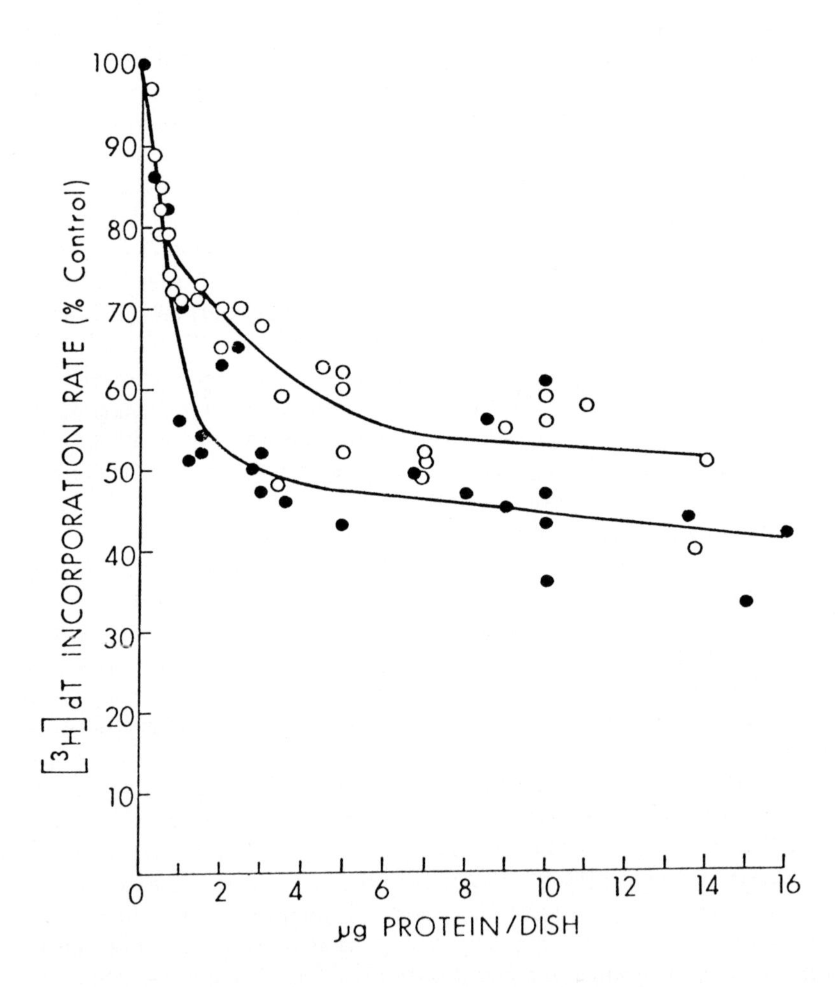

Fig. 6. Inhibition of growth of 3T3 cells by membranes. Sparse 3T3 cells were incubated with the indicated concentrations of plasma membranes from 3T3 cells (●), or an octylglucoside extract from such membranes (O) for 24 hours, after which the rate of thymidine incorporation into DNA was measured. Octylglucoside was removed from the extract by dialysis before addition of the cells. For details see Whittenberger *et al.* (1978). The rate of thymidine incorporation into DNA under these conditions is a measure of the number of cells in S phase of the cell cycle.

or become arrested. It should be realized that different cells, and cells grown under different conditions, will show various sensitivities to these signals and thus response to cell density or mitogens will differ with different cells and under different growth conditions.

Cells may not only respond to the presence of other cells by cessation of growth, but they may also initiate growth as a response to the presence of other cells. In one specific case, Schwann cells prepared from dorsal root ganglia by the procedure of Wood (1976), respond to the presence of neurites from dorsal root ganglia by initiating cell division. (Wood and Bunge, 1975). The signal appears to require contact between neurite and Schwann cells, inasmuch as separation of the neurite and Schwann cells by a collagen diaphragm, 6 micrometers thick, prevents the mitogenic effect (Salzer, J., unpublished observation).

The mitogenic effect of intact neurites can be reproduced, by the addition of neurite surface membranes to Schwann cells. The signal is absent from the neurite cytosol, is trypsin and heat sensitive, and appears to be due to the presence of one or more proteins on the surface of the neurites. The signal is highly specific and is absent from plasma membranes of a variety of cells, including neuronal cell lines (Salzer *et al.*, 1979).

These two systems, contact inhibition of growth and mitogenicity of neurites for Schwann cells, illustrate the fact that functional consequences of cell-to-cell adhesion can be reproduced when membrane vesicles are added to the appropriate target cells. The systems in which this has been successful are complex, in that we do not understand the factors that control progression of cells through the cell cycle and, therefore, cannot even make a very educated guess as to the nature of the transmembrane signals generated by cell-to-cell contact. While currently it would be tempting to suggest that specific protein phosphorylation or dephosphorylation may represent such a signal, it is clear that the number of other possibilities too extensive to mention may also apply. If the components involved can be purified in functional form, this may aid in the elucidation of the steps involved in growth control. It is not clear, for example, how many of the events associated with growth control are generated by the same signals. For example, the addition of plasma membranes to 3T3 cells inhibits growth and decreases the rate of transport of α-aminoisobutyric acid (Lieberman *et al.*, 1979), but we do not know with certainty whether these two events respond to the same surface signal; purification of membrane components may serve to resolve such problems.

The assay of functional consequences of cell adhesion, rather than cell adhesion per se, has certain practical and theoretical advantages. The practical advantage is that a binding assay for solubilized membrane components is not required and, as discussed above, may be difficult. The theoretical advantage is that cell-to-cell adhesion, measured in the laboratory, may be the fortuitous consequence of the presence of certain ligands on the cell surface. The observation of a defined physiological effect of cell adhesion, to a large extent, avoids this difficulty. This is particularly true of the mitogenic effects of neurites on Schwann cells, which clearly correlate with *in vivo* events. If it becomes possible to isolate in pure form the molecules responsible for this biological activity, it will be important to return to the intact cell system to prove that these molecules are, in fact, involved in the original cell-to-cell interaction rather than simply mimicking that event.

V. SUMMARY AND PERSPECTIVES

It is striking that in spite of many years of investigation in many laboratories, the precise mechanism by which any two cells specifically adhere to each other is not known. In different systems certain molecules have been identified that clearly represent components of the cell recognition system. For example, the lectins in slime molds, but we do not know enough in any system about the remaining components to be able to describe the adhesive events in detail. The impetus for the study of cell adhesion is very often the study of the precise interaction of cells during embryogenesis, for example synaptogenesis. These precise connections established in the embryo are in a sense microscopic events, in that each neuron connects with a limited repertoire of other cells. The methods for the study of cell adhesion are all macroscopic techniques, which examine the average properties of many cells.

It is possible, in fact, likely, that cell adhesion in different systems uses different molecules and that the adhesive events may involve cytoskeletal elements and complex rearrangements at the cell surface. This divergence between different cellular systems is likely to be even more striking when one examines adhesion under laboratory conditions with different assays.

No single approach or system is a priori more likely to be more fruitful than others, and novel approaches to deal with the purification of small quantities of intrinsic membrane proteins and their assay as components of cell-to-cell adhesion will clearly be required.

The approach that we favor at the moment is the use of functional assays of cell-to-cell adhesion where these are possible and it is hoped that studies of this type of system will become more popular. In embryonal systems these assays are less likely to be fruitful and cell adhesion will continue to be measured by current assays. Defined anatomical regions within the embryo are complex and contain many cell types. It would seem increasingly important to cease to treat each anatomical region as if it contained a uniform cell population and to place increasing emphasis on the preparation of uniform populations of differentiated cells of a given type. Specific antibodies are clearly one approach to this type of problem and may provide invaluable tools for the study of cell-to-cell adhesion if during cell fractionation the structure of the cell surface can be maintained.

The gap between the morphological and the chemical approaches remains, not by choice, but because of the quantitative aspects of the tools. The morphologist can examine a few cells in their natural environment, the chemist needs quantities of material several orders of magnitude in excess of what is required for morphology, and because of this gap, problems such as the precise chemical basis of neuronal connectivity continue to be outside the grasp of the chemist.

No attempt has been made to present a comprehensive review of the field, since several said reviews have been published recently, rather we have tried to highlight some of the major problems currently under investigation and cite some recent references.

ACKNOWLEDGEMENTS

The concepts presented have benefited from many discussions with D. Gottlieb, M. Lieberman, R. Merrell, F. Moya, G. Mintz, D. Raben, J. Salzer and B. Whittenberger, who are also responsible for any progress we have made in this field. In particular, I have benefitted from the collaboration with R. Bunge on the research on Schwann cells which originated in his laboratory. Work in my laboratory has been supported by grants GM 18405 and NSF PCM77-15972.

REFERENCES

Antoniades, H.N., Scher, C.D. and Stiles, C.D. (1979). *Proc. Nat. Acad. Sci. U.S.A.* **76**, 1809-1813.

Barbera, A.J., Marchase, R.B. and Roth, S. (1973). *Proc. Nat. Acad. Sci. U.S.A.* **70**, 2482-2486.

Barsh, G.S. and Cunningham, D.D. (1978). *J. Supramol. Struct.* **7**, 61-77.

Beug, H. and Gerisch, G. (1972). *J. Immun. Methods* **2**, 49-57.

Burger, M.M. and Jumblatt, J. (1977). *In* "Cell and Tissue Interaction" (J.W. Lash and M.M. Burger, eds.), pp. 155-172. Plenum Press, New York.

Dulbecco, R. (1970). *Nature* **227**, 802-806.

Folkman, J. and Moscona, A. (1978). *Nature* **273**, 345-349.

Frazier, W.A. and Glaser, L. (1979). *Ann. Rev. Biochem.* **48**, 491-523.

Gottlieb, D.I. and Arington, C. (1979). *Develop. Biol.*, in press.

Gottlieb, D.I. and Glaser, L. (1980). *Ann. Rev. Neurosciences* **3**, in press.

Gottlieb, D.I., Rock, K. and Glaser, L. (1976). *Proc. Nat. Acad. Sci. U.S.A.* **73**, 410-414.

Hausman, R.E. and Moscona, A.A. (1975). *Proc. Nat. Acad. Sci. U.S.A.* **72**, 916-920.

Hausman, R.E. and Moscona, A.A. (1976). *Proc. Nat. Acad. Sci. U.S.A.* **73**, 3594-3596.

Hausman, R.E. and Moscona, A.A. (1979). *Exp. Cell. Res.* **119**, 191-204.

Holley, R.W. (1975). *Nature* **258**, 487-490.

Jakoi, E.R. and Marchase, R.B. (1979). *J. Cell Biol.* **80**, 642-650.

Lieberman, M.A., Raben, D.M., Whittenberger, B. and Glaser, L. (1979). *J. Biol. Chem.* **254**, in press.

Marchase, R.B. (1977). *J. Cell Biol.* **75**, 237-257.

Martz, E. and Steinberg, H.S. (1972). *J. Cell Physiol.* **79**, 189-210.

McClay, D.R., Gooding, L.R. and Fransen, M.E. (1977). *J. Cell Biol.* **75**, 56-66.

McGuire, E.J. and Burdick, C.L. (1976). *J. Cell Biol.* **68**, 80-89.

Merrell, R., Gottlieb, D.I. and Glaser, L. (1976). *In* "Neuronal Recognition" (S.H. Barondes, ed.), pp. 249-273. Plenum Press, New York.

Mintz, G. and Glaser, L. (1978). *J. Cell Biol.* **79**, 132-137.

Moscona, A.A. (1965). *In* "Cells and Tissues in Culture" (E.N. Willmer, ed.), Vol, I, pp. 197-220. Academic Press, New York.

Moya, F., Silbert, D.F. and Glaser, L. (1979). *Biochim. Biophys. Acta* **550**, 485-494.

Muller, K. and Gerisch, G. (1978). *Nature* **274**, 445-449.

Muller, W.E.G., Zahn, R.K., Kurdee, B., Muller, I., Uhlenbreck, G. and Vaith, P. (1979). *J. Biol. Chem.* **254**, 1280-1287.

Obrink, B., Kuhlenschmidt, M.S. and Roseman, S. (1977). *Proc. Nat. Acad. Sci. U.S.A.* **74**, 1077-1081.

O'Neill, C.H., Riddle, P.N. and Jordan, P.W. (1979). *Cell* **16**, 409-418.

Porzig, E.F. (1978). *Develop. Biol.* **67**, 114-136.

Ross, R. and Vogel, A. (1978). *Cell* **14**, 203-210.

Roth, S. (1968). *Develop. Biol.* **18**, 602-613.

Roth, S. and Weston, J.A. (1967). *Proc. Nat. Acad. Sci. U.S.A.* **58**, 974-978.

Rutishauser, U., Thiery, J.P., Brackenburg, R., Sela, B.A. and Edelman, G.M. (1976). *Proc. Nat. Acad. Sci. U.S.A.* **73**, 577-581.

Rutishauser, U., Thiery, J.P., Brackenbury, R. and Edelman, G.M. (1978a). *J. Cell Biol.* **79**, 371-381.

Rutishauser, U., Gall. W.E. and Edelman, G.M. (1978b). *J. Cell Biol.* **79**, 382-393.

Salzer, J., Glaser, L. and Bunge, R. (1979). *J. Cell Biol.*, in press.

Sanes, J.R., Marshall, L.M. and McMahan, U.J. (1978). *J. Cell Biol.* **78**, 196-198.

Sankaran, L., Proffitt, R.T., Petersen, J.R. and Pogell, B.M. (1977). *Proc. Nat. Acad. Sci. U.S.A.* **74**, 4487-4490.

Santala, R. and Glaser, L. (1977). *Biochem. Biophys. Res. Commun.* **79**, 285-291.

Santala, R., Gottlieb, D. I., Littman, D. and Glaser, L. (1977). *J. Biol. Chem.* **252**, 7625-7634.

Schnaar, R.L., Weigel, P.H., Kuhlenschmidt, M.S., Lee, Y.C. and Roseman, S. (1978). *J. Biol. Chem.* **253**, 7940-7951.

Schubert, D., Harrimann, S., Carlisle, W., Tarikas, H., Kimes, B., Patrick, J., Steinbach, J.H., Culp, W. and Brandt, B.L. (1974). *Nature* **249**, 224-227.

Schur, B.D. and Roth, S. (1975). *Biochim. Biophys Acta* **415**, 473-512.

Steinberg, M.S. (1978). *In* "Cell-Cell Recognition." SEB Symp. 32 (A.S.G. Curtis, ed.), pp. 25-49. Cambridge University Press.

Stoker, M. and Piggott, D. (1974). *Cell* **3**, 207-215.

Todaro, G.J., Green, H. and Goldberg, B.D. (1963). *Proc. Nat. Acad. Sci. U.S.A.* **51**, 66-73.

Umbreit, J. and Roseman, S. (1975). *J. Biol. Chem.* **250**, 9360-9368.

Weigel, P.H., Schnell, E., Lie, Y.C. and Roseman, S. (1978). *J. Biol. Chem.* **253**, 330-333.

Whittenberger, B. and Glaser, L. (1977). *Proc. Nat. Acad. Sci. U.S.A.* **74**, 2251-2255.
Whittenberger, B. and Glaser, L. (1978). *Nature* **272**, 821, 823.
Whittenberger, B., Raben, D., Lieberman, M.A. and Glaser, L. (1978). *Proc. Nat. Acad. Sci. U.S.A.* **75**, 5457-5461.
Whittenberger, B., Raben, D. and Glaser, L. (1979). *J. Supramol. Struct.* **10**, 307-327.
Wood, P.M. (1976). *Brain Res.* **115**, 361-375.
Wood, P. and Bunge, R.P. (1975). *Nature* **256**, 662-664.

II. The Cell Surface and Early Development

Membrane Adhesions Between *Chlamydomonas* Gametes and Their Role in Cell—Cell Interactions

Ursula W. Goodenough

Department of Biology
Washington University
St. Louis, MO 63130

W. Steven Adair

Department of Anatomy and Neurobiology
School of Medicine,
Washington, University
St. Louis, MO 63110

I. INTRODUCTION

The mating reaction of *Chlamydomonas* gametes entails cell-cell interactions analogous, if not homologous, to those that may occur between embryonic cells. The two gamete types (*mt*+ and *mt*−) first recognize one another by specific molecules displayed on surface membranes, and cell-cell adhesions are established. These adhesions, in turn, stimulate pairs of interacting gametes to establish junction-like connections between specialized regions of their plasma membranes. Once the junctions form, the cells proceed to fuse together, and the initial adhesions then come apart. The circuit, in short, entails an initial

ISBN 0-12-612984-3

recognition, an "on" response to the recognition event, the formation of junctions, and an "off" response once the interaction is complete.

We have, during the past 3 years, had several opportunities to review studies of the *Chlamydomonas* mating reaction (Goodenough, 1977; Goodenough and Forest, 1978; Goodenough *et al.*, 1979; Goodenough, 1980). To minimize duplication, and to keep within the topic of this symposium, we focus here on the initial recognition/adhesion events, with some attention necessarily given to the "on" response they elicit.

II. THE DEVELOPMENT OF ADHESIVE MEMBRANES

Gametic differentiation in haploid *C. reinhardi* cells is easily synchronized (Kates and Jones, 1964; Martin and Goodenough, 1975) by suspending a synchronous liquid culture of vegetative (mitotic, nonsexual) cells into a nitrogen-free minimal medium (NFHSM). Adhesive flagellar membranes are ordinarily not produced by the cells until the end of the 10-12 hr differentiation period. At this time the cells withdraw their nonadhesive vegetative flagella and undergo one or, more commonly, two successive rounds of mitosis, just as in a vegetative mitosis (Johnson and Porter, 1968). The new flagella that grow out following these divisions are found to be adhesive when, for example, an *mt*+ culture is mixed with differentiated *mt*− gametes or the reverse.

Genetic control over the differentiation process is ultimately exerted by the *mt*+ and *mt*− loci, which segregate as a pair of alleles during meiosis (Smith and Regnery, 1950) and which control the expression of a number of gametic traits. The flagellar adhesiveness of *mt*+ gametes is influenced by 2 unlinked gene loci, designated *sag-1* (*s*exual *ag*glutination) and *sag-2* (Goodenough *et al.*, 1978). The *sag-1* locus is marked by 5 mutations (*imp-2, -5, -6, -7,* and *-9*) while the *sag-2* locus is to date marked by a single mutation (*imp-8*). Neither locus is linked to *mt*+, yet both require the presence of *mt*+ for their expression: an *mt*+ *sag-1* (or *sag-2*) cell undergoes a normal gametic *mt*+ differentiation but fails to develop adhesive flagella (Bergman *et al.*, 1975) while an *mt*− *sag-1* cell undergoes a normal *mt*− gametogenesis and displays normal *mt*− agglutinins on its flagella. The expression of the *sag-1* and *sag-2* loci, in other words, is said to be *sex-limited* to *mt*+ cells. The mutation *gam-1*, which affects the signaling phase of the adhesion reaction, is similarly found to be sex-limited to *mt*− cells (Forest and Togasaki, 1975; Forest *et al.*, 1978).

The phenotype of *sag-1* and *sag-2* mutants reveals that the development of sexual adhesiveness proceeds independently of other aspects of gametic differentiation. Specifically, gametogenesis is known to entail the acquisition of flagellar-tip agglutinability by Concanavalin A (Wiese and Shoemaker, 1979; McLean and Brown, 1974) and by anti-flagellar antiserum (Goodenough and Jurivich, 1978). It also entails the assembly of mating structures (Friedmann *et al.*, 1968), which participate in junction formation and which undergo activation in response to tip adhesions (Weiss *et al.*, 1977). All of these capacities differentiate in the *sag-1* and *sag-2* mutants as well. Indeed, the mutants can be induced to undergo normal mating and zygote formation (allowing genetic analysis) simply by substituting antibody-mediated adhesions for sexual adhesions (Goodenough and Jurivich, 1978). With the possible exception of *imp-7* (Goodenough and Jurivich, 1978), each mutant strain we have isolated appears to be defective in a single gametic phenotype. In other words, the phenomenon of pleiotropy, which plagues many genetic analyses of development, does not present a problem with the mutants at hand.

While means have been found to bypass the necessity for sexual adhesiveness during laboratory matings, the adhesins of *Chlamydomonas* are unquestionably the natural initiators of cell-cell interactions, and their specificity is quite remarkable. Different species of *Chlamydomonas* exhibit no sexual interactions with one another (Wiese, 1969). Moreover, antisera raised against *C. reinhardi* flagella have no agglutinative effects when presented to *C. moewusii* or to various multicellular Volvocales (e.g. *Pandorina, Endorina*) (A. Coleman and U. Goodenough, unpublished experiments). We can therefore envisage flagellar membrane surfaces as highly differentiated features of these organisms, a reasonable conclusion when it is recalled that their cell membranes are normally covered by glycoproteinaceous walls. The flagellar membranes, therefore, emerge as the focal surfaces for detecting and transmitting external information.

III. MAINTAINING ADHESIVE MEMBRANES

Four sets of observations are relevant to the question of how the adhesive state is maintained. 1) Vegetative and gametic *Chlamydomonas* cells are known to engage in continuous "blebbing" of small membrane vesicles from their flagellar tips. These vesicles can be harvested from the culture media and, when they derive from gametic cells, they are found to bear *mt*-specific agglutinins (Bergman *et al.*, 1975; Snell, 1976): mt^+ vesicles will cause mt^- gametes to isoagglutinate, and vice-versa, presumably because the vesicles are "multivalent." 2) This continuous shedding process must necessitate the continuous synthesis of fresh surface components, and the experiments of Solter and Gibor (1977, 1978) demonstrate the existence of a "pool" of agglutinins, persistent after a cycloheximide block, which undergoes normal assembly into a regenerating flagellum. 3) If sexual agglutination is followed by zygotic cell fusion, the flagella disadhere. The agglutinin pool is also inactivated by zygotic cell fusion, moreover, since deflagellated zygotes regenerate nonadhesive flagella. 4) The agglutination reaction between nonfusing *imp-1 mt*+ mutants and normal *mt*- gametes persists for many hours, indicating that in the absence of fusion, adhering cells continue to maintain their adhesive state. By contrast, if gametes of one *mt* are agglutinated with isolated flagella of opposite *mt*, the reaction ceases after about 15 min (Snell and Roseman, 1979): the cells remain responsive to freshly added flagella but the "spent" flagella have no adhesive properties when presented to fresh cells.

We can incorporate these observations into the following model. We visualize adhesive molecules as being continuously synthesized and then inserted into proximal sectors of the gametic flagellar surface, replacing molecules lost by vesiculation from the distal end. We propose that, during sexual agglutination, interacting *mt*+ and *mt*- adhesive molecules effectively neutralize one another. Since isolated flagella cannot replace such neutralized molecules by the synthesis/insertion pathway, they become "spent" by adhesive interactions. We further imagine that zygotic cell fusion brings about a change in the synthesis/insertion process: perhaps the *mt*+ and *mt*- "pool agglutinins", when brought together by cytoplasmic confluence, are able to neutralize one another, or perhaps the insertion process itself is blocked. As a consequence, neither the shed nor the neutralized agglutinins of fused cells are replaced at the base, flagellar disagglutination results, and a functional change is conferred upon the zygotic pool.

IV. FLAGELLAR MEMBRANE COMPOSITION

The foregoing model can be rigorously tested only after the adhesive molecules are identified biochemically. Moreover, the isolation and identification of the adhesins must precede any experimental studies of how the adhesins interact, when they are synthesized, and how *mt*+ and *mt*- adhesins differ. The pioneering work of Wiese and his colleagues (1969) established that the agglutinins are sensitive to proteolysis and suggested that they are glycosylated. More recent work in this laboratory, most notably by Ken Bergman, Steve Adair, and Brian Monk, and work in the laboratory of Bill Snell, has yielded the following results.

We first analyzed the glycopolypeptide composition of shed vegetative and gametic mt^+ and mt^- flagellar membrane vesicles (Bergman *et al.*, 1975; Snell, 1976) and determined that these membranes bear a single major PAS-positive species, with very slow electrophoretic motility. We concluded that this major component was indistinguishable in the four types of membrane preparations analyzed, a result consistent with at least 4 interpretations: 1) the agglutinins are such minor components of the vesicle membranes that they are not detected in SDS gels; 2) the agglutinins co-migrate in SDS gels with the "major membrane glycopolypeptide"; 3) the major membrane polypeptide of the vegetative flagellum is modified during gametic differentiation to bear agglutinative groups without changing its electrophoretic mobility; 4) the vegetative membrane polypeptide is biochemically identical to that in gametes but becomes an agglutinin as a consequence of some other change in gametic flagellar architecture (e.g. increased membrane "fluidity").

More recent studies of total flagellar membrane composition (Monk *et al.*, 1979) reveal that, in addition to the "major membrane glycopolypeptide", some 20 minor polypeptides of diverse electrophoretic mobilities are detected by ^{125}I-surface labeling techniques. Most of these polypeptides are not, however, included in the vesiculated material. In other words, only certain polypeptides, including the "major membrane glycopolypeptide" and the agglutinin, are shed from the flagellar tips. The mechanism for this selectivity is presently unknown.

If antisera are prepared against the 4 flagellar types (αV^+, αV^-, αG^+, αG^-), all four cell types are agglutinated by all 4 antisera, indicating the existence of common antigen(s). This conclusion is confirmed by indirect immunoautoradiography (Adair *et al.*, 1978), which reveals that all 4 antisera contain antibody directed against most if not all of the 20 ^{125}I-labeled species. Immunoabsorption studies (e.g. αG^+ exhaustively absorbed to mt^- vegetative or gametic cells, followed by testing with mt^+ gametes) fail to reveal *mt*-specific antibodies in these sera (unpublished experiments from this laboratory). This result would be expected if the anti-agglutinins are minor species that are nonspecifically lost during absorption and/or if the agglutinins are side groups of common antigenic species. It is also, of course, the expected result if the agglutinins are not antigenic, although experiments to be described in a later section indicate that this is not the case.

V. THE REQUIREMENT FOR A LIVING CELL

An intriguing property of the sexual agglutinins is that their activity cannot be detected unless at least one mating type in the assay is a living cell (Goodenough, 1977). Thus, for example, *mt*+ vesicles, flagella, or glutaraldehyde-fixed gametic cells fail to exhibit any adhesiveness to their *mt*- counterparts, but all are readily recognized and adhered to by living *mt*- gametes. This requirement for a living cell is not encountered if lectins or antisera are utilized: isolated flagella, for example, agglutinate readily in the presence of anti-flagellar antiserum. These observations can be interpreted to mean that some vital activity such as ATP generation must attend the sexual adhesive process. Alternatively, one can imagine that 2 membranes will stick together only if the agglutinating molecules "patch" together to form concentrated aggregates and that, for the sexual adhesins, such patches fail to form between two sets of nonliving membranes.

The requirement for a living cell becomes relevant in designing agglutinin-isolation schemes, since most detergents one might choose to solubilize the agglutinin are toxic to, or at least cause deflagellation of, living tester gametes. We have therefore utilized the detergent octylglucoside (OG), which is readily dialyzed, to extract the agglutinin from isolated flagella.

VI. ISOLATING AND IDENTIFYING THE AGGLUTININ POLYPEPTIDES

Described here is the standard protocol for isolating the *mt*- agglutinin, developed by Steve Adair and Carol Hwang (Adair *et al.*, 1979). An identical protocol allows the isolation of the *mt*+ species.

Isolated *mt*- flagella (Witman *et al.*, 1972) are extracted with 30 mM OG in 10 mM Tris, pH 7.4, containing 10^{-4} M EGTA and 0.1 mM dithiothreitol. Extracted axonemes are pelleted by high-speed centrifugation; the OG extract, which contains both membrane and liberated flagellar-matrix components, is dialyzed overnight against several changes of NFHSM. The dialyzed extract is most readily assayed for activity by allowing it to absorb to glass-wool fibers and presenting the fibers to various cell types. It is found that *mt*+ gametes agglutinate vigorously to the coated fibers, whereas no reaction occurs when the fibers are presented to *mt*- gametes, *mt*+ or *mt*- vegetative cells, or *mt*+ *sag-1* or *mt*+ *sag-2* mutant gametes. Moreover, such agglutination is found to trigger *mt*+ mating-structure activation (Mesland *et al.*, 1980), indicating that the OG extract retains full signaling ability.

We have taken several approaches to fractionating the OG extracts so as to selectively enrich for the agglutinins. One approach has involved immunoaffinity chromatography in which, for example, an *mt*+ OG extract is passed first through an αV^--Sepharose 4B column to remove "common antigens" and then exposed to an αG^+-Sepharose 4B column. When the material retained on this second column is eluted with potassium isothiocyanate, precipitated with acetone, and tested for activity, strong *mt*+ agglutinin activity is found; when displayed by SDS/PAGE, a band migrating in the "major membrane glycopolypeptide" region is heavily enriched. In overloaded gels, however, faint bands appear elsewhere as well; therefore, this approach is ultimately only able to indicate agglutinin *candidates*.

To determine which polypeptide(s) in OG extracts in fact carries agglutinin activity, we performed a modification of what we have come to call the "green-band experiment". In its original formulation, this experiment involved presenting an unstained SDS gel of *mt*⁻ flagellar polypeptides to living *mt*+ gametes and looking for the green cells agglutinating to a particular sector of the gel. When we realized that the heat-sensitive agglutinin would probably lose activity when subjected to electrophoresis at room temperature, we turned to lithium dodecyl sufate (LDS) (Delepelaire and Nam-Hai Chua, 1979), which can be used as the detergent in electrophoresis running buffers maintained at 4°C. We determined that if a dialyzed OG extract is acetone-precipitated, solubilized in cold LDS, and re-precipitated in acetone, *mt*-specific agglutinin activity is not lost.

The green-band experiment was therefore successfully performed by subjecting OG extracts to LDS/PAGE in preparative slab gels, cutting bands from the gels, eluting the material from the bands either by prolonged washing in Tris-glycine or by electrophoresis, acetone-precipitating the eluants, and testing their activity. For both *mt*+ and *mt*⁻ extracts, sex-specific agglutinin activity is indeed found to localize in the high-molecular-weight glycopolypeptide region enriched on the immunoaffinity chromatography columns. That activity resides in a single region of LDS gels indicates that the agglutinin is not dependent for its activity on its association with additional polypeptides having different electrophoretic mobilities. It is also clear that agglutinin activity is not dependent on membrane architecture.

With a bioassay now at hand, it is possible to analyze the candidate polypeptides biochemically. We will first determine, by N–terminal analysis, whether purified agglutinin bands carry one or several polypeptide species. If several species are present, we will attempt to separate these and determine which carry agglutination activity. We will next subject "active polypeptides" to limited proteolytic cleavage to ascertain whether smaller peptides can be isolated which retain activity. Finally, we will subject *mt*+ and *mt*⁻ material to biochemical analysis (e.g. carbohydrate and amino-acid composition) and will ask whether the active (poly) peptides interact *in vitro*. Parallel studies will be made of preparations from vegetative flagella and from the nonagglutinating mutants.

We have been able to simplify the OG extracts by dialyzing them against distilled water: most of the extract components precipitate out, leaving a "dialyzed OG supernatant" (DOGS) which is greatly enriched in the major membrane polypeptide relative to the other polypeptides in the extract. DOGS preparations from *mt*+ and *mt*⁻ flagella exhibit excellent agglutinin activity; they are devoid of activity when derived from vegetative flagella, from *sag-1* or *sag-2* mutant flagella, or from flagella obtained from *mt*+ gametes pre-treated with trypsin to inactivate their agglutinability (Wiese, 1969). SDS/PAGE of the various DOGS preparations has to date revealed no detectable electrophoretic differences between the major membrane polypeptide regions of gametic *mt*+, *mt*⁻, *sag-1*, *sag-2*, or trypsinized samples. A lower electrophoretic mobility of this polypeptide region has, on the other hand, been observed for vegetative DOGS preparations; whether this is due to different growth conditions or to the acquisition of agglutinability is presently being explored.

VII. HOW DO ADHESIONS SIGNAL?

Flagellar adhesions between *mt*+ and *mt*⁻ gametes unquestionably bring cells of the correct species and mating type into close proximity, and one can design fertilization schemes in which achieving such proximity represents the sole "purpose" of agglutination. We have found, however, that for *Chlamydomonas*, the adhesions also trigger the subsequent steps of the mating reaction.

When living gametic flagellar surfaces are cross-linked by either sexual agglutinins, anti-flagellar antibody, or Concanavalin A, the cells respond by moving the complexes out to the flagellar tips (Goodenough and Jurivich, 1978; Mesland and van den Ende, 1979). This "tipping" reaction is not performed by vegetative cells and is blocked by colchicine and vinblastine. Dick Mesland and collaborators have recently discovered that tipping is accompanied by the accumulation in the flagellar tips of a dense substance (fibrous tip material or FTM) beneath the flagellar membrane, by an elongation of flagellar tip microtubules, and by signal transmission to the cell body (Mesland and Goodenough, 1979; Mesland *et al.*, 1980). Cell–wall lysis and mating-structure activation occur in response to this signal and, at the time of disadhesion, flagella return to their non-activated state, with FTM disappearing and microtubules resuming their original length. When cycles of adhesion/disadhesion are induced by presenting successive samples of isolated flagella to cells of opposite mating type, parallel cycles of tip activation/deactivation are observed.

That it is the adhesions themselves that elicit the signal is most convincingly shown by the following series of observations (Mesland *et al.*, 1980). 1) If an EM grid surface is coated with a Formvar film and then with polylysine, and *mt*+ gametes stick by their flagella to this film, the cells do not activate their flagellar tips nor their mating structures. Flagellar immobilization *per se*, in other words, is not sufficient for the response. 2) If the polylysine-tethered cells are presented with a drop of *mt*⁻ flagella, tip and mating-structure activation occur normally. The polylysine is not, therefore, inhibiting cellular responsiveness *per se*. 3) If a Formvar-coated grid is instead covered with a layer of dialyzed *mt*⁻ OG flagellar extract and then with a monolayer of *mt*+ gametes, flagella and mating structures activate normally. In this situation, neither *mt*⁻ cells, flagella, nor membranes are present, nor do the *mt*+ cells associate with one another *via* some isoagglutinating agent. We therefore conclude that an individual cell can autonomously generate a full sexual response to agents that generate flagellar membrane adhesions.

We presently have no idea how the above described changes in flagellar structure might serve to elicit cellular responses. Since colchicine and vinblastine have no effect on flagellar adhesiveness but block tip activation, however, it is clear that adhesions are necessary but not sufficient to trigger the subsequent steps in the reaction. Additional cellular activities, sensitive to the antitubulin drugs, couple agglutination with tip signaling in some fashion.

VIII. SPECULATIONS ON THE ROLE OF ADHESION IN METAZOAN CELL-CELL INTERACTIONS

Most examples of cellular adhesions in the Metazoa (reviewed by Frazier and Glaser, 1979) involve cells of apparently identical phenotype:

slime-mold amoeba or sponge cells associate with their genetically identical counterparts; tissues are formed by adhesions between clonally derived sister cells. Such adhesions are in three respects different from those described here. 1) They are stable rather than short-lived and reversed. 2) They are mediated or at least stabilized to a large extent by extracellular materials (e.g. lectins or glycocalyx components) and not by membrane-membrane interactions alone. 3) The adhesive molecules are typically found to be effective in agglutinating nonliving cells (i.e. the "requirement for a living cell" is not operative).

Are there, then, adhesive interactions during embryogenesis or histogenesis in which *un*like cells make brief, reversible, membrane-membrane contacts with one another which require that at least one cell be alive? To our knowledge, this question cannot as yet be answered, although the opportunity for such contacts certainly arises during inductive encounters between differentiating cells. It is intriguing to learn that most if not all early embryonic cells, including those of mesodermal origin, bear short, nonmotile cilia of undefined function (Sorokin, 1968; Rash *et al.*, 1969). There are, therefore, at least morphological grounds for imagining that cilia may play role(s) in embryogenesis, a speculation that is given impetus by recent evidence that mammalian syndromes characterized by mutant cilia may include aberrant embryological development (Afzelius, 1976; Arden and Fox, 1979).

What kinds of insights, then, might the *Chlamydomonas* mating reaction provide in seeking to understand other kinds of cell-cell interactions? 1) It may become fruitful to consider that operative interactions are mediated by small portions of the cell surface rather than the entire surface. 2) It may be important that interacting molecules become concentrated by some sort of cross-linking mechanism and that they migrate to a particular sector of the cell [e.g., the region overlying the centriole, as in the capping response (Schreiner and Unanue, 1976)]. 3) A clone of cells may display different agglutinins at different developmental stages, each responsive to different cell types but all perhaps triggering a common signal mechanism. 4) With *Chlamydomonas,* it is possible to stimulate the same mating response using either Concanavalin A, antibody, or sexual agglutinins as the cross-linking agents. It may be possible to play similar "tricks" on embryonic cells under appropriate experimental conditions. 5) The entire adhesion/signal/response/disadhesion cycle in *Chlamydomonas* requires only 30 sec. Such rapid initial kinds of interactions might miss detection in more complex situations, particularly if they are followed by slower and more conspicuous forms of associative behavior such as junction formation or the secretion of extracellular adhesins.

ACKNOWLEDGEMENTS

The experiments from this laboratory described in this review are the result of a most enjoyable and fruitful collaboration involving, at present, Jackie Hoffman, Dennis Hourcade, Carol Hwang, Dick Mesland, and Brian Monk and, in the recent past, Ken Bergman, Eve Caligor, Charlene Forest, Don Jurivich, Jane Warren, and Dick Weiss. We are supported by grants from the NIH and NSF.

REFERENCES

Adair, W.S., Jurivich, D. and Goodenough, U.W. (1978). *J. Cell Biol.* **79**, 281-285.
Adair, W.S., Hwang, C. and Goodenough, U.W. (1979). *J. Cell Biol.*, **83**, 75a.
Afzelius, B.A. (1976). *Science* **193**, 317-319.
Arden, G.B. and Fox, B. (1979). *Nature* **379**, 534-536.
Bergman, K., Goodenough, U.W., Goodenough, D.A., Jawitz, J. and Martin, H. (1975). *J. Cell Biol.* **67**, 606-622.
Delepelaire, P. and Nam-Hai Chua. (1979). *Proc. Nat. Acad. Sci. U.S.A.* **76**, 111-115.
Forest, C.L. and Togasaki, R.K. (1975). *Proc. Nat. Acad. Sci. U.S.A.* **72**, 3652-3655.
Forest, C.L., Goodenough, D.A. and Goodenough, U.W. (1978). *J. Cell Biol.* **79**, 74-84.
Frazier, W. and Glaser, L. (1979). *Ann. Rev. Biochem.* **48**, 491-523.
Friedmann, L., Colwin, A.L. and Colwin, L.H. (1968). *J. Cell Sci.* **3**, 113-128.
Goodenough, U.W. (1977). *In* "Microbial Interactions," Receptors and Recognition, Series B, (J.L. Reissig, ed.) **3**, 323-350, Chapman and Hall, London.
Goodenough, U.W. (1980). Symp. Soc. Gen. Microbiol., in press.
Goodenough, U.W. and Forest, C.L. (1978). *In* "The Molecular Basis of Cell-Cell Interaction" (R.A. Lerner and D. Bergsma, eds.) pp. 429-438. Alan R. Liss, Inc., New York.
Goodenough, U.W. and Jurivich, D. (1978). *J. Cell Biol.* **79**, 680-693.
Goodenough, U.W., Hwang, C. and Warren, A.J. (1978). *Genetics* **89**, 235-243.
Goodenough, U.W., Adair, W.S., Caligor, E., Forest, C.L., Hoffman, J.L., Mesland, D.A.M. and Spath, S. (1980). *In* "Membrane-Membrane Interactions" (N.B. Gilula, ed.), Raven Press, New York.
Johnson, U.G. and Porter, K.R. (1968). *J. Cell Biol.* **38**, 403-425.
Kates, J.R. and Jones, R.F. (1964). *J. Cell. Comp. Physiol.* **63**, 157-164.
Martin, N.C. and Goodenough, U.W. (1975). *J. Cell Biol.* **67** 587-605.
McLean, R.J. and Brown, R.M. Jr. (1974). *Develop. Biol.* **36**, 279-285.
Mesland, D.A.M. and van den Ende, H. (1979). *Protoplasma* **98**, 115-129.
Mesland, D.A.M., and Goodenough, U.W., (1979). *J. Cell Biol.*, **83**, 182a.
Mesland, D.A.M., Hoffman, J.L., Caligor, E. and Goodenough, U.W. (1980). *J. Cell Biol.*, in press.
Monk, B., Adair, W.S., Hoffman, J.L. and Goodenough, U.W. (1979). *J. Cell Biol.*, **83**, 182a.
Rash, J.E., Shay, J.W. and Biesle, J.J. (1969). *J. Ultrastr. Res.* **29**, 470-484.
Schreiner, G.F. and Unanue, E.R. (1976). *Adv. Immunol.* **24**, 37-165.
Smith, G.M. and Regnery, D.C. (1950). *Proc. Nat. Acad. Sci. U.S.A.* **36**, 246-248.
Snell, W.J. (1976). *J. Cell Biol.* **68**, 48-69.

Snell, W.J. and Roseman, S. (1979). *J. Biol. Chem,* **254**, 10820-10829.
Solter, K.M. and Gibor, A. (1977). *Nature* **265**, 444-445.
Solter, K.M. and Gibor, A. (1978). *Exp. Cell Res.* **115**, 175-181.
Sorokin, S.P. (1968). *J. Cell Sci.* **3**, 207-230.
Weiss, R.L., Goodenough, D.A. and Goodenough, U.W. (1977). *J. Cell Biol.* **72**, 144-160.
Wiese, L. (1969). Algae. *In* "Fertilization: Comparative Morphology, Biochemistry, and Immunology.", Vol. 2., (C.B. Metz and A. Monroy, eds.) pp. 135-188. Academic Press, New York.
Wiese, L. and Shoemaker, D.W. (1970). *Biol. Bull.* **138**, 88-95.
Witman, G.B., Carlson, K. Berliner, J. and Rosenbaum, J.L. (1972). *J. Cell Biol.* **54**, 507-539.

Physiology of Fertilization and Fertilization Barriers in Higher Plants

H. F. Linskens

Department of Botany
Section Molecular Developmental Biology
University of Nijmegen
The Netherlands

I. STAGES OF THE PROGAMIC PHASE AND OF CELL SURFACE CONTACT

Fertilization in higher plants is a complex phenomenon consisting of a series of physiological and biochemical events which results in the formation of the zygote from the union of the sexually differentiated gametic nuclei.

In higher plants, e.g. angiosperms, fertilization is mediated by the controlled processes of the progamic phase which lies between pollination, the transfer of the male gametophyte from the anthers to the perceptive organs, and syngamy, the fusion of the gametes in the process of double fertilization.

ISBN 0-12-612984-3

The first contacts between the sexes are made during the progamic phase. Following a certain sequence of cytological and biochemical events, the pollen tube's contents are injected into the egg cell, leading to the definitive stage of fusion.

The mediating role of the pollen tube in the progamic phase can be separated into the following stages (Linskens, 1975b), each of which has its own molecular control:

(1) pollen *germination* and initiation of tube formation.
(2) *penetration* of the pollen tube into the female tissue of the stigma.
(3) directed *growth of the pollen tube* within the style, with simultaneous transfer of the male nuclei into the ovary.
(4) *opening of the pollen tube tip* to enable the release of the male material into the egg apparatus.

There are different sites of cell surface contacts during the progame phase. During pollen germination the outer wall of the pollen grain and the stigmatic surface which is mostly covered with a stigmatic fluid, a secretion product of the stigmatic tissue, come into contact. Throughout the penetration stage the pollen tube walls are in close contact with either the transmitting tissue of the solid style or with the inner surface of the stylar canal, also covered with a slimy secrete of the glandular tissue covering the cavity. The most intimate contact between the cell surfaces occurs during syngamy when the plasmatic content of the male gametophyte is released into the egg apparatus. These complicated interactions precede the fusion of the male and female nuclei.

II. FERTILIZATION AS A RECOGNITION PROCESS

The achievement of a successful fertilization in higher plants is, in many cases, a casual event: if the male gametophyte is transported by wind in anomophilic species it is only by chance that the right species of pollen reaches the receptive surface of the stigma; in cases of other vectors, e.g. insects, the visits of the congenial stigma also depend on many factors, although by ecological arrangements the chances of legitimate combinations are improved.

Thus, the first and decisive filter, the one determining whether or not the right partners are brought together, is the contact surface between pollen and stigma. Two different and opposing events occur on the stigma: On one hand the stigma must be made interested enough to catch the pollen in order to provide the chance for fertilization, on the other hand illegitimate unwanted pollen must be prevented from participating in the fertilization process.

The catching is improved by the stickiness of the stigma exudate. The selection is based on recognition processes which are based on surface determinates. Male-female recognition in flowering plants is induced by a mutual contact of pollen and stigma surface compounds. The chemical compounds involved are mainly glycoproteids (carbohydrate-protein-complexes (Linskens, 1956, 1958), complex mixtures of proteins, glycoproteins and glycolipids (Clarke *et al.*, 1979).

The stigma and pollen surfaces contain all the components of an ideal adhesive. Recently it has been shown, that the lectin Concanavalin A (Con A) is able to bind specifically to stigma surface receptors (Knox *et al.*, 1976). There is evidence that this binding is responsible for the reduction of the adhesive capacity for pollen proteins. Its capacity is restored when Con A is applied in the presence of its complementary ligand suggesting that specific binding of Con A alters the topography of the receptive surface in a way which "results in a less ideal contact surface for general adhesion. This implies that the stigma surface must be an efficient adhesive as well as carrying receptors for pollen recognition" (Clarke *et al.*, 1979).

Evidence also exists which indicates that the recognition reactions between both animal and plant cells, respectively as in flowering plants, are based on similar types of molecular interactions (Clarke and Knox, 1978). There is also a parallel between the recognition process in fertilization and the host-pathogen interaction (Linskens, 1968 a,b, 1976; Pegg, 1977).

The research into recognition processes in fertilization of higher plants is based on the problem of fertilization barriers. Therefore discussing the guards which make the selection for the all-or-not success of the pollen tube — stigma encounter is the best way to provide information of the nature of the interaction, which results in discrimination (Heslop-Harrison, 1978).

III. FERTILIZATION BARRIERS IN ANGIOSPERMS

At all stages of the progamic phase we find such barriers. These result in either complete sterility, which is the non-occurrence of the ultimate goal of sexuality, the fusion of the sexual cells, or a selection among the gametes which results in a disruption of random Mendelian segregation. These deviations of the target of the sexuality are genetically controlled.

One can distinguish 4 different forms of fertilization barriers in higher plants (Linskens, 1969).

A. *Gamete Competition*

This case is especially well known as embryo sac concurrence and therefore is also called gone concurrence. In hybrid plants the different embryo sacs of one ovary have unequal chances to be fertilized. Their position in the ovary depends on gametic constitution and plant development. The unequal frequency of the various egg types can be determined by the polarized position of the meiotic spindle. As a consequence, certain chromosome combinations are always directed toward the micropyle while others are orientated to the chalaza. Polarity phenomena during the ontogeny of the female gametophyte are related to nutritional supply and hormonal induction from the ovule (Noher de Halac and Harte, 1977). The position of the megaspore within the ovary determines the chance to be fertilized. Certain embryo sacs are preferred by the arriving pollen tubes (Harte, 1958).

B. *Selective Fertilization*

The term selective fertilization covers all processes which lead to a situation where the relative frequency with which the expected diploid genotypes are formed by fertilization deviates from the probability with which they are expected to form according to random meetings of male and female gametes (Harte, 1975). If the meeting is determined by the genetic constitution of the two gametes and the random distribution of the fertilization product is selectively disturbed, we are then encountered with a preferential or selective fertilization process. It results from an unequal attractive force between the sexually differentiated gametes, the embryo sac and the pollen tubes. The frequency of combination of two different types of gametes depends likewise on their genetic background. The intensity of affinity and the velocity of reaction depend on phenotypic and genotypic factors (Haustein, 1955; Glenk, 1964).

The biochemical background of both the gamete competition and the selective fertilization is not yet well understood. Apparently there exists differences in the chemotropic attractiveness of the various ovaries in one bud. *In vitro* differences in chemotropic activity have been demonstrated. Ovaries of different chemotropic activity differ not only in concentration but also in composition of the released chemical substances that attract the pollen tubes. Optimal selective fertilization can only take place if the gametes encounter each other in a stage of specific maturity. It is species-specific. Orientating chemical analyses have demonstrated that a mixture of sugars, amino acids and peptides are responsible for the selective fertilization. The differences in affinity of tubes for various types of ovaries is determined by quantitative differences in qualitatively identical mixtures. Up until now no specific attracting substances have been demonstrated.

Both these fertilization barriers, gamete competition, as well as selective fertilization, are only *discriminating barriers;* they do not prevent the process. Fertilization does take place, but the results are abnormal and do not conform to the Mendelian law of free combination of genes.

The other two types of fertilization barriers to be described, incongruity and incompatibility, are *total barriers.* They have mechanisms for non-functioning of the intimate partner relationship of the fertilization.

C. *Incongruity*

This is defined as an outbreeding mechanism which maintains the species character by preventing interpopulation matings. It is based on a lack of genetic information in one partner about the other (Hoogenboom, 1973). For example, during the progamic phase the pollen tubes need many enzymes for penetrating into the pistil and for orientating its growth down the style. These enzymes prepare the way, dissolve material, produce substrates for energy metabolism, and build up the tube wall material. If the adequate enzymes are not synthesized because of the lack of genetic information, the pollen tubes are not able to penetrate the female organ in order to fulfill the later stages of the progamic phase.

These barriers are the result of missing coordinated information. As a result the pollen tube will stop somewhere between pollination and syngamy during the progamic phase.

Incongruity as a consequence of non-matching genic systems in pistil and pollen results from a failure in the co-evolution of pollen and pistil. It may serve as a general basis for the explanation of breeding experiences in interspecific crosses. The principal of co-evolution is a mutually dependent development of two connected but unrelated taxa (Hoogenboom, 1975) which become more and more closely adapted to each other. Incongruity in a sexual relationship is quite similar to the relationship in a host-parasite system. In both cases the barriers exist because of a mistake in the chain of events in both partners. In other words, lack of coordination and interaction between pollen/pollen tube and pistil during the progamic phase.

From the genetic point of view, incongruity can be defined as the inability of genetically different gamete systems to fuse. Called on occasion heterogenic incompatibility, it shows up in higher plants usually as a unilateral incompatibility (Esser, 1976). It is sometimes called interspecific incompatibility but this term should be avoided because its genetic mechanism is completely distinct from self-incompatibility. The physiological mechanisms of incongruity is not yet understood but the barrier in fertilization is caused by genetic differences. Incongruity is the rule and congruity the exception.

D. *Incompatibility*

This is an inbreeding mechanism which prevents the mating in monoecious systems of individuals with identical S-genes. Incompatibility is a physiological barrier to fertilization within a species which prevents the fusion of the male and the female gametes even though they both are fully fertile.

Generally known as self-incompatibility, para-sterility or homogenic incompatibility is a mechanism which prevents inbreeding. It has been known about for more than 200 years (reviews: Linskens and Kroh, 1967; Townsend, 1971; de Nettancourt, 1977) and occurs in about 3000 species from more than 50 families.

From the developmental point of view, one can distinguish between *sporophytically* determined and *gametophytically* determined systems. In plants with sporophytic incompatibility, the gene action of the pollen tube and the stylar tissue, respectively, are determined by the diploid sporophyte; in plants with gametophytic incompatibility, each incompatibility gene acts independently in the haploid pollen and in the diploid stylar tissue. The gametophytic system is the most frequently occurring system of incompatibility in angiosperms.

An interesting correlation exists between the incompatibility system, pollen cytology and the site of the fertilization barrier in homomorphic plants (Brewbaker, 1957). Gametophytic incompatibility is observed primarily in species with binucleate pollen where inhibition occurs during the pollen tube growth in the style, whereas sporophytic incompatibility is linked with trinucleate pollen grains and inhibition of pollen germination, or the disturbance of the early penetration processes into the stigma. This almost perfect *correlation between pollen cytology and the site of the incompatibility barrier* exists for 42 genera from 23 families. It has been suggested that this association might be the result of a difference in the amount of available metabolites in the bi- vs. trinucleate pollen grains at anthesis. The hypothesis is that the division of the generative cell into two sperm cells consumes the valuable metabolites required for penetration of the stigma. Recently this hypothesis was strengthened by Hoekstra (1973), who found that the respiration rate in trinucleate pollen is about 3 times higher during the first phase of development. This can also be linked to the rapid decrease of its germination capacity.

The *localization of the incompatibility barrier* can occur in different parts of the plant as well as during different stages of the progamic phase:
(a) on the *surface of the stigma.* Here either pollen germination is inhibited or the pollen tube is unable to penetrate the stigma. This type of incompatibility is found within the families of the Cruciferae, Papilionaceae, Gramineae, Compositae and others.
(b) during the *growth of the pollen tube through the style,* either in the transmitting tissue or the stylar canal. This type of incompatibility barrier is found in many Solanaceae, as well as in Trifolium and Abutilon.
(c) *in the ovary, the ovules or the embryo sac.* In some cases the pollen tubes of incompatible crosses are able to grow as rapidly as compatible ones and so reach the base of the style. In some lilies, (such as Hemerocallis), in Gasteria and Theobroma cacao, the incompatibility reaction results in the non-fusion of the sperm cells with the egg apparatus. Such unfertilized ovules degenerate. It is interesting to mention that most of the species with the incompatibility barrier localized in the ovules are characterized by hollow styles. It seems that intimate contact between pollen tubes and the conducting tissue of the style is a precondition for the occurrence of the inhibition reaction (Brewbaker, 1957). The observation that the oxygen tension in the style is the control mechanism for the bursting of the pollen tube tips in connection with the release of the tube content into one of the synergids can be linked to these findings (Linskens and Schrauwen, 1966; Stanley and Linskens, 1967).

From the experimental point of view, the stigma and style, as sites of the incompatibility reaction, have attracted the most attention during the last 25 years. The advantages of such studies are a strict localization of the physiological processes and a relatively easy method for obtaining material for electronmicroscopical and biochemical analysis from vegetatively propagated genetically homogenous plant material with known S-allele structures.

From the investigations of many researchers it became evident that it was necessary to distinguish two steps in the *incompatibility* barrier, which in fact is a process of self-recognition with regards to a certain S-gene: *recognition* and *rejection.* Recognition has to precede rejection but both work together to form the barrier which prevents self-fertilization. Whereas in former years most of the attention was directed to the sequence of events which results in the prevention of the fusion process, during the last years the interest has been directed more towards the molecular events of the recognition reaction.

1. *Stigmatic inhibition.* The excellent work of Heslop-Harrison and his associates (Heslop-Harrison, 1978) over the last 10 years has elucidated the recognition site on the stigmatic papillae in the external protein pellicle. This coating, which covers the cutinised outer layer of the wall, is the site of recognition, but also has properties which function in the capture and hydration of the pollen grains (Mattsson *et al.*, 1974; Heslop-Harrison, 1975, 1978; Heslop-Harrison and Heslop-Harrison 1975; Heslop-Harrison *et al.*, 1974, 1975). This is true for the so called "dry" stigma surfaces, which have no visible secretion exudate or liquid cuticle. This layer can be observed both by optical and electron optical methods and is disrupted by pronase but not be lipase. It is a proteinaceous layer with specific structures for the passage of water into the pellicle from the protoplasm of the papillar cells.

The counterpart in the recognition reaction is the exineborne fraction. From the first attempts to identify the precise constitution of this material, it appeared that the diffusates of pollen (Brassica oleracea) from parent plant homozygotes for a known S-allele induced formation of callose after implantation into the stigma of the same S-genotype. This reaction is typical for an incompatible combination. The effective molecules are proteins or glycoproteins (Heslop-Harrison *et al.*, 1975; Knox *et al.*, 1975). These experiments also demonstrated sporophytic control of the incompatibility reaction: That pollen coat material tryphine is in fact derived from the exine and is, therefore, a deposit of the tapetum of the parental anther (Dickinson and Lewis, 1973).

The pollen-stigma-interaction can be described as a dialogue between the pollen grain and the stigmatic papilla (Heslop-Harrison *et al.*, 1975): Immediately after attachment, some water molecules diffuse to the pollen grain. They cause pollen hydration and the elution of proteins and other substances which are localized in the spaces of the submicroscopical structure of the exine (perhaps also from the intine). These signal substances react with the preformed extracuticular protein film from the pellicle of the stigma. Whether or not the signal is set on "stop," that is "incompatible," or on "go," that is "compatible," cutinases are released from the pollen grains (Linskens and Heinen, 1962). They break down the cuticular layer of the stigmatic cell so the tube can enter.

All those processes occur on the stigma surface during the first minutes after pollination (30 seconds to 30 minutes). In many cases the incompatibility barrier of the sporophytic type can be by-passed by bud-pollination, a stage at which apparently the recognition system is inactive or has not yet been fully developed. By mechanical disruption of the stigma surface and inplantation of the pollen grains into the stylar transmitting tissue, successful pollination with incompatible pollen is possible. This demonstrates that stigmatic inhibition is strictly localized to the surface papillae.

The specificity of the proteins involved in the pollen stigma incompatibility barrier has also been demonstrated (Nasrallah and Wallace, 1967; Nasrallah *et al.*, 1969, 1973; Hinata and Nishio, 1978; Nishio and Hinata, 1978): Each S-allele codes for a specific protein. Apparently the specific proteins differ from one another by amino acid substitutions.

2. *Stylar inhibition.* In the gametophytic system, with the exception of the grasses and Oenothera, the incompatibility barrier is localized in either the style or the ovary. The ultimate degree of the inhibition of pollen tube growth varies from species to species, depending not only on the S-alleles involved, but also on the general vegetative condition of the receptor plant and ambient temperature (Lewis, 1942). In other words, in the case of the gametophytic system, the incompatibility barrier acts, after pollen germination, in the style (stylar reaction) and no doubt very early during the contact between pollen tube and stylar tissue (Linskens, 1975a). Among the first events is a differential nucleic acid synthesis (van der Donk, 1975). This RNA and protein synthesis begins at a later stage of the pollen-style interaction after compatible rather than after incompatible pollination. If RNA with messenger activity is injected into the egg cells of *Xenopus laevis,* active proteins are produced which are effective after re-injection into styles.

One should clearly distinguish the *recognition reaction* from the *rejection reaction* which can include a chain of biochemical events such as an increase of the respiration level, changes of the substrate pool, differentiation in the growth hormone balance (Barendse *et al.*, 1970), an increase in enzyme activities (Schlösser, 1961; Roggen, 1967) and altered protein patterns (Linskens, 1955; Desborough and Peloquin, 1968). The presence of concentric endoplasmic reticulum in inhibited pollen tubes was observed as a constant feature of the rejection reaction (de Nettancourt *et al.*, 1974) as well as deviations of the cell wall structure (van der Pluijm and Linskens, 1966). They all reflect the abnormal metabolism that occurs after recognition which contributes to the building up of the barrier, preventing the pollen tubes from reaching the ovary.

3. *Distance effects.* The inhibition barrier can be considered as a short-distance effect of the recognition reaction. But in addition to this, there is long-distance information going out from the site of recognition. Long before the pollen tube tips are half way to the ovule, activation of the synthetic processes in the ovaries has differentially occurred (Linskens, 1973a; Deurenberg, 1977). differences can be detected in protein

metabolism after cross- and self-pollination. Also the flow pattern, that is the redistribution of the soluble material in the flower, is different after cross-and self-pollination, and noticeable before the pollen reach their ultimate goal (Linskens, 1974, 1975a). Consequently, long-distance translocation phenomena are adequate indicators of the rejection reaction. The wilting phenomenon can also be considered a long-distance indicator for the incompatibility barrier (Gilissen, 1978). It is an open question which type of signal transfers the information through the style down to the ovules and provides the information whether or not they can expect to be fertilized compatibly or incompatibly. There are some indications that this transfer of information is an electrical stimulus (Linskens and Spanjers, 1973; Spanjers, 1978).

4. *Breakdown of incompatibility.* The incompatibility barrier is one of the important breeding barriers in cultivated plant species. Plant breeders, therefore, are constantly searching for methods which enable them to make a controlled use of the barrier in experimental crosses (Linskens, 1973). Many methods of modifying and transforming incompatibility barriers have been worked out (de Nettancourt, 1977). In addition to manipulating genetic changes leading to a state of compatibility (mutations, modification of the genetic background, polyploidy), physiological methods are also promising, e.g., bud-pollination and old, wilted flowers which are in some cases unable to maintain the barrier. Depending on the species and the S-allele, the strength of the incompatibility reaction alters with the flowering season (Linskens, 1977), a phenomenon called end-season-fertility. Chronic and acute radiation, as well as mentor pollen, can sometimes be used to lower the incompatibility barriers but have to be worked out for a species to species combination. Pistil and stem grafting in practice is complicated but electrically aided pollination (Roggen *et al.*, 1972) can be useful in some of the Cruciferae. For laboratory purposes, it is also possible to use successfully in vitro fertilization: The incompatible pollen is deposited in the neighbourhood of the ovule so that selfed seeds are received (Rangaswamy and Shivanna, 1972).

The fertilization barriers on higher plants are of scientific interest because they make it possible to follow gene action during a developmental process. Equally, they are also economically very important since plant breeders can manipulate them effectively thus improving food-production (Cresti *et al.*, 1978).

IV. POLLEN TUBE DISCHARGE

A. *Role of Synergids*

The last and most decisive step of fertilization in higher plants is the release of the pollen tube contents into the embryo sac. In angiosperms both sperm cells are released into one of the so-called synergids, which have a characteristic structure in the filiform apparatus (van Went and Linskens, 1967). The pollen tube tip opens, triggered by oxygen pressure at the top of the egg apparatus and discharges the nuclei in such a rapid manner that it has not yet actually been observed by electron microscopy. The specific interaction between synergid and pollen tube cytoplasm is apparently a prerequisite of the tube tip entrance and discharge into the embryo sac, and for the male gamete transfer to the egg and central cell (van Went and Linskens, 1967; Jensen, 1972; Mogensen, 1978). But one has to confess that the details of the interaction between pollen tube and synergid are still unsolved.

The transport of the male or sperm cells into the egg and the central cell respectively, a process which is called double fertilization and is characteristic for angiosperms, is still far from being elucidated (Jensen, 1976). Also the meeting of the cell surfaces, the plasma membranes of the egg and central cell with the sperm cells is still unobserved.

B. *Fusion of Sexual Nuclei*

From electron microscopical observations, however, it can be seen that nuclear fusion takes place in the following way (Linskens, 1968b; van Went, 1971). The joining nuclei touch each other and locally become flattened. At several places fusions of the outer nuclear membranes take place. The space between both nuclear membranes disappears and they come into contact, resulting in a spreading-out over a large area. Immediately thereafter the same type of fusion takes place between the inner nuclear membranes. In this way bridges between the nuclei result at several sites which remain at least until the membranes of both the nuclei are completely unified (Jensen, 1964; van Went, 1971). In higher plants the mediator effect of the cell surfaces in the fertilization process is evident, though far from recognized in detail.

V. SUMMARY

Pistils of angiosperms are highly selective and discriminating organs which determine the success of sexual relationships. There are two different mechanisms available to the plant:

(1) The outbreeding mechanism, which maintains the species character by preventing interpopulation matings. It can be called *incongruity* (cross-incompatibility) and is based on a lack of genetic information in one partner about the other.

(2) The inbreeding mechanism, which prevents the mating of individuals with identical S-genes. It is called *incompatibility* (self-incompatibility, parasterility).

The incompatibility mechanisms can be effected on the stigma surface, in the conducting tissue of the style, within the stylar canal, or in the ovary. If the genome of the pollen-producing plant determines the reaction, we call it *sporophytic* incompatibility. It was found that this form of incompatibility is mostly linked with the trinuclear type of pollen grains.

The other type, *gametophytic* incompatibility, is present if the pollen genome is the determinant. It is mostly linked with binuclear pollen bearing plants. The incompatibility reaction can be divided into two different steps: *recognition* and *rejection.*

In recognition, specific substances of glycoproteinaceous character are involved. S-allele specific polypeptides of the pollen induce qualitatively different protein synthesis in the stigma a short time after the first contact of the pollen coat with the stigma surface. This can be interpreted as a biochemical dialogue between both partners of the mating process. Presumably the primary products of the S-alleles have a regulating effect on gene expression.

The succeeding rejection reaction includes changes in the enzyme pattern, respiration pathway, permeability and formation of inhibitory substances. The end result is inhibition of pollen germination and/or pollen tube growth so that no fusion of the male nuclei with the female nuclei can take place. Whereas the mediating function of the cell surface in the recognition reaction of fertilization barriers is evident, the interrelationship of membranes in the later stages of syngamy in higher plants is still unsolved.

REFERENCES

Barendse, G. W. M., Rodrigues Pereira, A. S., Berkers, P. A., Driessen, F. M., van Eyden-Emons, A. and Linskens, H. F. (1970). *Acta Bot. Neerl.* **19**, 175-186.

Brewbaker, J. L. (1957). *J. Hered.* **48**, 217-277.

Clarke, A.E. and Knox, R. B. (1978). *Quart. Rev. Biol.* **53**, 3-28.

Clarke, A. E., Gleeson, P., Harrison, S. and Knox, R. B. (1979)., in preparation
Cresti, M., Donini, B. and Devreux, M. (1978). *Riv. Ortoflorofrutticol. Ital.* **4**, 330-349.
Desborough, S. and Peloquin, S. J. (1968). *Theor. Appl. Genet.* **38**, 327-331.
Deurenberg, J. J. M. (1977). Thesis, Univ. Nijmegen.
Dickinson, H. G. and Lewis, D. (1973). *Proc. Roy. Soc.* (London) B **184**, 149-165.
van der Donk, J. A. W. M. (1975). Thesis, Univ. Nijmegen.
Esser, K. (1976). *In* "Conservation of Threatened Plants" (Simons, J.B., Beyer, R.I., Brandham, P.E., Lucas, G.L. and Parry, V.T.H., eds.) pp. 185-197, Plenum Press, New York.
Gilissen, L. (1978). Thesis, Univ. Nijmegen.
Glenk, H. O. (1964). *In* "Pollen Physiology and Fertilization" (H. F. Linskens, ed.), pp. 170-179, North Holland Publ. Co., Amsterdam.
Harte, C. (1958). *Z. Vererbungsl.* **89**, 473-496, 497-507, 715-728.
Harte, C. (1975). *In* "Gamete Competition in Plants and Animals" (D. L. Mulcahy, ed.), p. 31-41, North Holland Publ. Co., Amsterdam-Oxford.
Haustein, E. (1955). *Z. Bot.* **43**, 253-261.
Heslop-Harrison, J. (1975). *Ann. Rev. Plant Physiol.* **26**, 403-425.
Heslop-Harrison, J. (1978). *In* "Cellular Recognition Systems in Plants." Studies in Biology no. 1000. Arnold, London.
Heslop-Harrison, J. and Heslop-Harrison, Y. (1975). *Ann. Bot.* **39**, 163-165.
Heslop-Harrison, J., Knox, R. B. and Heslop-Harrison, Y. (1974). *Theor. Appl. Genet.* **44**, 133-137.
Heslop-Harrison, J., Heslop-Harrison, Y. and Barber, J. (1975). *Proc. Roy. Soc.* (Lond.) B **188**, 287-298.
Hinata, K. and Nishio. T. (1978). *Heredity,* **41**, 93-100.
Hoekstra, F.A. (1973). *Incompatibility Newsletter* **3**, 52-54.
Hoogenboom, N. G. (1973). *Agricult. Res. Rep.* 804.
Hoogenboom, N. G. (1975). *Proc. Roy. Soc.* (Lond.) B **188**, 361-375.
Jensen, W. A. (1964). *J. Cell Biol.* **23**, 669-672.
Jensen, W. A. (1972). *In* "The Embryo Sac and Fertilization in Angiosperms". *H. L. Lyon Arboretum Lecture* no. **3**, 1-32.
Knox, R. B., Heslop-Harrison, J. and Heslop-Harrison, Y. (1975). *Biol. J. Linnean Soc.* **7**, suppl. 177-188.
Knox, R. B., Clarke, A. E., Harrison, S., Smith, P. and Marchalonis, J. J. (1976). *Proc. Nat. Acad. Sci.* U.S.A. **73**, 2788-2792.
Lewis, D. (1942). *Proc. Roy. Soc.* (Lond.) B **131**, 13-26.
Linskens, H. F. (1955), *Z. Bot.* **43**, 1-44.
Linskens, H. F. (1956). *Ber. dtsch. bot. Ges.* **69**, 353-360.
Linskens, H. F. (1958). *Ber. dtsch. bot. Ges.* **71**, 2-10.
Linskens, H. F. (1968a). *Neth. J. Plant Pathol.,* **74**, (suppl. 1), 1-8.
Linskens, H. F. (1968b). *Quad. Acc. Nat. Lincei* (Roma) **104**, 47-56.
Linskens, H. F. (1969). *In* "Fertilization" (C. B. Metz and A. Monroy, eds.)., Vol. II, pp. 189-254, Academic Press, New York.
Linskens, H. F. (1973a). *Caryologia* **25**, suppl. 27-41.
Linskens, H. F. (1973b). *Fizil. Rast.* **20**, 192-203.
Linskens, H. F. (1974). *In* "Fertilization in Higher Plants" (Linskens, H. F., ed.), pp 285-292, North Holland Publ. Co., Amsterdam-Oxford.
Linskens, H. F. (1975a). *Proc. Roy. Soc.* (Lond.) B **188**, 299-311.
Linskens, H. F. (1975b). *Biol. J. Linnean Soc.* **7**, suppl. 143-152.
Linskens, H. F. (1976). *In* "Specificity in Plant Diseases" (R. K. S. Wood and A. Graniti, eds.). *NATO Adv. Study Inst.,* Ser. A. **10**, 311-325.

Linskens, H. F. (1977). *Acta Bot. Neerl.* **26**, 411-415.
Linskens, H. F. and Heinen, W. (1962). *Z. Bot.* **59**, 338-347.
Linskens, H. F. and Kroh, M. (1967). *Encycl. Plant Physiol.* (W. Ruhland, ed.) **18**, 506-630.
Linskens, H. F. and Schrauwen, J. (1966). *Planta* **71**, 98-106.
Linskens, H. F. and Spanjers, A. W. (1973). *Incompatibility Newsletter* **3**, 81-85.
Mattsson, O., Knox, R. B., Heslop-Harrison, J. and Heslop-Harrison, Y, (1974). *Nature* **274**, 298-300.
Mogensen, H. L. (1978). *Amer. J. Bot.* **65**, 953-964.
Nasrallah, M. E. and Wallace, D. H. (1967). *Heredity* **22**, 519-527.
Nasrallah, M. E., Barber, J. T. and Wallace, D. H. (1969). *Heredity* **24**, 23-27.
Nasrallah, M. E., Wallace, D. H. and Savo, R. M. (1973). *Genet. Res.* **20**, 151.
de Nettancourt, D. (1977). *In* "Incompatibility in Angiosperms." Springer, Berlin-Heidelberg-New York.
de Nettancourt, D., Devreux, M., Lanieri, U., Cresti, M., Pacini, E. and Sarfatti, G. (1974). *Theor. Appl. Genet.* **44**, 278-288.
Nishio, T. and Hinata, K. (1978). *Jap. J. Genet.* **53**, 27-33; *Jap. J. Genet.* **54**, 307-311.
Noher de Halac, I. and Harte, C. (1977). *Plant Syst. Evol.* **127**, 23-28.
Pegg, G. F. (1977). *In* "Cell Wall Biochemistry Related to Specificity in Host-Plant Pathogen Interactions" (B. Solheim and J. Raa, eds.). Universitetsforlaget, Tromsø .
van der Pluijm, J. E. and Linskens, H. F. (1966). *Theor. Appl. Genet.* **36**, 220-222.
Rangaswamy, N. S. and Shivanna, K. R. (1972). *Phytomorphology* **21**, 284-289.
Roggen, H. P. J. R. (1967). *Acta Bot. Neerl.* **16**, 1-31.
Roggen, H. P., van Dijk, A. J. and Dorsman, C. (1972). *Euphytica* **21**, 181-184.
Schlösser, K. (1961). *Z. Bot.* **49**, 266-288.
Spanjers, A. W. (1978). *Experientia* **34**, 36.
Stanley, R. G. and Linskens, H. F. (1967). *Science* **157**, 833-834.
Townsend, C. E. (1971). *In* "Pollen Development and Physiology," (J. Heslop-Harrison, ed.), pp. 281-309, Butterworth, London.
van Went, J. L. (1971). Thesis, Univ. Nijmegen.
van Went, J. L. and Linskens, H. F. (1967). *Theor. Appl. Genet.* **37**, 51-62.

Molecular Alterations in Gamete Surfaces During Fertilization and Early Development

Bennett M. Shapiro, Robert W. Schackmann, Christopher A. Gabel, Charles A. Foerder, and Martha L. Farance

Department of Biochemistry

E. M. Eddy

Department of Biological Structure

S. J. Klebanoff

Deparment of Medicine

University of Washington
Seattle, Washington 98195

ISBN 0-12-612984-3

I. INTRODUCTION

The early events of fertilization involve many changes at the surfaces of both gametes. These include events that prepare the sperm for fusion with the egg, fusion of sperm and egg to initiate development of a new individual, and the ensuing changes in the egg requisite for normal development. Many of the mechanisms used by gametes during the events of fertilization are similar to those used by other cell types for their specific functions, and often gametes constitute the most convenient system for studying general problems of membrane related events in cell biology. For example, there are several membrane fusion events in fertilization, some of which involve exocytosis, and invertebrate gametes provide a rich source of material for studying the general properties of such events.

Over the past several years, work in our laboratory has concentrated on some molecular mechanisms of fertilization and early development, that will be summarized here. We have been examining properties of the sperm as it undergoes the acrosome reaction, a prerequisite to fertilization, and have found that sperm behave like many other excitable cells, in altering ion fluxes in response to specific stimuli. We have examined the fate of the sperm surface after fertilization, and have found some unusual properties of this physiologically important membrane fusion system that suggest that the sperm surface may play a subsequent role in development. We have also studied the barrier that is erected by the egg at the time of fertilization to prevent polyspermy and to protect the new embryo. This phenomenon, that occurs as a result of the cortical reaction, a massive change in the properties of the egg plasma membrane and cell coat, includes the activation of a complex peroxidative system that has been highly conserved throughout evolution. Most of our experiments have been done with gametes of the sea urchin, *Strongylocentrotus purpuratus,* but many of the results appear to be applicable to gametes from other types of animals, and to involve phenomena that occur in other cellular systems.

II. TRIGGERING OF THE ACROSOME REACTION

The acrosome reaction of sperm is an absolute requirement for fertilization. In echinoderm sperm, it is elicited by contact with the egg surface, and is expressed as exocytosis of the acrosomal granule contents and polymerization of subjacent globular actin into an acrosomal filament (Fig. 1). Our work on the acrosome reaction of sperm from *S. purpuratus* uses inhibitors of the reaction to dissect the mechanism of the process, including the sequence of ionic movements associated with

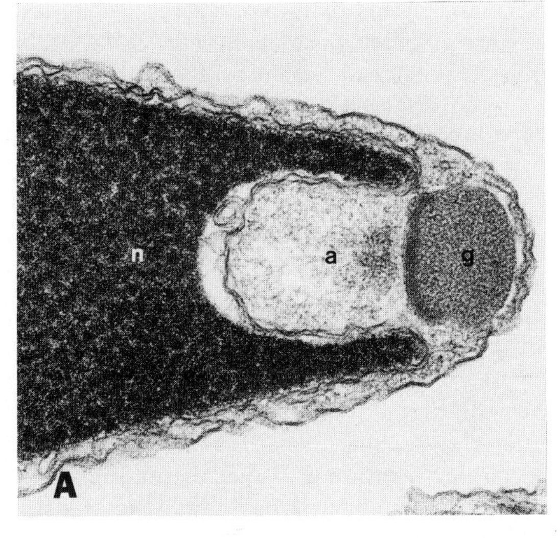

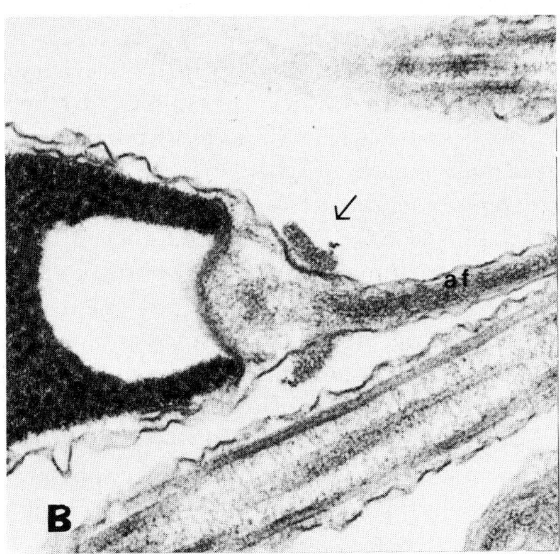

Fig. 1 Acrosome reaction of sperm of the sea urchin S. *purpuratus*. 1A) shows sperm prior to contact with egg jelly coat. The acrosomal granule (g) remains intact and the periacrosomal region contains globular actin (a). The nucleus is designated by n; 1B) shows a sperm that has undergone the acrosome reaction, triggered by egg jelly. The actin has polymerized to form the acrosomal filament (af) and the contents (arrow) of the acrosomal granule have been exposed by exocytosis.

membrane fusion, exocytosis, and filament extension. We have found that several inhibitors of ion movements prevent the acrosome reaction, that specific alterations in ion gradients or membrane permeabilities can induce the reaction, and that induction of the reaction is associated with alterations in sperm membrane permeability to Ca^{2+}, K^{+}, Na^{+}, and H^{+}. Each of these topics will be dealt with individually below. These studies, which represent extensions of work initiated by Jean Dan and her collaborators (1954, 1964) on the importance of Ca^{2+} in the acrosome reaction, have led us to hypothesize that triggering of the acrosome reaction occurs as a result of multiple, interrelated alterations in the permeability of the sperm plasma membrane due to sperm contact with a component (or components) of the egg surface referred to as "jelly."

A. *Calcium Requirement*

Dan (1954) was the first to demonstrate that the acrosome reaction required calcium; this appears to be the only requirement for extracellular Ca^{2+} in fertilization (Takahashi and Sugiyama, 1973). The requirement for Ca^{2+} is supported by inhibitor studies: for example, La^{3+} and the local anesthetics, procaine and xylocaine, agents known to block Ca^{2+} fluxes in other biological systems, block the acrosome reaction (Decker *et al.*, 1976; Collins and Epel, 1977). We have found that verapamil or D 600, drugs that block excitable Ca^{2+} currents in heart (Kohlhardt *et al.*, 1972), are nearly 20 times more potent than xylocaine in inhibiting the acrosome reaction (Schackmann *et al.*, 1978).

When the acrosome reaction is triggered, Ca^{2+} uptake is stimulated immediately, although we have not yet been able to resolve whether uptake occurs prior to acrosomal filament extension or after the reaction is completed. However, since Ca^{2+} uptake occurs into reacted sperm regardless of the triggering method, and since an ionophore, A23187, that allows for Ca^{2+} influx, triggers the acrosome reaction (Decker *et al.*, 1976; Collins and Epel, 1977), Ca^{2+} influx is probably an initial step in the reaction mechanism. Inhibition of 80%-90% of the Ca^{2+} uptake by uncoupling agents, such as FCCP (carbonyl cyanide, p-trifluoromethoxyphenylhydrazone) and localization of the Ca^{2+} by electron probe microanalysis (Cantino, unpublished data) show that most of the measured Ca^{2+} accumulation is into the mitochondria. The data suggest that an increase in the sperm plasma membrane Ca^{2+} permeability is closely associated with the acrosome reaction, and that the mitochondria serve as a sink for the Ca^{2+} that has entered the sperm. The large amount of mitochondrial Ca^{2+} uptake makes it difficult to determine the amount of Ca^{2+} influx necessary for triggering the reaction but implies that sufficient Ca^{2+} enters the sperm to allow binding to material with affinities on the order of the mitochondrial uptake system, 2–4 μM (Bygrave *et al.*, 1971). Since sea urchin sperm contain substantial quantities of calmodulin (Jones *et al.*, 1978), a Ca^{2+} binding regulatory protein with a dissociation constant of less than 1 μM (Wolfe *et al.*, 1977), the Ca^{2+} influx may alter the amount of Ca^{2+} bound to this protein and thereby effect either the acrosome reaction or the cortical reaction that occurs upon subsequent fertilization. In the latter case, when egg and sperm fuse, the sperm could provide a source of Ca^{2+} perhaps bound to calmodulin, and thus trigger the Ca^{2+} stimulated exocytosis of the cortical reaction (see below). In other words, by providing a bolus of Ca^{2+} to the egg, the sperm may act as a "calcium bomb."

With regard to possible roles for Ca^{2+} and calmodulin in fertilization, suggestive evidence that such a process may occur comes from some preliminary experiments with trifluoperazine (TFP), a phenothiazine derivative that binds to the Ca^{2+}–calmodulin complex (Levin and Weiss, 1977) and interferes with processes regulated by calmodulin. Fertilization is 90% inhibited at 10 μM TFP. This is probably due to inhibition of the jelly triggered acrosome reaction, which is half maximally inhibited at 2 μM TFP and almost completely blocked at 10 μM TFP. When sperm are acrosome reacted by treatment with egg jelly, then rapidly transferred to a suspension of eggs, TFP still prevents fertilization although 50% inhibition requires higher levels of the drug (9 μM). Although these data are compatible with the involvement of a Ca^{2+}–calmodulin complex in the acrosome and cortical reactions, they are far from compelling, since TFP binds nonspecifically to many proteins (Levin and Weiss, 1977), and thus may be inhibiting other cellular processes.

Another feature of the Ca^{2+} requirement has emerged from a study of inhibitors of the acrosome reaction. Most of these block the jelly triggered Ca^{2+} uptake (Table I): verapamil, tetraethylammonium (TEA), reduced pH, and elevated K^+ (30 mM) prevent both $^{45}Ca^{2+}$ uptake and the acrosome reaction. However, if we reduce the [Na^+] to less than 10 mM, by substitution with choline, Ca^{2+} accumulation occurs whether or not jelly is present, although filament extension is prevented (Table I). Thus, although Ca^{2+} uptake is necessary for the acrosome reaction, it is not sufficient. Additionally, if the Na^+ concentration is increased to 25 mM, there is triggering of the acrosome reaction even without the addition of egg jelly; if the [Na^+] is increased above 70 mM, there is no spontaneous triggering (the [Na+] in sea water is about 460 mM). The spontaneous triggering in the presence of low [Na+], as well as the requirement for Na+ for any reaction to occur, implies that altered Na+ movement is a distinct event in the triggering pathway.

TABLE I

Effect of Inhibitors of the Acrosome Reaction on Ca^{2+} Uptake[a]

Conditions	Initial Uptake Rate (nmole $Ca^{2+}/10^8$ sperm-min)	
	- jelly	+ jelly
ASW	0.1	2.3
ASW (pH 6.8)	0.1	0.1
ASW (30 mM K+)	0.1	0.1
NaFSW (5 mM Na+)	0.9	0.6
ASW + TFP (10 μM)	0.1	0.1
Low NaSW (25 mM Na+)	5.3	X

[a]Sperm were suspended under the various conditions to give 1–5 x 10^8 sperm/ml in artifical seawater containing 10 mM $CaCl_2$ (4 μCi/ml $^{45}Ca^{2+}$), 50 mM $MgCl_2$, 10 mM KC1, 5 mM Tris–2.5 mM Hepes, pH 8.0 and 360 mM NaCl (ASW) (except as noted). Egg jelly was added and Ca^{2+} uptake monitored as described previously (Schackmann, *et al.*, 1978). NaFSW designates solutions in which choline chloride was substituted on a molar basis for all but 5 mM NaCl. Low NaSW has choline chloride substituted for all but 20 mM–30 mM NaCl. Addition of jelly caused 80% of the sperm to undergo the acrosome reaction in normal ASW. Low NaSW is the only condition that causes sperm to react spontaneously without egg jelly. Under no other conditions did more than 1% of the sperm react spontaneously.

B. *Relationship with Other Ions*

In support of the involvement of Na+ in the acrosome reaction is the finding that when the acrosome reaction is triggered with jelly, Na+ is accumulated and acid is released, with the same kinetics as for extension of the acrosomal filament; all are essentially complete within 15–20 sec. Using estimates made from measurement of $^{22}Na^+$ uptake, and the pH shift of the medium in which the reaction is triggered, some 23 nmole Na+ are taken up while 20 nmole H+ equivalents are released for 10^8 sperm reacted. The close correlation, both temporally and quantatively, suggests that a Na+:H+ exchange is associated with and required for the acrosome reaction (Schackmann, unpublished data). All inhibitors of jelly induced triggering prevent both Na+ uptake and H+ efflux. In particular, removal of Ca^{2+} blocks the Na+:H+ exchange, suggesting that a Ca^{2+} requiring step occurs first.

Since the normal acrosome reaction is inhibited by low concentrations of TEA (1 mM-5 mM) that also inhibit excitable potassium channels (Hille, 1967), we thought that potassium might be involved in regulating the acrosome reaction. Compatible with this is the finding that the acrosome reaction is inhibited completely merely by increasing the K^+ from 10 mM to 20 mM. These results suggest that K^+ movement, or perhaps a K^+ dependent membrane potential change, is involved in induction of the reaction.

In pursuing the role of K^+, we have observed, using both $^{42}K^+$ and a K^+ selective electrode, that jelly treatment elicits K^+ efflux from sperm. However, when careful measurements were made with the K^+ electrode, the efflux clearly began approximately 15 sec after jelly addition, after the acrosome reaction was completed, suggesting that the K^+ loss is a result of the reaction and not directly involved in the triggering mechanism. Likewise, if TEA is added after the sperm have reacted, there is no effect on the K^+ efflux; thus, the K^+ efflux is not occurring through TEA sensitive K^+ channels. Therefore, the bulk of the K^+ efflux is not related to the triggering mechanism *per se*, but occurs as a late event. The involvement of altered K^+ permeability in triggering is likely to be an early event in the process, as suggested by the data with TEA and with slightly elevated K^+ concentrations.

By analyzing the kinetics, stoichiometries, and interrelationships of the ion movements described above, and the effects on them by treatment with inhibitors, we have arrived at the tentative partial sequence presented below.

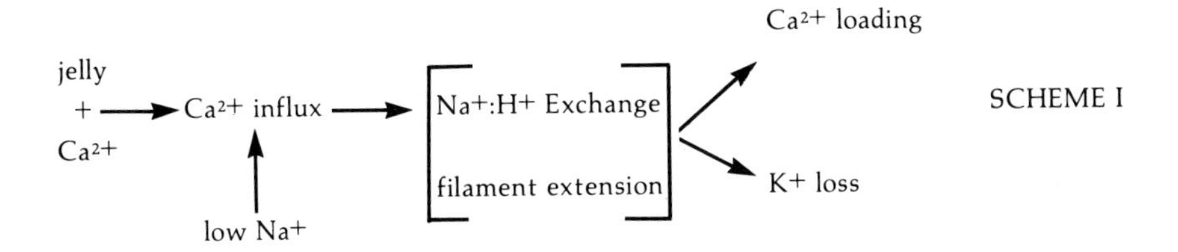

Jelly plus Ca^{2+} lead to an initial Ca^{2+} dependent step, presumably involving Ca^{2+} uptake, that is followed by the $Na^+:H^+$ exchange and filament extension. After the reaction is complete, additional Ca^{2+} accumulation occurs and the sperm loses its ability to retain K^+. While this sequence shows one interpretation of the data, at this stage of our knowledge alternative explanations are undoubtedly possible, and the scheme remains a working hypothesis rather than a formal model.

III. SPERM-EGG FUSION

Following the acrosome reaction, the sperm is poised for binding to the egg and initiating development. Electron microscopic studies of fertilization in invertebrates (Colwin and Colwin, 1963) and mammals (Barros and Franklin, 1968) have indicated that the plasma membrane of the fertilizing sperm fuses with that of the egg to form a continuous, mosaic zygote surface. Following this fusion, sperm cytoplasmic organelles (i.e., pronucleus, mitochondria, centriole, and axoneme) are found in the egg cytoplasm (reviewed in Austin, 1968; Longo, 1973), but the distribution of sperm surface components within the zygote is unknown. On the basis of what we know about membrane fluidity in other cell types (Edidin *et al.*, 1976; Schlessinger *et al.*, 1976) and in artificially fused heterokaryons (Frye and Edidin, 1970), one might predict that sperm surface components would rapidly intermix with those of the egg, to form a uniform zygote surface. The fertilized egg, however, is known to possess special properties that could affect the distribution of the sperm surface components. For example, the surface of the fertilized sea urchin egg retains a topographic mosaicism (Eddy and Shapiro, 1976) and is much more rigid than its unfertilized counterpart (Hiramoto, 1974). In addition, whereas lipids and antigens of the unfertilized mouse egg surface have mobilities characteristic of other cells, their mobilities decrease by as much as 100-fold following fertilization (Johnson and Edidin, 1978). The significance of this apparent "rigidification" of the zygote cortex, and its effect on the newly incorporated sperm surface, remain unclear.

As there are no distinct morphological features of the sperm membrane, it is not possible to follow the sperm surface by electron microscopy after fertilization. The sperm surface does possess antigens not found on the unfertilized egg, but attempts at visualizing these antigens in the newly fertilized egg using immunocytochemical techniques have failed (Artzt *et al.*, 1973; Menge and Fleming, 1978), although such antigens were detected later in development. The approach we have taken for tracing the fate of sperm surface components involves treating the sperm prior to fertilization with small vectorial probes. The sperm are subsequently used to fertilize eggs and the fate of the modified components can be followed in the early embryo both morphologically and biochemically, by following the fate of the covalently attached probe.

A. *Labeling of the Sperm Surface*

Since the modified sperm must be capable of effecting fertilization, neither the reagent nor the conditions used during the labeling reaction should cause a loss of viability. It had previously been reported that treatment of mammalian sperm with fluorescein isothiocyanate (FITC) did not affect their motility (Mellish and Baker, 1970). We found that

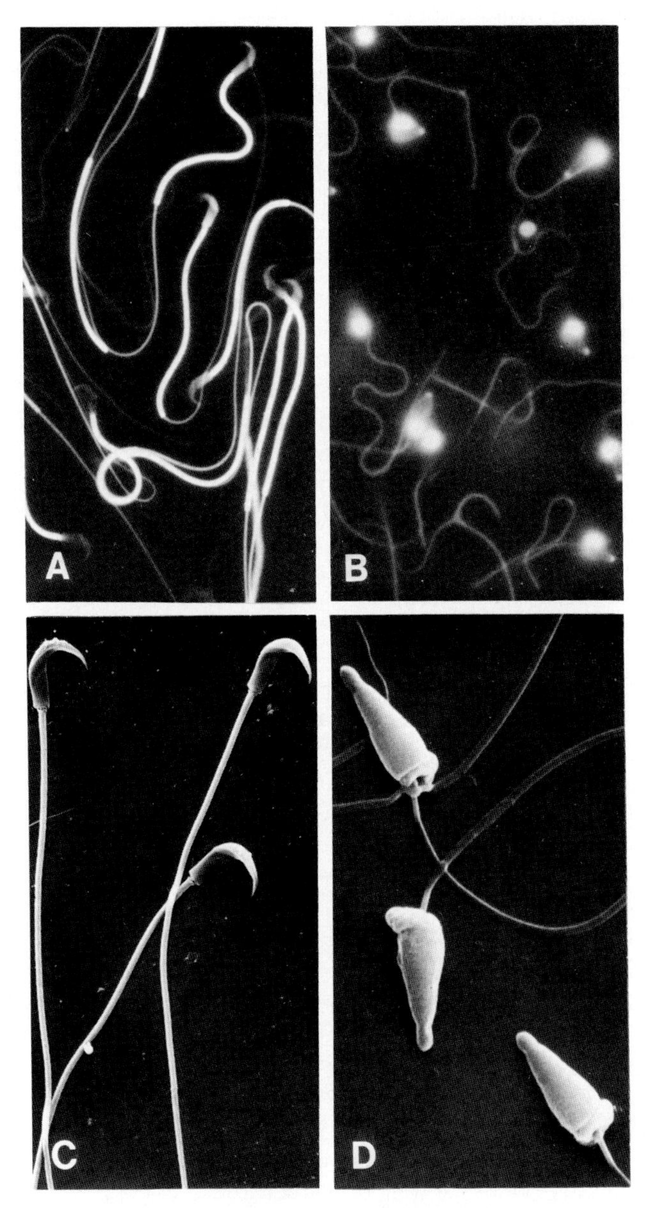

Fig. 2. Fluorescent labeling of sperm. Fluorescence micrographs of hamster (A) and sea urchin (B) sperm treated with FITC. The scanning electron micrographs show the cellular morphology for hamster (C) and sea urchin (D) sperm. See Gabel *et al.*, 1979 b, c.

FITC labels not only mammalian sperm but also spermatozoa from sea urchins and fish (Gabel *et al.*, 1979a). In all cases, labeling is nonuniform, with the highest fluorescense intensity occurring over the sperm midpiece (Fig. 2). In addition to the fluorescent probe, sperm may also be labeled with a radioactive derivative of FITC, ^{125}I-diiodofluorescein isothiocyanate (^{125}IFC), to identify the labeled molecular components (Gabel *et al.*, 1979a) and to estimate the amount of label bound.

The fluorescein derivatives are thought to be labeling components of the sperm surface for the following reasons: 1) FITC does not penetrate the plasma membrane of viable muscle cells and ^{125}IFC is excluded from red blood cells, where labeling is restricted to a single protein of the membrane (Gabel and Shapiro, 1978); 2) IFC-labeled sea urchin sperm are agglutinated in the presence of an antibody against the chromophore whereas unlabeled sperm are not (Gabel *et al.*, 1979a); 3) Triton X-100 extraction of ^{125}IFC-labeled sea urchin sperm removes 96% of the bound radioactivity (Gabel *et al.*, 1979a) under conditions previously shown to remove the plasma membrane but does not disrupt the head or tail elements (Gibbons and Gibbons, 1972); 4) the locus of binding for IFC to hamster sperm is at the midpiece surface as indicated in an indirect immunoperoxidase electron microscopic assay (Gabel *et al.*, 1979a), in agreement with the localization by fluorescence.

B. *Transfer of the Labeled Sperm Components to the Egg*

^{125}IFC-labeled sea urchin sperm quantitatively transfer the bound radioactivity to the zygote at fertilization (Gabel *et al.*, 1979b). In addition, as shown in Fig. 3, the sperm-derived radioactivity persists with the developing embryos at a ratio of approximately one sperm equivalent per embryo; a result consistent with monospermic fertilization. Homogenization of these embryos and subsequent centrifugation results in approximately 70% of the radioactivity being associated with the particulate fraction. It is, therefore, unlikely that the radioactivity of the embryo is due to free ^{125}IFC, or to ^{125}IFC-amino acids derived from degradation of sperm proteins. In addition, if the embryos were to degrade the sperm-derived peptides to their constituent amino acids, it is unlikely that the ^{125}IFC-labeled residues would be reincorporated into new proteins because of the high specificity of the protein synthesizing machinery (von Ehrenstein, 1970). Thus it appears that the ^{125}IFC-labeled sperm polypeptides have unusually slow turnover rates within the embryo; the extent to which these proteins are modified by the embryo is currently under investigation.

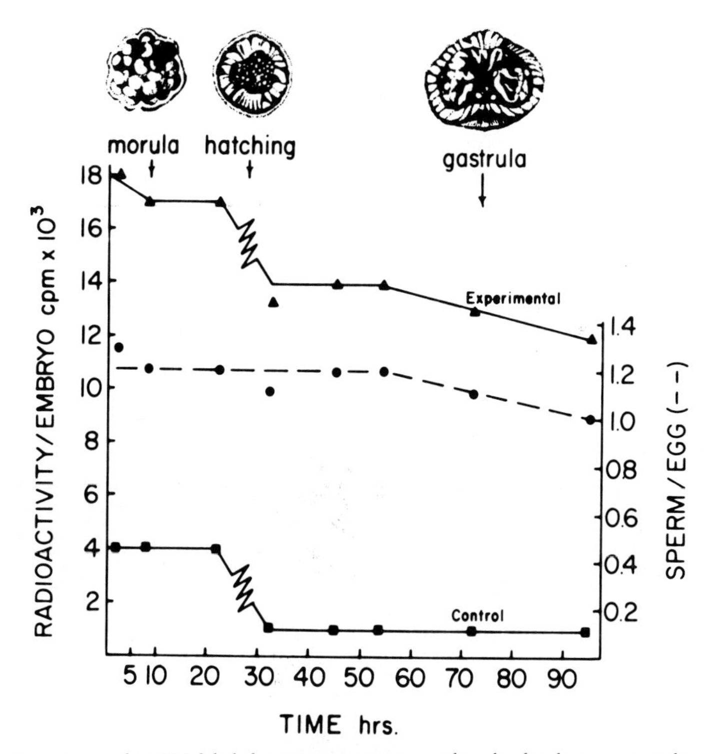

Fig. 3. Persistence of ^{125} IFC-labeled sperm components within the developing sea urchin embryo. ^{125}IFC-labeled sperm were added to unfertilized eggs and, after 7 min., the eggs were diluted into a large volume of Millipore filtered seawater (MSW). When the eggs settled, the supernatant was removed and the eggs were resuspended in fresh MSW. The washing was repeated four times after which the eggs were resuspended in MSW containing 50 μg/ml of penicillin and streptomycin and allowed to develop at 12° C. At the indicated times, aliquots of the culture were removed and the embryos were killed by the addition of CN^-. The dead embryos were collected by hand centrifugation, resuspended in a small volume of MSW, layered onto 10 ml of 5% Ficoll in MSW, and again collected by centrifugation. The final embryo pellet was washed once with MSW, then aliquots were removed and counted to determine the radioactivity present/embryo (—▲—▲—). For the control curve (—■—■—) the eggs were fertilized with unlabeled sperm and, after 7 min., ^{125}IFC-labeled sperm, equivalent to the amount used in the experimental culture, were added and the eggs treated as described above. The sperm/egg curve (—•—•—), was calculated by subtracting the control curve from the experimental and dividing by the specific radioactivity of the sperm solution (1.1 x 10^{-2} cpm/sperm). From Gabel *et al.*, 1979b.

C. *Distribution of the Labeled Components in the Embryo*

The labeled sperm components may be followed directly after fertilization by fluorescence microscopy (Gabel *et al.*, 1979c). Fig. 4A is a bright field micrograph of an 8-cell sea urchin embryo and Fig. 4B the corresponding fluorescence micrograph of this embryo. A single fluorescent patch is associated with one of the blastomeres. At the 16-cell stage the fluorescent patch is randomly associated with the three cell types; this is consistent with fertilization being random with respect to the animal-vegetal axis of the unfertilized sea urchin egg. As development continues, the patch becomes increasingly more difficult to detect, possibly due to diffusion of the labeled components (and thus, loss of the signal due to background fluorescence), or because of the difficulty in detecting the small patch in a group of embryos of random orientation.

The persistence of the labeled sperm components as a patch also occurs in mammalian embryos. After fertilization fluorescent mouse sperm components remain as a patch that is distributed asymmetrically during the early cleavages. This is shown for a two-cell mouse embryo in Fig. 4C.

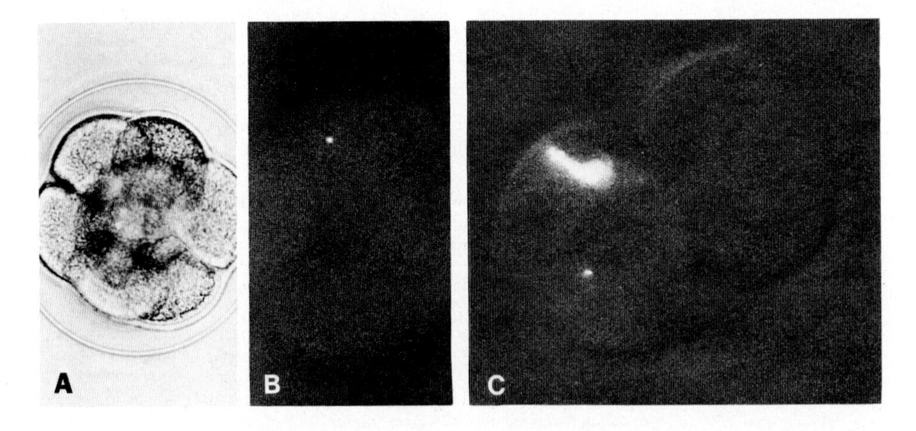

Fig. 4. Fluorescence localization of labeled sperm surface components in the early embryo. A) A bright field micrograph of an 8-cell sea urchin embryo derived from an egg fertilized with an FITC-labeled sperm. B) Fluorescence micrograph of the same sea urchin embryo. C) Fluorescence micrograph of a 2-cell mouse embryo recovered 24 hours post-insemination from a mouse artificially inseminated with tetramethylrhodamine isothiocyanate-labeled sperm.

The existence of the labeled sperm surface components as a patch within the embryo and the persistence of these components throughout early development are striking phenomena, but the significance of these observations is unclear. The fluorescent patch of labeled sperm components may simply reflect the rigidification of the egg cortex that occurs post-fertilization (Johnson and Edidin, 1978; Hiramoto, 1974). Alternatively, the asymmetric distribution of the sperm components may serve as a developmental signal. There is growing evidence that although early blastomeres are totipotent (Driesch, 1900; Kelly, 1977) there are molecular differences between them. For example, early mouse blastomeres are heterogeneous for the distribution of a surface antigen detected with a monoclonal antibody (Willison and Stern, 1978). Twins that are obtained by separating the blastomeres after the first cleavage in the sea urchin are non-identical (Marcus, 1979). In addition, the asymmetric distribution of sperm surface components that we find to occur within the embryo similarly distinguishes the early blastomeres, and such differences may affect future differentiation. Further analysis of this interesting group of heterogeneously distributed sperm proteins by using the fluorescent vectorial reagents and immunological techniques for purifying the labeled components (Gabel *et al.*, 1979a) should allow an assessment of their role in development.

In addition to following the fate of the labeled components, the fluorescent sperm may be used to study the process of sperm–egg fusion. It has previously been reported that treatment of invertebrate eggs with cytochalasin B (CB) prevents sperm pronuclear incorporation (Longo, 1978; Gould-Somero *et al.*, 1977), but it is not known to what extent sperm–egg fusion occurs in cytochalasin B treated eggs. We have found (G. Gundersen, unpublished) that greater than 50% of *S. purpuratus* eggs inseminated with FITC-labeled sperm in the presence of CB contain a patch of fluorescence closely associated with them. Although the patches in the majority of these eggs are within the hyaline layer (associated with the egg plasma membrane), a number of patches have been observed between the hyaline and the elevated fertilization membrane. These data indicate that an exchange of labeled sperm surface components with the egg may occur and that CB does not prevent sperm–egg fusion, but only the subsequent incorporation of sperm components, such as the pronucleus.

IV. THE CORTICAL REACTION

The cortical reaction of fertilization begins at the point of sperm entry and sweeps over the egg surface (Runnstrom, 1966; Anderson, 1968; Guidice, 1973). The reaction seems to be mediated by a wave of Ca^{2+} release from intracellular sites; as suggested above, this exocytosis may be initiated by the Ca^{2+} that is introduced into the egg by the sperm. Following the elevation of the concentration of free Ca^{2+}, cortical granules that underlie the plasma membrane undergo exocytosis, releasing a protease (Runnstrom, 1966; Vacquier *et al.*, 1972; Schuel *et al.*, 1973; Carroll and Epel, 1975), a "structural" protein (Bryan, 1970 a, b), a hyaline protein (Bryan, 1970a; Kane, 1970), a β-1, 3-glucanase (notably in *S. purpuratus;* Epel *et al.*, 1969; Muchmore *et al.*, 1969: Schuel *et al.*, 1973; Detering *et al.*, 1977; Schuel, 1978), and a peroxidase (Katsura and Tominaga, 1974; Foerder and Shapiro, 1977; Hall, 1978). Several of these components interact with the vitelline layer, a mucopolysaccharide-protein complex (Ito, 1969), to form the fertilization membrane (Runnstrom, 1966). At this time the plasma membrane is topographically mosaic, due to the incorporation of the membranes of the cortical granules; the microvilli, with which the plasma membrane of the egg is

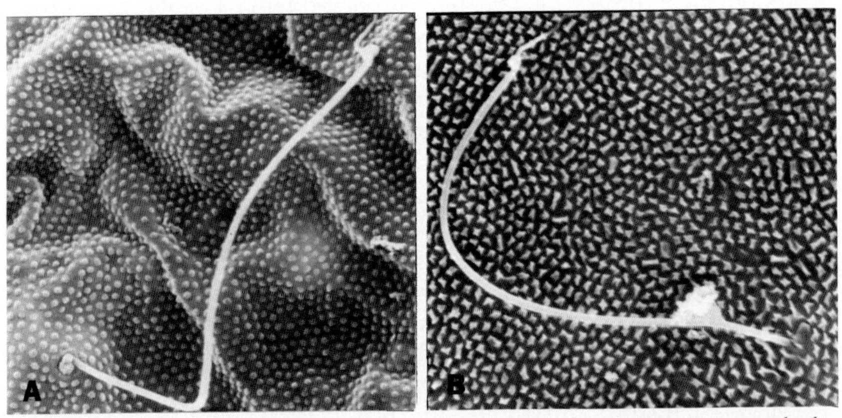

Fig. 5. A) The soft fertilization membrane containing igloo-shaped projections. Sixty seconds after insemination, the elevated fertilization membrane has collapsed onto the surface of the egg during fixation. The tail of the fertilizing sperm protrudes from the fertilization membrane. One can clearly see the uniform dome-shaped projections which occur on the fertilization membrane before it hardens and the soft, flexible nature of that structure. Magnification 6200x. B) The hardened fertilization membrane containing tent-shaped projections. Three minutes after fertilization, the fertilization membrane no longer collapses onto the surface of the egg during fixation, and all of the projections occur in tent form. The tail of the fertilizing sperm projects from the fertilization membrane. Magnification 5,750x. See Veron *et al.*, 1977.

covered, elongate (Eddy and Shapiro, 1976) to accommodate the increased membrane surface area. The conversion of the vitelline layer to a hardened, impermeable fertilization membrane occurs as a series of steps (Veron *et al.*, 1977) in a defined sequence. Sixty seconds after fertilization the fertilization membrane, now elevated from the surface of the egg, is "soft" (i.e., can be dissolved by protein denaturants) and has igloo-shaped (I form) projections upon it; these appear to be imprints of the microvilli (Fig. 5A). By 180 seconds the fertilization membrane has been altered (by a glycine ethyl ester-inhibited step; Lallier, 1971) into a structure that is still soft but now has tent-shaped (T-form) projections (Fig. 5B). In the next 30 seconds the membrane becomes "hard," i.e., does not dissolve readily (Runnstrom, 1966; Paul and Epel, 1971; Shapiro, 1975). Between 5–7 minutes after fertilization, the fertilization membrane develops a decreased permeability to substances such as Concanavalin A (Veron and Shapiro, 1977).

A. *The Peroxidative System of Fertilization*

The hardening of the fertilization membrane, that makes it so resistant to all but the most vigorous attempts to solubilize it (Runnstrom, 1966; Carroll and Baginski, 1978), appears to be mediated by an ovoperoxidase (OPO)-induced linkage of tyrosyl groups producing dityrosine (Foerder and Shapiro, 1977) as shown in Fig. 6. The evidence for a peroxidative system being activated at fertilization, to cross-link the fertilization membrane with dityrosine crosslinks, is several-fold (Foerder and Shapiro, 1977; Foerder *et al.*, 1978; Hall, 1978; Klebanoff *et al.*, 1979).

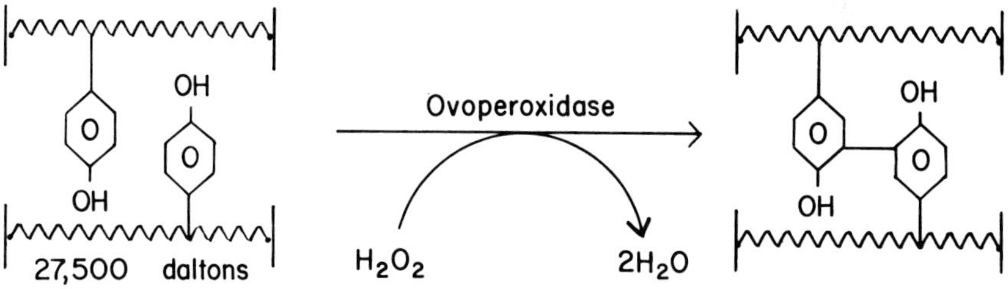

Fig. 6. Mechanism of hardening.

Dityrosine residues are found in hardened fertilization membranes, at about 1 residue per 50,000 daltons of fertilization membrane protein (Foerder and Shapiro, 1977). A peroxidase is released from the cortical granules at fertilization; some of the ovoperoxidase is found in the sea water surrounding the eggs, and some is incorporated into the fertilization membrane (Foerder and Shapiro, 1977; Klebanoff *et al.*, 1979). The localization of the ovoperoxidase in eggs before and after fertilization is shown in Fig. 7; note that it resides in the lamellae of the cortical granules of *S. purpuratus* before fertilization and within the structure of the fertilization membrane afterwards. Inhibitors of hardening, as determined from classical studies on fertilization, inhibit the ovoperoxidase (Motomura, 1954; Lallier, 1971);when known peroxidase inhibitors were tested for their effects on fertilization, they inhibited hardening (Foerder and Shapiro, 1977).

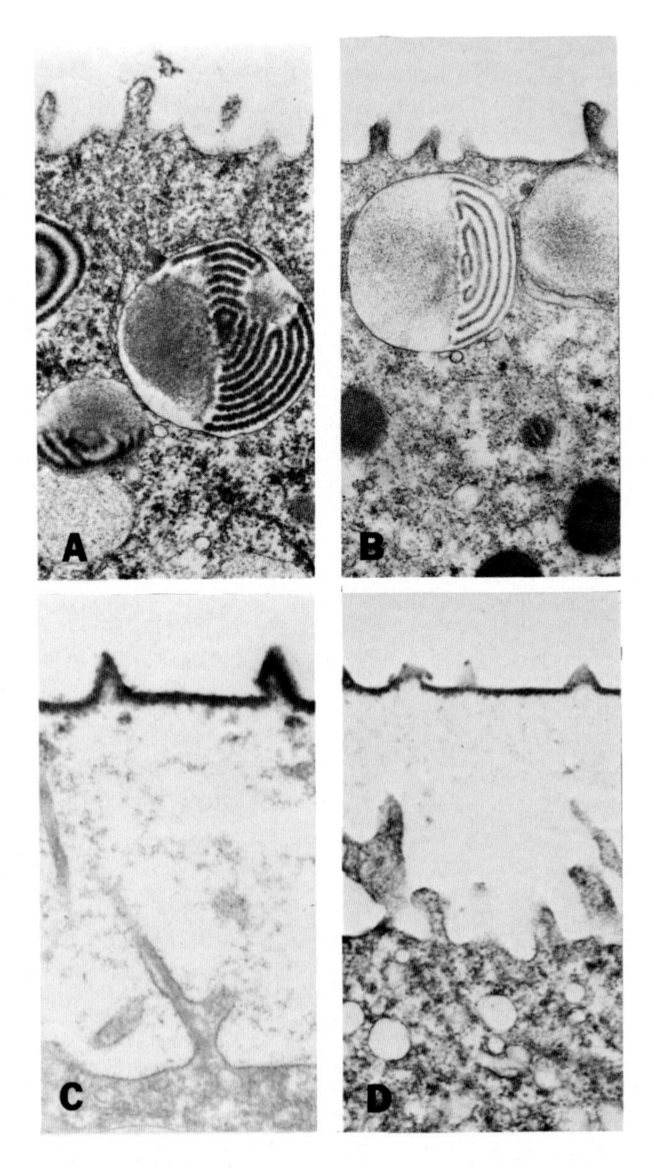

Fig. 7. Electron microscopic localization of ovoperoxidase in unfertilized and fertilized eggs. Living eggs were incubated with diaminobenzidine and hydrogen peroxide before being fixed with osmium tetroxide (Klebanoff *et al.*, 1979). The reaction product indicative of peroxidase activity is present in the lamellae of the cortical granules of the unfertilized eggs (A) and in the hardened fertilization membrane of fertilized eggs (C). Addition of aminotriazole (a peroxidase inhibitor) during incubation resulted in cortical granules (B) and hardened fertilization membrane (D) similar in density to those of eggs not reacted for peroxidase (data not shown). Magnification 20,000x. From Klebanoff *et al.*, 1979.

The ovoperoxidase has been highly purified from the material released by the egg at fertilization (the "fertilization product"). In these experiments, the vitelline layers were treated with trypsin, to prevent formation of the fertilization membrane, and the cortical granules were stimulated to release their contents by fertilization. The most highly purified preparation of ovoperoxidase is some 200-fold purified from the fertilization product. The enzyme behaves as a species of apparent molecular weight 50,000 upon gel filtration on Sephacryl 200. When examined by SDS-polyacrylamide gel electrophoresis, the principal protein components accounting for over 90% of the material on the gel consist of a doublet of polypeptides of 45,000 and 48,000 daltons in approximately equivalent concentrations. We are presently exploring which of these polypeptides is responsible for the ovoperoxidase activity, or whether one is related to the other.

By forcing eggs treated with the ovoperoxidase inhibitor, aminotriazole, through a Nitex mesh screen, the soft fertilization membranes can be removed and isolated from eggs (L. Ballou, unpublished data). Soft fertilization membranes isolated in this manner are useful for studying the mechanism of assembly of the fertilization membrane, a structure that arises from components of the vitelline layer and the cortical granule exudate. Soft fertilization membranes are not crosslinked, and thus can be disaggregated in order to determine their composition. When one examines the proteins of the cortical granule exudate and the composition of the soft fertilization membrane, several striking observations can be made. In the first place, only the 48,000 dalton component of the ovoperoxidase doublet is found in the soft fertilization membrane, as determined by SDS polyacrylamide gel electrophoresis. Several other cortical exudate components can be detected, especially several high molecular weight proteins, as can some polypeptides from the vitelline layer. One mechanism for stabilizing the structure of the soft fertilization membrane appears to be via ionic interactions. If the soft fertilization membrane preparation is diluted into either distilled water or into dilute buffer containing no divalent cations, its light scattering properties decrease enormously, and when examined under the phase microscope it appears much more "ghost-like" than normal. This change in the soft fertilization membrane does not occur if it is diluted into seawater, or into dilute buffer containing 5 mM Ca^{2+}, 40 mM Mg^{2+} or 460 mM Na^{+} (the concentrations of cations found in seawater); in fact, the concentration of magnesium required to preserve stability of the soft fertilization membrane is on the order of 5 mM. When the soft fertilization membranes become "ghost-like," after dilution in distilled water, they can regain their former properties, both their increased refractility, as determined by phase microscopy, and their increased light scattering, by subsequent re-addition of 50 mM Mg^{2+}. Thus, one of the interactions stabilizing the fertilization membrane as it assembles is an ionic one mediated by the cations in the seawater (B. Shapiro, unpublished data), and such interactions probably play a role in the initial assembly process.

Additionally, the soft fertilization membranes can be hardened merely by washing away the aminotriazole used to isolate them, and by adding hydrogen peroxide; i.e., the soft fertilization membranes contain a still active peroxidase than can crosslink the structures to make them resistant to dissolution in strong denaturants, e.g. 6 M guanidine with 30 mM dithiothreitol. This *in vitro* hardening reaction, although dependent upon the addition of hydrogen peroxide, does not require the re-addition of soluble ovoperoxidase. When the preparations are examined before and after hardening, by SDS polyacrylamide gel electrophoresis, two high molecular weight proteins of the soft fertilization membrane (that were contributed by the cortical granule exudate) disappear, and an increased amount of material is excluded from the gel (L. Ballou, unpublished data) as hard fertilization membrane (Shapiro, 1975). We interpret this to mean that the proteins contributed by the cortical granule exudate are somehow involved in the final crosslinked structure. However, crosslinking does not involve all of the proteins of the soft fertilization membrane; for example, the band corresponding to the ovoperoxidase does not itself disappear, suggesting that it is not crosslinked into the structure, although it is associated with it by noncovalent interactions. We are just at the beginning of understanding the mechanism of assembly of this extracellular barrier, but it is a fascinating structure, assembled from two distinct cellular compartments.

B. *Hydrogen Peroxide Production and Other Similarities to Phagocytosis*

The ovoperoxidase requires H_2O_2 as a substrate, which is produced by eggs at the time of the cortical reaction. The increased O_2 consumption at fertilization, first reported by Warburg (1908), appears to be in large part converted to H_2O_2 (Foerder *et al.*, 1978). H_2O_2 production occurs with approximately the same kinetics as O_2 uptake and ⅔ of the O_2 uptake can be accounted for by the production of H_2O_2.

The ovoperoxidase is involved in the production of chemiluminescence in fertilized eggs (Foerder *et al.*, 1978); thus, the sea urchin egg "flashes" when activated, just as it gasps. Inhibitors of the ovoperoxidase inhibit light emission. H_2O_2 production and chemiluminescence are two of many similarities between the events following fertilization in the sea urchin egg and those following phagocytosis in polymorphonuclear leukocytes (PMN). In the case of human PMNs, some of the chemiluminescence depends on myeloperoxidase activity; the remainder requires superoxide (see Rosen and Klebanoff, 1976 for futher references). Thus, as shown in Table II, both sea urchin eggs and human PMN's generate activated oxygen species and utilize them in subsequent peroxidative reactions. In most cases, more is known about human PMNs in this regard, and we have yet to unravel the nature of the oxidase in the egg, or whether activated oxygen species other than H_2O_2 exist. The peroxidative system of the egg may contribute to the block to polyspermy. Foerder and Shapiro (1977) have suggested that OPO and H_2O_2 could kill sperm in the vicinity of the fertilized egg; H_2O_2 is toxic to echinoderm sperm (Evans, 1947), the peroxidative system of PMNs (myeloperoxidase + H_2O_2 + halide) is toxic to bovine spermatozoa as are comparable systems employing lactoperoxidase or rat uterine fluid peroxidase (Smith and Klebanoff, 1970), and lactoperoxidase activity kills the sperm of *S. purpuratus* (B. Shapiro, unpublished data). However, the importance of any *in vivo* effect of this system is not yet known.

TABLE II
Similarities Between Fertilization and Phagocytosis

OXIDASE: O_2 ⟶ Activated Oxygen; PEROXIDASE: Activated Oxygen ⟶ Chemical modification

OXIDASE	Fertilization	Phagocytosis
respiratory burst	+	+
NAD(P)H as reductant	?	+
increased hexose monophosphate shunt	+	+
H_2O_2 production	+	+
superoxide production	?	+
PEROXIDASE	**Ovoperoxidase**	**Myeloperoxidase**
located in cytoplasmic granule	+	+
released upon specific stimulus	+	+
chemiluminescence	+	+
iodination	+	+
spermicidal	?	+
deiodination	+	+
estradiol binding	+	+
dityrosine synthesis	+	?

A remarkable feature of this peroxidative mechanism is the extent to which it is conserved, having many features in common in both the invertebrate egg and the human white blood cell. For example, they both have an oxidative reaction, with a burst of oxygen consumption (Warburg, 1908; Sbarra and Karnovsky, 1959) and H_2O_2 production (Iyer *et al.*, 1961; Foerder *et al.*, 1978), that may be related to the increased activity of the hexose monophosphate shunt pathway that is common to the two systems (Iyer *et al.*, 1961; Guidice, 1973). Both have peroxidases that are located in cytoplasmic granules (Agner, 1941; Katsura and Tominaga, 1974; Foerder and Shapiro, 1977) and released upon stimulation, and both exhibit chemiluminescence (Allen *et al.*, 1972; Foerder *et al.*, 1978) and other model reactions characteristic of peroxidative systems (Rosen and Klebanoff, 1977; Klebanoff *et al.*, 1979).

However, the PMN is a highly specialized cell without the capacity to divide, its whole purpose being to ingest and kill invading organisms. The egg has a quite different role; being totipotent, it will give rise to all cells of the adult organism, yet it has a similar peroxidatic system in every aspect seriously examined to date. We think that in the egg too, the peroxidatic system may play a protective role, by killing excess sperm. Additionally, the egg utilizes the peroxidatic system in another way, by mediating a complex extracellular synthesis, in which a highly targeted and carefully placed enzyme (the ovoperoxidase) is ideally located (within the fertilization membrane) to effect the biosynthetic redox reaction that results in the formation of a stable, covalent dityrosine bond. Whether such a capacity for responding to stimuli, by activating a peroxidative system is a general property of cells remains to be seen, but it is certainly suggested by the work on sea urchin eggs and human PMN's.

V. SUMMARY AND CONCLUSIONS

In our analysis of several membrane-related events of fertilization, we have been struck repeatedly by the observation that gametes perform activities characteristic of many specialized cell types. The movement of ions in the sperm as it undergoes the acrosome reaction is reminiscent of phenomena seen in excitable cells like muscle and nerve. The early changes that occur in the sperm, with an apparent rapid change in the membrane permeability to several ions, suggest that one of the first triggering mechanisms may involve alteration in the membrane potential of the cell. Upon fusing with the egg, the sperm membrane does not diffuse rapidly, as would be expected from studies of other cells, but rather exists as a patch throughout early development. Such an unusual patch may play some role in morphogenesis, but as yet there is no specific evidence for the hypothesis, it being based on inferences taken from some of the classic literature in developmental biology (e.g., the role of the sperm as a partial determinant of the grey crescent; reviewed by Brachet, 1977). The egg responds to the sperm by altering its surface, both by inserting some 10,000 cortical granules into the plasma membrane and by having a concomitant, and perhaps related, increase in the rigidity of the cortical zone. The egg additionally dramatically changes the properties of its glycocalyx, in assembling a complex fertilization membrane. This structure then hardens under the influence of a coordinated peroxidative system, ideally suited for extracellular biosyntheses, that is strikingly similar in many of its properties to a system in the human phagocytic white blood cell. In all of this work, we see the guiding hand of evolution, where similar mechanisms are preserved in different cell types from different species. In this way, a process once established can be modified slightly to effect different events, certainly much more efficient a mechanism than starting over from scratch. Gametes may be particularly striking in showing us inferences about the generality of mechanisms between cells, for they are at the beginning of the path that will ultimately express itself in all the functions that are accessible to the living state.

ACKNOWLEDGEMENTS

This work was supported by grants from the NSF (PCM 7720472) and NIH (GM 23910) to B.M.S. and NIH grant HD 02266 (to S.J.K.) C.A.G. was supported by a National Research Award from the National Institutes of Health (GM 07270) and C.A.F. by Training Grant GM 00052. We are grateful to Ms. Abby A. Dudley for expeditiously typing and assembling this manuscript.

REFERENCES

Agner, K. (1941). *Acta Physiol. Scand.* **2**, Suppl. 8.

Allen, R.C., Stjernholm, R.C. and Steele, R.H. (1972). *Biochem. Biophys. Res. Commun.* **47**, 678-684.

Anderson, E. (1968). *J. Cell Biol.* **37**, 514-539.

Artzt, K., Dubois, P., Bennett, D., Condamine, H., Babinet, C. and Jacob, F. (1973). *Proc. Nat. Acad. Sci. U.S.A.* **70**, 2988-2992.

Austin, C.R. (1968). "Ultrastructure of Fertilization." Holt, Rinehart and Winston, New York.

Barros, C. and Franklin, L.E. (1968). *J. Cell Biol.* **37**, c13-c18.

Brachet, J. (1977). *Curr. Top. Develop. Biol.* **11**, 133-186.

Bryan, J. (1970a). *J. Cell Biol.* **44**, 635-644.
Bryan, J. (1970b). *J. Cell Biol.* **45**, 606-614.
Bygrave, F.L., Reed, K.C. and Spencer, T.L. (1971). *Nature, New Biol.* **230**, 8.
Carroll, Jr., E.J. and Epel, D. (1975). *Develop. Biol.* **44**, 22-32.
Carroll, Jr., E.J. and Baginski, R.M. (1978). *Biochemistry* **17**, 2605-2612.
Collins, F. and Epel, D. (1977). *Exp. Cell Res.* **106**, 211-222.
Colwin, L.H. and Colwin, A.L. (1963). *J. Cell Biol.* **19**, 501-518.
Dan, J.C. (1954). *Biol. Bull.* **107**, 335-349.
Dan, J.C., Ohori, Y. and Kushida, H. (1964). *J. Ultrastruct. Res.* **11**, 508-524.
Decker, G.L., Joseph, D.B. and Lennarz, W.J. (1976). *Develop. Biol.* **53**, 115-125.
Detering, N.K., Decker, G.L., Schmell, E.D. and Lennarz, W.J. (1977). *J. Cell Biol.* **75**, 899-914.
Driesch, H. (1900). *Arch. f. Entw. Mech.* **10**, 361-410.
Eddy, E.M. and Shapiro, B.M. (1976). *J. Cell Biol.* **71**, 35-48.
Edidin, M., Zagyansky, Y. and Lardner, T.J. (1976). *Science* **191**, 466-468.
Epel, D., Weaver, A.M., Muchmore, A.V. and Schimke, R.T. (1969). *Science* **163**, 294-296.
Evans, T.C. (1947). *Biol. Bull.* **92**, 99-109.
Foerder, C.A. and Shapiro, B.M. (1977). *Proc. Nat. Acad. Sci. U.S.A.* **74**, 4214-4218.
Foerder, C.A., Klebanoff, S.J. and Shapiro, B.M. (1978). *Proc. Nat. Acad. Sci. U.S.A.* **75**, 3183-3187.
Frye, L.D. and Edidin, M. (1970). *J. Cell Sci.* **7**, 319-335.
Gabel, C.A. and Shapiro, B.M. (1978). *Anal. Biochem.* **86**, 396-406.
Gabel, C.A., Eddy, E.M. and Shaprio, B.M. (1979a). *J. Cell Biol.*, **82**, 742-754.
Gabel, C.A., Eddy, E.M. and Shapiro, B.M. (1979b). Mini Symposium on the Spermatozoon. (D. Fawcett, ed.), Urban and Schwartzenberg, Maryland.
Gabel, C.A., Eddy, E.M., and Shapiro, B.M. (1979c) *Cell*, **18**, 207-215.
Gibbons, B.H. and Gibbons, I.R. (1972). *J. Cell Biol.* **54**, 75-97.
Giudice, G. (1973). *In* "Developmental Biology of the Sea Urchin Embryo," Academic Press, New York.
Gould-Somero, M., Holland, L. and Paul, M. (1977). *Develop. Biol.* **58**, 11-22.
Hall, H.G. (1978). *Cell* **15**, 343-355.
Hille, B. (1967). *J. Gen. Physiol.* **50**, 1287-1302.
Hiramoto, Y. (1974). *Exp. Cell Res.* **89**, 320-326.
Ito, S. (1969). *Fed. Proc.* **28**, 12-35.
Iyer, G.Y.N., Islam, D.M.F. and Quastel, J.H. (1961). *Nature* **192**, 535-541.
Johnson, M. and Edidin, M. (1978). *Nature* **272**, 448-450.
Jones, H.P., Bradford, M.M., McRorie, R.A. and Cormier, M.J. (1978). *Biochem. Biophys. Res. Commun.* **82**, 1264-1272.
Kane, R.E. (1970). *J. Cell Biol.* **45**, 615-622.
Katsura, S. and Tominaga, A. (1974). *Develop. Biol.* **40**, 292-297.
Kelly, S.J. (1977). *J. Exp. Zool.* **200**, 365-376.
Klebanoff, S.J., Foerder, C.A., Eddy, E.M. and Shapiro, B.M. (1979). *J. Exp. Med.* **149**, 938-953.
Kohlhardt, M., Bauer, B., Krause, H. and Fleckenstein, A. (1972). *Pflugers Archiv.* **335**, 309-322.
Lallier, R. (1971). *Experientia* **27**, 1323-1324.
Levin, R.M. and Weiss, B. 1(977). *Molecular Pharmacol.* **13**, 690-697.
Longo, F.J. (1973). *Biol. Reprod.* **9**, 149-215.
Longo, F.J. (1978). *Develop. Biol.* **67**, 249-265.
Marcus, H.N. (1979). *Develop. Biol.* **70**, 274-277.

Mellish, K.S. and Baker, R.D. (1970). *J. Anim. Sci.* **31**, 917-922.
Menge, A.C. and Fleming, C.H. (1978). *Develop. Biol.* **63**, 111-117.
Motomura, I. (1954). *Sci. Rep. Tohoku Univ.*, Series 4, **20**, 219-225.
Muchmore, A.V., Epel, D., Weaver, A.M. and Schimke, R.T. (1969). *Biochim. Biophys. Acta* **178**, 551-560.
Paul, M. and Epel, D. (1971). *Exp. Cell Res.* **65**, 281-288.
Rosen, H. and Klebanoff, S.J. (1976). *J. Clin. Invest.* **58**, 50-60.
Rosen, H. and Klebanoff, S.J. (1977). *J. Biol. Chem.* **252**, 4803-4810.
Runnstrom, J. (1966). *Adv. Morphogen.* **5**, 221-325.
Sbarra, A.J. and Karnovsky, M.L. (1959). *J. Biol. Chem.* **234**, 1355-1361.
Schackmann, R.W., Eddy, E.M. and Shapiro, B.M. (1978). *Develop. Biol.* **65**, 483-495.
Schlessinger, J., Koppel, D.E., Axelrod, D., Jacobson, K., Webb, W.W. and Elson, E.L. (1976). *Proc. Nat. Acad. Sci. U.S.A.* **73**, 2409-2413.
Schuel, H. (1978). *Gamete Res.* **1**, 299-382.
Schuel, H., Wilson, W.L., Chen, K. and Lorand, L. (1973). *Develop. Biol.* **34**, 175-183.
Shapiro, B.M. (1975). *Develop. Biol.* **46**, 88-102.
Smith, D.C. and Klebanoff, S.J. (1970). *Biol. Reprod.* **3**, 229-235.
Takahashi, Y.M. and Sugiyama, M. (1973). *Develop. Growth Different.* **15**, 261-267.
Vacquier, V.D., Epel, D. and Douglas, L.A. (1972). *Nature* **237**, 34-36.
Veron, M. and Shapiro. B.M. (1977). *J. Biol. Chem.* **252**, 1286-1292.
Veron, M., Foerder, E., Eddy, E.M. and Shapiro, B.M. (1977). *Cell* **10**, 321-328.
Von Ehrenstein, G. (1970). In "Aspects of Protein Biosynthesis" (C.B. Anfinsen, Jr., ed.), Academic Press, New York.
Warburg, O. (1908). *Hoppe Seylers Z. Physiol. Chem.* **57**, 1-16.
Willison, K.R. and Stern, P.L. (1978). *Cell.* **14**, 785-793.
Wolff, D.J., Poirer, P.G., Bronstrom, C.O. and Bronstrom, M.A. (1977). *J. Biol. Chem.* **252**, 4108-4117.

The Adhesion of Sperm to Sea Urchin Eggs

Victor D. Vacquier

Marine Biology Research Division, A-002
Scripps Institution of Oceanography,
University of California, San Diego,
La Jolla, California 92093

I. INTRODUCTION

For the past one hundred years the gametes of sea urchins have been the favorite material for studies of the fertilization process. This intense interest in such a fundamental process as fertilization in this one animal group has resulted in the fact that we now know more about the mechanism of fertilization in sea urchins than in any other animal group. There are many reasons why sea urchins are so popular for fertilization

ISBN 0-12-612984-3

research: The adults are readily obtainable at marine laboratories and from biological supply companies, the sexes are separate, the gametes are extruded from gravid adults as pure populations of either sperm or eggs, fertilization is external occurring in sea water, the events of sperm-egg interaction occur with great synchrony in a time span of seconds and large quantities of gametes can be obtained inexpensively and with relative ease.

Two of the most intensely studied problems in sea urchin fertilization research are the adhesion-detachment cycle of sperm to eggs (Fig. 1; Epel

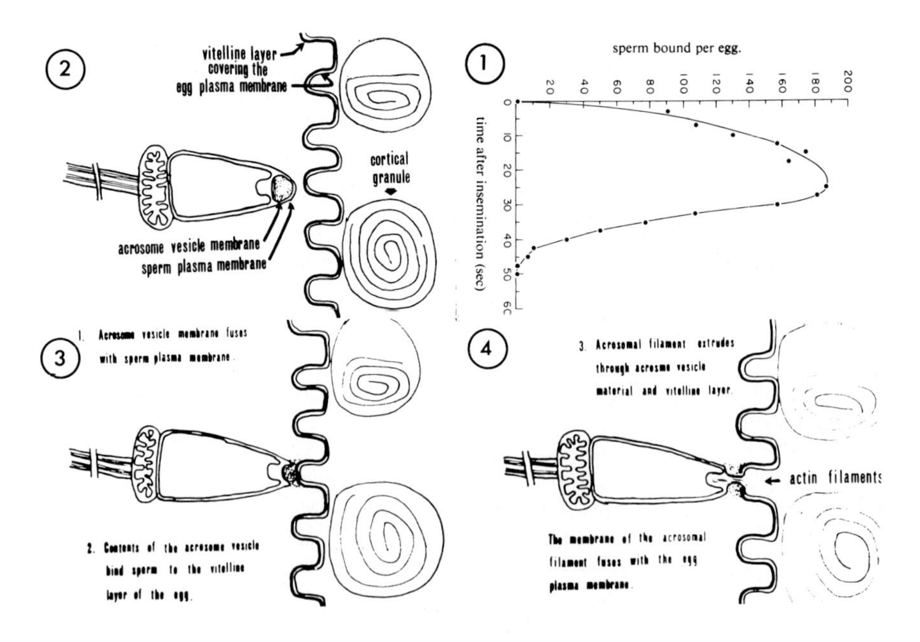

Fig. 1. The adhesion-detachment cycle of *S. purpuratus* sperm to eggs. Sperm adhere to eggs during the first 25 sec following insemination. The release of the fertilization protease from cortical granules at 25 to 50 sec destroys the sperm–egg bond (Vacquier *et al.*, 1973; Schuel, 1979).

Fig. 2. The interacting surfaces of sea urchin gametes. Cortical granules lie beneath the egg plasma membrane which is overlain by the vitelline layer. Acrosome granule (vesicle) is situated in the anterior apex of sperm.

Fig. 3. The acrosome granule (vesicle) fuses with the sperm PM releasing the contents of the granule which bind to the egg vitelline layer.

Fig. 4. The acrosome process passes through the mass of acrosome granule contents and fuses with the egg.

and Vacquier, 1978) and the metabolic activation of the egg immediately after fusion with a sperm (Epel, 1978). The attachment and fusion of sperm with egg comprises at least three separate events involving cell surface recognition. The first is the interaction of the egg jelly coat fucose sulfate polymer with the sperm plasma membrane. This induces the sperm acrosome reaction which consists of the exocytosis of the acrosome granule and the extrusion of the finger-like acrosome process (SeGall and Lennarz, 1979). The second is the attachment (binding, adhesion) of sperm by the acrosome process to the egg vitelline layer (VL). Third is the fusion of egg and sperm plasma membranes, a process about which we know virtually nothing (Figs. 2-4). These three recognition events must surely involve protein components on the surfaces of the interacting gametes (Metz, 1978; Vacquier, 1979).

The interaction of sperm and eggs exhibits a high degree of homologous species preference which is most often termed species specificity. When foreign sperm are mixed with eggs the sperm usually fail to undergo the acrosome reaction, so consequently they also fail to bind to or fuse with the egg. This species specificity is not absolute since interspecies, intergeneric and even interphyletic hybrid zygotes can be achieved. For this reason it may be more correct to use the term "self species preference" in place of "species specific" when referring to the process of sperm-egg interaction. The phenomenon of self species preference in gamete interaction has long fascinated those who have studied fertilization. Jacques Loeb, one of the founders of fertilization research, was probably the first to review the question of interspecific sperm-egg interaction and to speculate that preference for self species must involve proteins on the gamete surfaces. It should be remembered that at the time of Loeb's intuitive suggestion essentially nothing was known about the cell surface, the composition of cell membranes or extracellular membrane proteins (Loeb, 1916). Today, we recognize that probably all specific intercellular interactions are mediated through the binding of complementary cell surface proteins and glycoproteins.

The pioneering work of the late Jean C. Dan on the ultrastructural aspects of the invertebrate sperm acrosome reaction and her demonstration that the acrosome reaction was required to make the sperm capable of attachment to and fusion with the egg presented the idea that the acrosome reaction exposed a new membrane that was different from the rest of the sperm surface (Dan, 1967). Her descriptions of the golgi-produced acrosome granule (AG; also termed vesicle or vacuole) in the anterior apex of intact echinoderm sperm, the exocytosis of the AG and the extension of the acrosome process coated with the contents of the AG were some of the great steps forward in our knowledge of the fertilization process (Dan, 1967; Metz, 1978).

The implication from the early transmission electron microscopy of Dan (1967), and also the Colwins (1967), was that the exocytosed content of the AG was the substance responsible for the attachment of the sperm acrosome process to the egg VL (Figs. 2–4). With improved fixation and sectioning methods Summers and Hylander (1974) showed conclusively for the sand dollar *E. parma,* that the contents of the AG are present at the site of the bond of sperm to egg (Summers *et al.,* 1975). Following this, Hylander and Summers (1977) elegantly demonstrated that in the clam, *Chama,* the AG contents bind to filamentous material carried on the tips of microvilli that protrude through the egg VL.

Our interest in the biochemistry of gamete interaction, especially sperm-egg recognition and adhesion, prompted us to attempt to isolate the adhesive substance of the sperm AG and also to identify the egg surface component to which it bound. This article reviews the evidence that the major component of the sea urchin sperm acrosome granule is a protein we have called bindin (short for sperm borne binding protein) which has species preferential affinity for a bindin receptor glycoprotein found on the VL of the unfertilized egg.

II. THE ISOLATION OF THE ACROSOME GRANULE OF SEA URCHIN SPERM AND IDENTIFICATION OF BINDIN AS THE MAJOR COMPONENT

The AG of *S. purpuratus* sperm is a bullet-shaped, phase dense granule about 0.25 μm in diameter forming the anterior tip of the cell (Fig. 5). The AG was isolated by suspending the sperm in CA^{2+}–free sea water (pH 5.8) containing 5 mM sodium acetate, 5 mM EGTA (to chelate Ca^{2+}), 2.5% Triton X–100, 0.05% sodium azide and ovomucoid trypsin inhibitor at 0.1 mg/ml (Vacquier and Moy, 1977). The suspension was shaken by hand for 1 min (Fig. 6) which demembranated the cells, dissolved the midpiece and freed the AG. The shakate, composed of sperm nuclei, flagella and AGs was centrifuged 30 min at 1000 xg to sediment the nuclei. The supernate, containing AGs and broken flagella was carefully decanted and poured through a bed of glass fiber (8 μm diameter) which traps the flagella but allows the spherical AGs to pass through (Fig. 6). The liquid passing through the glass fiber was centrifuged 50 min at 13,000 xg to sediment the AGs. The AGs form a thin film on the tube bottom. When fixed, embedded, sectioned and viewed in the electron microscope this film is composed of AGs that have lost their limiting membranes (Fig. 7), but seem to be about the same granular texture as the AG of the intact cell (Fig. 5). When the AG film is dissolved in sodium dodecyl sulfate and electrophoresed in polyacrylamide gels, the AG material migrates as a single major protein-staining component of 30,500 Daltons apparent molecular weight (Fig. 8). The material also migrates as

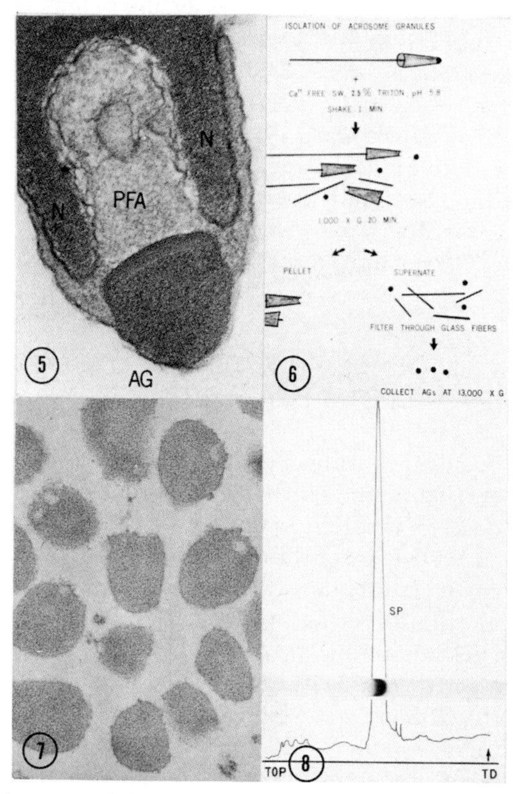

Fig. 5. Thin section of anterior end of an *S. purpuratus* sperm. The acrosome granule (AG) protrudes from the cell. N = nucleus; PFA = profilamentous actin. x86,000.

Fig. 6. Scheme for the isolation of acrosome granules devoid of limiting membranes. The spun glass was the type used for Christmas Tree decoration. (Vacquier and Moy, 1977).

Fig. 7. Thin section of pellet obtained in last step of isolation scheme. The spherical granular material is composed of demembranated acrosome granules. x50,000.

Fig. 8. 12.5% polyacrylamide gel containing 0.1% sodium dodecylsulfate on which 50 μg acrosome granule pellet was loaded. The gel was stained with Coomassie Blue and scanned at 550 nm. The granules are composed almost entirely of a single protein of 30,500 Daltons apparent mol. wt. We call this protein bindin for sperm borne binding protein.

a single band in acetic acid-urea gels and it has a single N–terminal amino acid which is tyrosine. By dry weight it is 100% protein as measured against a bovine serum albumin standard and it does not stain for carbohydrate. It is also negative for carbohydrate by both the phenol-sulfuric acid and the anthrone colorimetric tests (Vacquier and Moy, 1977).

Although the electrophoretic analysis shows one major band, it is very possible that the intact AG contains other important proteins that are solubilized by the Triton X–100 and high ionic concentration of the isolation medium. For example, data from other laboratories show that exocytosis of the AG exposes a trypsin-like protease (Levine *et al.*, 1978), a chymotrypsin (Csernansky *et al.*, 1979) and a phospholipase (Metz, 1978). Comparing Fig. 5 to Fig. 7 leaves little doubt that bindin is the major component of the sea urchin sperm AG. To this date bindin has been isolated by this method from three sea urchin species; *Strongylocentrotus purpuratus, S. franciscanus* and *Arbacia punctulata* (Glabe and Lennarz, 1979).

III. EVIDENCE THAT BINDIN MEDIATES SPERM-EGG ADHESION

Two lines of evidence support the hypothesis that bindin is the adhesive protein bonding sperm to egg. The first involves the immunoperoxidase localization of bindin on the surfaces of acrosome reacted sperm and in the site of the sperm-egg bond, and the second is the interaction of particulate bindin with unfertilized eggs.

Antibody to bindin was prepared by injecting unstained polyacrylamide gel containing the center of the bindin band (Fig. 8) into rabbits. The antiserum to this band showed a single line of identity when diffused against a PBS extract of whole sperm and purified bindin. The bindin is sparingly soluble after sonication in PBS and we had no difficulty obtaining precipitin lines in 1% agar double diffusion plates. Fixed sperm and eggs were reacted with rabbit anti-bindin, washed in PBS and then reacted with horseradish peroxidase-conjugated swine anti-rabbit serum. The cells were washed extensively and then reacted with diaminobenzidine (DAB), washed, fixed in OsO_4, embedded in epon and prepared for electron microscopy (Moy and Vacquier, 1979).

Sperm in which the acrosome reaction had been induced by raising the pH to 9.2 with NH_4OH showed a dense DAB deposit covering the acrosome process and the first 1 to 2 μm of plasma membrane running posteriorly from the base of the acrosome process (Figs. 9, 10). The

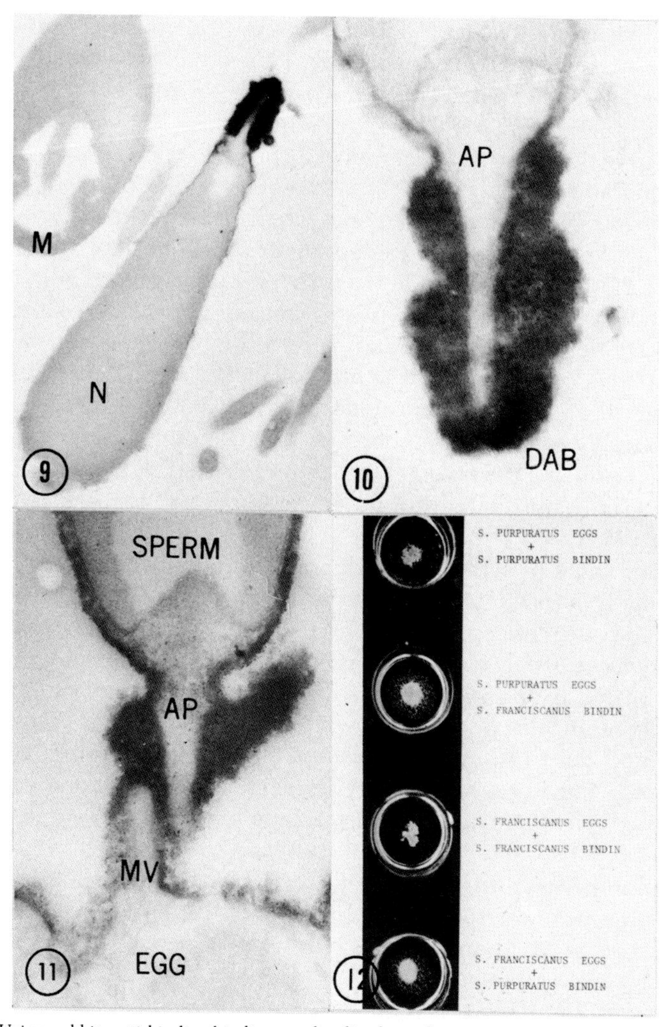

Fig. 9. Using rabbit anti-bindin, bindin was localized on the acrosome process and membrane of anterior tip of acrosome reacted sperm by the immunoperoxidase procedure. x25,000; M = mitochondrion midpiece; N = nucleus.

Fig. 10. The immunoperoxidase technique shows that bindin completely coats the acrosome process (AP). The membrane of the AP is destined to fuse with the egg plasma membrane. x125,000; DAB = diaminobenzidine precipitate.

Fig. 11. Immunoperoxidase localization of bindin at the site of sperm to egg bond. In this electron micrograph the sperm acrosome process is bound to the side of a microvillus (MV) of the egg. x75,000 (Moy and Vacquier, 1979).

Fig. 12. The agglutination of *S. purpuratus* and *S. franciscanus* eggs by bindin is species specific. Starting from the top, wells 1 and 3 show agglutination and 2 and 4 nonagglutination (Glabe and Vacquier, 1977a).

interesting finding is that the antibody localizes bindin as a thick covering along the acrosome process. The acrosome process membrane is the membrane destined to fuse with the egg. This coat of bindin on the fusing membrane suggests that bindin may also play a role in the process of membrane fusion. Control sperm, which had not undergone the acrosome reaction, did not react with either the primary or secondary serum (Moy and Vacquier, 1979).

When sperm bound to eggs were reacted with the antibody to bindin the DAB precipitate localized the bindin at the site of the sperm to egg bond (Fig. 11). We can conclude from these studies that bindin is at the correct place at the correct time to be the mediator of gamete adhesion. In regard to species cross-reactivity of the antibody to *S. purpuratus* bindin, we found that every echinoid (sand dollar, sea urchin, heart urchin) tested regardless of geographical location would cross-react. However, sperm of closely related Echinoderm classes (Ophiuroids, Holothuroids, Asteroids) would not cross-react showing that echinoid bindin is immunologically unrelated to other Echinoderm bindins (Moy and Vacquier, 1979).

The second line of evidence that bindin is the sperm to egg adhesive is the finding that particulate bindin is a species specific agglutinin of unfertilized sea urchin eggs (Fig. 12). In the experiment shown here, eggs of *S. purpuratus* or *S. franciscanus* were mixed with bindin from either species and the culture dishes agitated by rotary motion. The particulate bindin is a strong agglutinin of only its own species of eggs, and cross species agglutination does not occur regardless of the concentration of bindin (Glabe and Vacquier, 1977a). The same phenomenon has been quantitatively demonstrated using eggs and bindin of *A. punctulata* and *S. purpuratus* (Glabe and Lennarz, 1979). The importance of the phenomenon of bindin-induced agglutination of eggs is that it was the first demonstration of the interaction of a protein isolated from sperm with the surface of an animal egg.

IV. COMPARISON OF BINDINS FROM *S. purpuratus* (Sp) AND *S. franciscanus* (Sf)

The species specific agglutination of eggs with bindin isolated from these two species led us to perform a preliminary analysis of the two bindins to determine their degree of similarity. Both migrated on SDS gels with an apparent molecular weight of 30,500 Daltons. Isoelectric focusing gels showed the pI of Sp bindin to be 6.62 and of Sf bindin 6.59

(Bellet *et al.*, 1977). Amino acid analysis (Table I) of the two bindins showed they are composed of 40% nonpolar, 47% polar and 13% positively charged residues. The bindins are almost identical in composition for 16 amino acids (285 residues per molecule) with significant species differences in only two: Pro and Asx (Table I). Two dimensional tryptic peptide fingerprints also show a similarity: of the 37 spots for Sp and 34 for Sf bindin, 24 spots are superimposable (Fig. 13; Bellet *et al.*, 1977). We have determined the amino acid sequence of Sp and Sf bindins for the first 45 residues. Comparison of the sequences shows 31 out of 45 residues to be in the same position (Fig. 14). Both bindins are uncharged to at least this point suggesting that the N-terminal region of the molecule may be responsible for the hydrophobic properties of bindin.

V. EVIDENCE FOR BINDIN RECEPTORS ON THE VITELLINE LAYER OF EGGS

Since bindin is a species specific agglutinin of unfertilized eggs the molecular differences between any two sea urchin bindins such as Sp and Sf (Figs. 12–14; Table I) must be only half the ultimate story. The egg surface must possess specific bindin receptors that should show species differences in their affinities for bindin.

The idea that the vitelline layer (VL) of sea urchin eggs contains a "sperm receptor" substance has been around for many years (Lillie, 1919). Modern evidence for this hypothesis comes from a variety of indirect experiments only a few of which are listed below. First, it was found by several groups that trypsinization of eggs decreases fertilizability (Aketa *et al.*, 1972). This effect may result from the loss of a "sperm receptor" substance (bindin receptor) or by loss of the acrosome reaction inducing component of the egg jelly coat (Schmell *et al.*, 1977; Metz, 1978; SeGall and Lennarz, 1979; Vacquier *et al.*, 1979). Second, when eggs are treated with the protease which is released from eggs at fertilization, the VL, although still present, looses its capacity to bind sperm (Vacquier *et al.*, 1973; Carroll and Epel, 1975; Schuel, 1979). Third, when VLs isolated from eggs are returned to sea water and sperm added, the sperm attach only to the outer VL surface (Glabe and Vacquier, 1977b).

Many intercellular recognition and adhesion systems are based on the binding of a protein on one cell to a glycoprotein receptor on an opposing cell (Marchesi *et al.*, 1978). Our data that bindin appears to be a

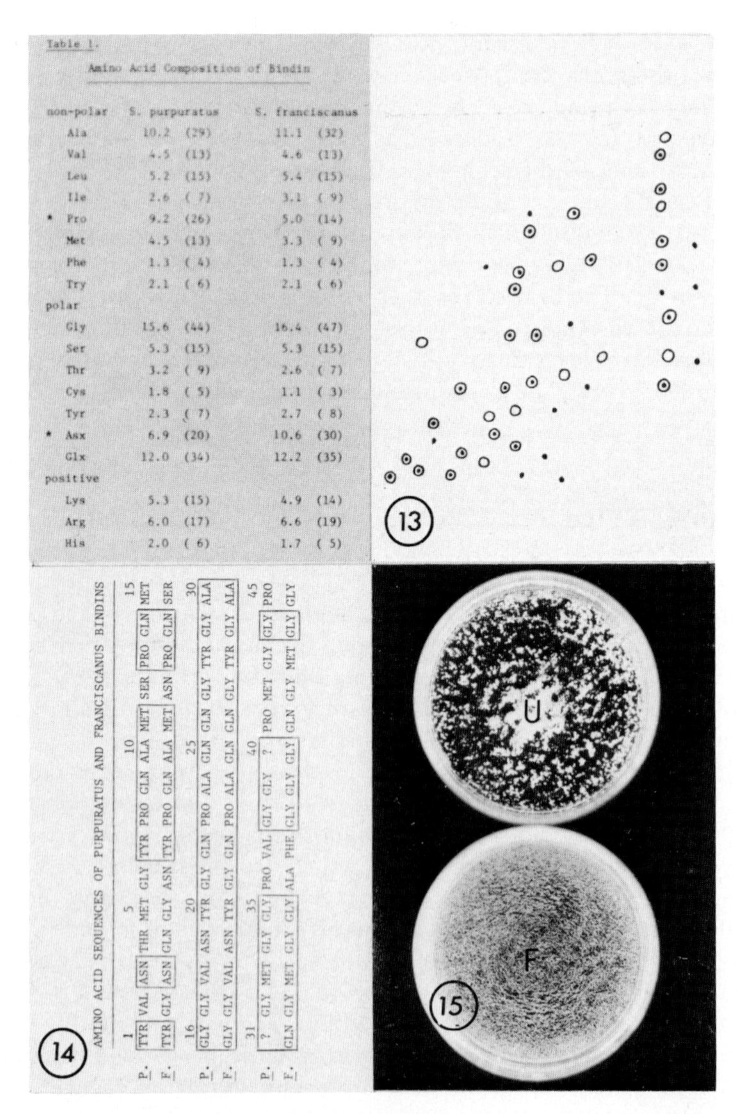
Table 1.

Amino Acid Composition of Bindin

	S. purpuratus	S. franciscanus
non-polar		
Ala	10.2 (29)	11.1 (32)
Val	4.5 (13)	4.6 (13)
Leu	5.2 (15)	5.4 (15)
Ile	2.6 (7)	3.1 (9)
* Pro	9.2 (26)	5.0 (14)
Met	4.5 (13)	3.3 (9)
Phe	1.3 (4)	1.3 (4)
Try	2.1 (6)	2.1 (6)
polar		
Gly	15.6 (44)	16.4 (47)
Ser	5.3 (15)	5.3 (15)
Thr	3.2 (9)	2.6 (7)
Cys	1.8 (5)	1.1 (3)
Tyr	2.3 (7)	2.7 (8)
* Asx	6.9 (20)	10.6 (30)
Glx	12.0 (34)	12.2 (35)
positive		
Lys	5.3 (15)	4.9 (14)
Arg	6.0 (17)	6.6 (19)
His	2.0 (6)	1.7 (5)

Fig. 13. Tryptic peptide maps of the two bindins *S. purpuratus* ●, *S. franciscanus* ○. Of the 37 spots for Sp and 34 for Sf, 24 spots are superimposable. In general the patterns of the two maps are quite similar. Ten spots are unique to Sf and 13 to Sp (Bellet *et al.*, 1977).

Fig. 14. Amino acid sequence analysis of Sp (P) and Sf (F) bindins shows the N-terminal regions to be somewhat related with 31 of the first 45 residues in the same position. Both bindins have no net charge to the 45th residue.

Fig. 15. Unfertilized eggs (U, upper dish) and fertilized eggs (F, lower dish) were mixed with particulate bindin (Fig. 7). Fertilized eggs do not agglutinate, presumably because the fertilization protease has removed the bindin receptor glycoprotein.

nonglycosylated protein and the data from Aketa's laboratory (Tsuzuki *et al.*, 1977) showing that the egg VL contains a "sperm receptor glycoprotein" led us to speculate that bindin might have affinity for a specific terminal oligosaccharide of a "bindin receptor glycoprotein" located on the VL. Support for this hypothesis was the finding that trypsin digestion of unfertilized egg surfaces released a mixture of glycopeptides that blocked the agglutination of eggs by bindin. Also, the trypsin-treated eggs failed to be agglutinated by bindin (Vacquier and Moy, 1977). The conclusion drawn from these results was that trypsin was removing the recognition oligosaccharide from the native bindin receptor which was then binding to a specific active site on the bindin molecule and thus blocking its interaction with the native receptor on the surface of nontrypsin-treated eggs (Vacquier and Moy, 1977). The trypsinized eggs fail to be agglutinated by bindin supposedly because the receptor oligosaccharide has been removed. Additional evidence that bindin interacts with egg surface carbohydrate was the finding that eggs fixed in formaldehyde or glutaraldehyde were still agglutinated by bindin. Also, the agglutination of the fixed eggs can be blocked by metaperiodate oxidation of the eggs prior to bindin addition (Vacquier and Moy, 1977; Glabe, 1978).

One problem with the glycopeptide inhibition of bindin induced egg agglutination is that the glycopeptide digest does not show species specificity. Digests of Sf and Sp eggs are equally effective in blocking the agglutination of Sp eggs by Sp bindin and the reciprocal experiment yields the same result. This problem has recently been given a theoretical analysis (Glabe, 1979) and may result from the possibility that the small bindin receptor oligosaccharide units are quite similar in structure, but the native, intact receptor glycoprotein may be structurally very different in different species. A similar situation is true for the fucose sulfate polysaccharide inducer of the acrosome reaction. Chemical analysis of the polysaccharide from several species shows almost identical composition, yet these molecules can show absolute species specificity in triggering the physiological responses resulting in the sperm acrosome reaction (SeGall and Lennarz, 1979).

During the first 25 seconds following the insemination of eggs, sperm continuously attach to the VL (Fig. 1). During this time the egg fuses with one or more sperm (Epel and Vacquier, 1978; Epel, 1978). From 25-50 seconds the egg releases the contents of its cortical granules one component of which is a trypsin-like protease. The egg protease functions in the detachment of the VL from the egg plasma membrane and in destroying the sperm binding capacity of the VL (Schuel, 1979). If the protease removes the bindin receptor one would assume that fertilized eggs, lacking the bindin receptor, would not be agglutinated by isolated bindin. Eggs from the same female were split into two lots and one lot was fertilized. The eggs were then washed in sea water containing soybean trypsin inhibitor (SBTI, 0.2 mg/ml) and particulate bindin (Fig. 7) added to both lots. Only the unfertilized eggs agglutinated (Fig. 15) regardless of the amount of bindin added.

VI. IDENTIFICATION OF A BINDIN RECEPTOR GLYCOPROTEIN ON THE EGG SURFACE

The observation that fertilized eggs are not agglutinated by bindin (Fig. 15) gave us the rationale for the isolation of a putative bindin receptor (BR) glycoprotein. The method used began with the ^{125}I-labelling of unfertilized eggs by either the lactoperoxidase or chloramine-T procedure (Glabe and Vacquier, 1978). After washing the eggs to remove free ^{125}I, the eggs were parthenogenetically activated using the Ca^{2+}ionophore A23187 (Steinhardt and Epel, 1974) in sea water containing 0.2 mg/ml SBTI. The idea was that the egg protease might cleave the radioactively labelled BR from the surface of the elevating VL (Fig. 16). The protease would then be inactivated by the SBTI. After allowing the activated eggs to settle the supernate was removed and centrifuged at 8,000 xg. The resulting supernate, containing about 10% of the total ^{125}-I bound to the eggs was designated "crude BR." A simple glass fiber filter disc assay was developed to test for complex formation between bindin and a BR, taking advantage of the fact that all the particulate bindin is retained on Whatman GF/C filters (Fig. 17; Glabe and Vacquier, 1978). The crude BR was mixed with isolated bindin, the tube incubated for 10 minutes and then the mixture diluted with sea water and poured in a filter cone. The filters were washed with sea water, dried and radioactivity per filter determined. Experiments showed that some component of the crude BR had saturable affinity for bindin (Fig. 18). When bindin was present in excess, 15 to 20% of the macromolecular ^{125}I had affinity for bindin. Kinetic data showed that the radioactivity accumulating on filters resulted from the formation of bindin-BR complexes with one-half maximum complex formation in 18 seconds (Glabe and Vacquier, 1978).

Species preference of the homologous BR for bindin was shown by competition experiments. Bindin and ^{125}I crude BR from *S. purpuratus* were mixed in quantities producing 70% saturation of the affinity sites on bindin. Aliquots of this mixture were added to tubes containing varying amounts of unlabeled crude BR from *S. purpuratus, S. franciscanus,* whole egg protein, bovine serum albumin (BSA) or SBTI. The results (Fig. 19) show that only unlabeled *S. purpuratus* crude BR was an effective competitor for the bindin- ^{125}I-BR complex. At 320 μg of added competitor protein unlabeled Sp crude BR dissociated 87% of the original complexes, whereas for Sf, crude BR 27% and for egg protein 30% was observed (Fig. 19). The proteins BSA and SBTI were noncompetitive showing that nonspecific electrostatic interaction was not responsible for the formation of bindin-BR complexes.

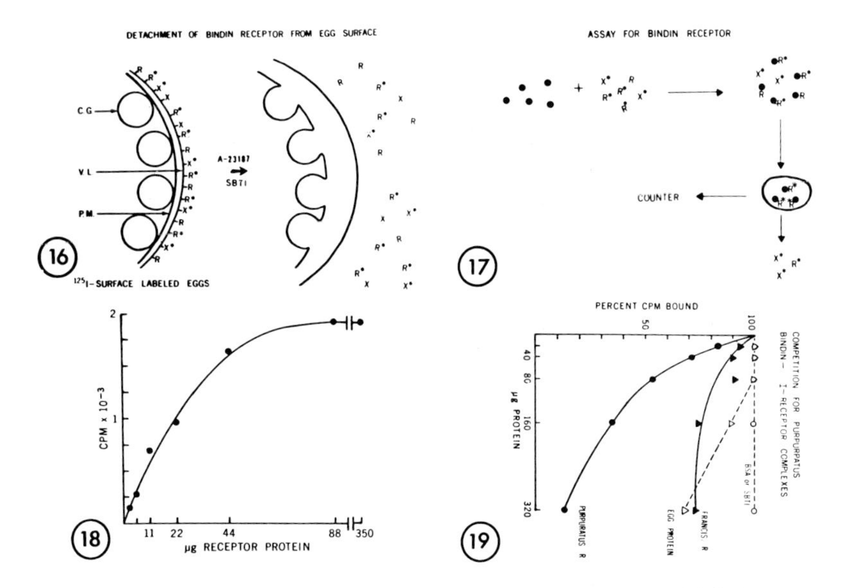

Fig. 16. Detachment of the bindin receptor from the egg surface. ^{125}I-labeled eggs were parthenogenetically activated by the ionophore A23187 in the presence of soybean trypsin inhibitor (SBTI). The exocytosis of the cortical granules released the fertilization protease which cleaved off about 10% of the macromolecular label before it was inactivated by the SBTI. CG, cortical granule; VL, vitelline layer; PM, plasma membrane; R&R*, unlabeled and labeled bindin receptor (BR); X&X*, non BR surface components.

Fig. 17. Scheme of the assay for bindin receptor activity. Particulate bindin (••) was mixed with surface released crude ^{125}I bindin receptor preparation (R,R*, X,X*). All the bindin is trapped on a GFC filter disk. Complexes formed between bindin and a labeled macromolecule were therefore retained on the filter which was then placed in a scintillation counter (Glabe and Vacquier, 1978).

Fig. 18. The binding of ^{125}I-crude bindin receptor (BR) to bindin with BR in excess. 1.0 ml of BR (0.8 mg; 620 K cpm/mg) was diluted in sea water and mixed with 0.05 ml bindin suspension (1 mg/ml). Ten min later the mixture was filtered and the cpm per filter related to the amount of BR protein. Bindin is saturated with the BR (Glabe and Vacquier, 1978).

Fig. 19. Species-preferential affinity of bindin for crude BR as shown by competition analysis with unlabeled crude BR from *S. franciscanus* and *S. purpuratus*. Complexes of *S. purpuratus* bindin- ^{125}I-BR were prepared and aliquots placed in tubes containing various amounts of unlabeled *S. purpuratus* and *S. franciscanus* crude receptor, whole fertilized egg protein, SBTI or BSA. The data show that *S. purpuratus* crude BR is the most effective competitor (Glabe and Vacquier, 1978).

VII. FRACTIONATION AND PARTIAL CHARACTERIZATION OF THE BINDIN RECEPTOR

Gel filtration of the ^{125}I crude BR on Biogel A5m fractionated the counts into three peaks (Fig. 20) consisting of the void volumn ($\geq 5 \times 10^6$

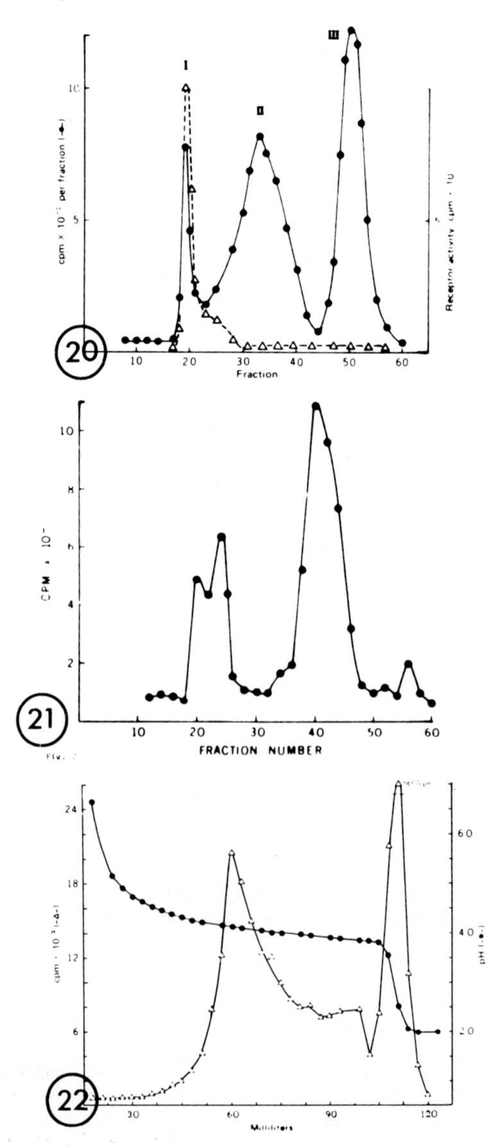

Daltons), an included peak of unknown size and a trailing peak of free ^{125}I. Each fraction was tested for BR activity (Fig. 17). The BR activity was found only in the peak I representing the void volumn fractions (Fig. 20). This showed that the BR activity was in a very large, aggregated form. To test the obvious hypothesis that the egg protease destroys the BR, we incubated the fractions of peak I (Fig. 20) with egg protease and found that all BR activity was indeed lost. Rerunning this sample on the same A5m column showed most of the counts ran well into the column (Fig. 21) which was a more definite indication that the BR activity was cleaved by the enzyme preparation (Glabe, 1978).

About 41% of the total macromolecular ^{125}I in the peak I void volume fractions of the A5m column (Fig. 20) has affinity for bindin when it is present in excess. Further purification of the BR activity was accomplished by preparative isoelectric focusing in a pH 3.5-5.0 gradient which separated peak I (Fig. 20) into two components of pI 4.02 and 2.5 (Fig. 22; Glabe and Vacquier, 1978). The pI 2.5 peak is residual egg jelly which is irreversibly precipitated in the acid reservoir (soluble jelly does not have BR activity). This residual jelly is only released from the egg surface by the exocytosis of the cortical granules at fertilization (Vacquier *et al.*, 1979). After dialysis of the pI 4 peak (Fig. 22) into sea water, 64% of the radioactivity had affinity for bindin when bindin was present in excess. By weight, the pH 4 peak fractions are 34% galactose and mannose in a 5:1 molar ratio and 4% sulfate. Sialic acid was not detected (Glabe and Vacquier, 1978).

Fig. 20. Gel filtration of ^{125}I-labeled crude receptor on Bio-Gel A5m. Two milliliters of crude receptor (0.35 mg of protein per ml, 817,000 cpm/mg) was fractionated in sea water. Three peaks of radioactivity were obtained (•). Peak I is the void volume, peak II is included material of undetermined size, and peak III is the included volume as determined by mercaptoethanol. One half milliliter of each fraction was mixed with 0.05 ml of bindin (2 mg/ml), incubated for 10 min, and then filtered. The bindin receptor activity (▽) was found only the peak I fractions, indicating a molecular weight $\geq 5 \times 10^{\circ}$. (Glabe and Vacquier, 1978).

Fig. 21. The peak I fractions from the A5m column (Fig. 20) were digested with the fertilization protease and then rerun on the same column and the BR activity of each fraction determined. All BR activity was lost and the radioactivity separated into 4 peaks, the one at fraction 40 containing most of the radiolabel (Glabe, 1978).

Fig. 22. Isoelectric focusing of peak I from the A5m column. Peak I fractions (Fig. 20) were focused in a pH 3.5–5.0 gradient. The material separated into two components of pI 4.02 and 2.5. The pI 2.5 peak is egg jelly. The pI 4.02 peak is believed to be the bindin receptor. Before isoelectric focusing, 41% of the cpm bound to bindin. After isoelectric focusing, the pI 4.02 peak fractions were dialyzed into sea water and affinity to bindin was determined. Sixty-four percent of this material bound bindin when bindin was present in excess (Glabe and Vacquier, 1978).

VIII. ARE SPERM BINDINS PRESENT IN OTHER SPECIES?

Self species preferential adhesion of sperm to eggs also occurs in mammals (Yanagimachi, 1978) and other invertebrates (Colwin and Colwin, 1967; Summers and Hylander, 1976). It is reasonable to speculate that such adhesions are mediated by proteins similar to sea urchin bindin. Recently, we isolated the acrosome granule from sperm of the oyster *Crassostrea gigas.* These AGs are composed of an insoluble material which is 85% protein, stains positively for carbohydrate and separates into two major components of 65,000 and 53,000 Daltons when electrophoresed on SDS-PAGE (Brandriff *et al.*, 1978). The oyster AG glycoprotein agglutinates oyster eggs and this agglutination is blocked by glycopeptides digested from egg surfaces. If this material is oyster bindin it is indeed very different from sea urchin bindin.

IX. SIGNIFICANCE OF SPERM BINDIN AND THE EGG SURFACE BINDIN RECEPTOR GLYCOPROTEIN

Model systems for studying the molecular basis of recognition and adhesion between two cell types would be ones with these attributes: (1) The biological significance of the intercellular interaction should be well established; (2) the cells should be homogeneous populations of single cells, readily distinguishable and separable from each other; (3) the interaction of the two cell types should be rapid and synchronous occurring in a time span of seconds; (4) milligram quantities of the interacting cell surface macromolecules should be easily obtainable in fairly pure form; and (5) it should be possible to demonstrate affinity of the interacting molecules for each other after their removal from the cell surface. The interaction of sea urchin sperm and eggs during fertilization possesses these five attributes.

Our hypothesis is that sperm binding is a lectin-ligand interaction in which bindin is the lectin which binds to glycosyl ligands on the outer surface of the vitelline envelope. We propose this is the general mechanism of sperm-egg binding in animals. This type of mechanism, in which one cell contains a protein and another cell a carbohydrate receptor, appears to be a general mechanism of recognition among cells (Marchesi *et al.*, 1978).

X. CONCLUSION

To our knowledge this is the first example where both surface components from two interacting cells have been isolated and their interaction demonstrated under *in vitro* conditions. Bindin is also the first protein to be isolated in pure form and milligram quantities that mediates a specific intercellular surface recognition and adhesion in a metazoan. Bindin could be considered one of the most important proteins of sexually reproducing species because it mediates the union of gametes which initiates embryonic development.

ACKNOWLEDGEMENTS

Much of the data presented here represents the research of my colleagues Neal F. Bellet, Brigitte Brandriff, Charles G. Glabe and Gary W. Moy and was supported by NIH Grant HD 12986.

REFERENCES

Aketa, K., Onitake, K. and Tsuzuki, H. (1972). *Exp. Cell Res.* **72**, 27-32

Bellet, N.F., Vacquier, J.P. and Vacquier, V.D. (1977). *Biochem. Biophys. Res. Commun.* **79**, 159-165.

Brandriff, B., Moy, G.W. and Vacquier, V.D. (1978). *Gamete Res.* **1**, 89-99.

Carroll, E.J. and Epel, D. (1975). *Develop. Biol.* **44**, 22-32.

Colwin, L.H. and Colwin, A.L. (1967). *In* "Fertilization I" (C.B. Metz and A. Monroy, eds.), pp. 295-367. Acad. Press, New York.

Csernansky, J.G., Zimmerman, M. and Troll, W. (1979). *Develop. Biol.* **70**, 283-286.

Dan, J.C. (1967). *In* "Fertilization I" (C.B. Metz and A. Monroy, eds.), pp. 237-293. Acad. Press, New York.

Epel, D. (1978). *Curr. Top. Develop. Biol.* **12**, 186-242.

Epel, D. and Vacquier, V.D. (1978). *In* "Cell Surface Reviews V" (G. Poste and G.L. Nicolson, eds.), pp. 1-63. Elsevier, North Holland.

Glabe, C.G. (1978). Ph.D. Thesis in Zoology, Univeristy of California, Davis.

Glabe, C.G. (1979). *J. Theor. Biol.* **78**, 1-7.

Glabe, C.G. and Lennarz, W.J. (1979). *J. Cell Biol.*, in press.

Glabe, C.G. and Vacquier, V.D. (1977a). *Nature* **267**, 836-838.

Glabe, C.G. and Vacquier, V.D. (1977b). *J. Cell Biol.* **75**, 410-421.

Glabe, C.G. and Vacquier, V.D. (1978). *Proc. Nat. Acad. Sci. U.S.A.* **75**, 881-885.

Hylander, B.L. and Summers, R.G. (1977). *Cell Tiss. Res.* **182**, 469-489.

Levine, A.E., Walsh, K.A. and Fodor, E.J.B. (1978). *Develop. Biol.* **63**, 299-306.

Lillie, F.R. (1919). "Problems of Fertilization." University of Chicago Press.

Loeb, J. (1916). "The Organism as a Whole." pp. 71-94. G.P. Putnam's Sons, New York.

Marchesi, V.T., Ginsburg, V., Robbins, P.W. and Fox, C.F. (1978). "Cell Surface Carbohydrates and Biological Recognition." Vol. 23, Progress in Clinical and Biological Research. Alan R. Liss Inc., New York.

Metz, C.B. (1978). *Curr. Top. Develop. Biol.* **12**, 107-147.
Moy, G.W. and Vacquier, V.D. (1979). *Curr. Top. Develop. Biol.* **13**, 31-44.
Schmell, E., Earles, B.J., Breaux, C. and Lennarz, W.J. (1977). *J. Cell Biol.* **72**, 35-46.
Schuel, H. (1979). *Gamete Res.* **1**, 299-382.
SeGall, G.K. and Lennarz, W.J. (1979). *Develop. Biol.* **71**, 33-48.
Summers, R.G. and Hylander, B.L. (1974). *Cell Tiss. Res.* **150**, 343-368.
Summers, R.G. and Hylander, B.L. (1976). *Exp. Cell Res.* **100**, 190-194.
Summers, R.G., Hylander, B.L., Colwin, L.H. and Colwin, A.L. (1975). *Amer. Zool.* **15**, 523-551.
Steinhardt, R.A. and Epel, D. (1974). *Proc. Nat. Acad. Sci. U.S.A.* **71**, 1915-1919.
Tsuzuki, H., Yoshida, M., Onitake, K. and Aketa, K. (1977). *Biochem. Biophys. Res. Commun.* **76**, 502-511.
Vacquier, V.D. (1979). *Amer. Zool.*, in press.
Vacquier, V.D. and Moy, G.W. (1977). *Proc. Nat. Acad. Sci. U.S.A.* **74**, 2456-2460.
Vacquier, V.D., Tegner, M.J. and Epel, D. (1973). *Exp. Cell Res.* **80**, 111-119.
Vacquier, V.D., Brandriff, B. and Glabe, C.G. (1979). *Develop. Growth Differen.* **21**, 47-60.
Yanagimachi, R. (1978). *Curr. Top. Develop. Biol.* **12**, 83-105.

Experimental Analysis of the Role of Intracellular Calcium in the Activation of the Sea Urchin Egg at Fertilization

David Epel

Hopkins Marine Station
Department of Biological Sciences
Stanford University
Pacific Grove, California 93950

I. INTRODUCTION

The problem I shall consider in this article concerns the means by which the brief interaction between sperm and egg initiates development. The system described will be that of sea urchin fertilization which is the best understood experimental model. This is in large part because prodigious amounts of sperm and eggs can be obtained and simply mixing the two together in seawater results in highly synchronous fertilization and development.

The vantage point from which I begin is a descriptive one, which is a necessary prelude to the subsequent dissection of events I will elucidate.

ISBN 0-12-612984-3

This "taxonomy" has resulted in the realization that a rigid program or sequence of events is triggered by the sperm. The current sequence, shown in Fig. 1, is presented with the caveat that there is much serendipity in this description and therefore that numerous changes occur which probably have not yet been described. Although some of these unknown changes may be critical for understanding activation, it does appear that several of the major regulatory events are now known and that our understanding of how these changes are controlled should result in considerable insights into the activation process.

The initial categorization of these changes indicated that the post-fertilization events could be divided into two temporally distinct events. There are "early" events beginning with sperm-egg binding and ending about 60 to 90 seconds later and "late" events beginning at about 5

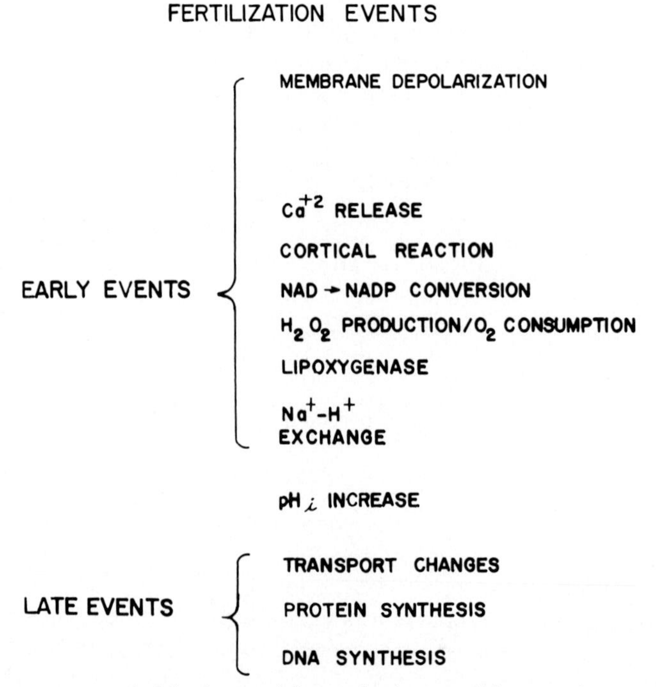

Fig. 1. The program or schedule of events following fertilization of the sea urchin egg. The "early events" are all initiated by 60 seconds after insemination; the "late events" begin about four — five minutes later.

minutes after insemination (see Epel, 1978a for review). To briefly review these changes, the earliest one following sperm-egg binding is a membrane depolarization which has been referred to as the "fertilization potential" or "activation potential." This apparently ensues from a sodium influx (Steinhardt *et al.*, 1971) which may be amplified by a stimultaneous influx of calcium ion (Hagiwara and Jaffe, 1979). This depolarization also results in the block to polyspermy by a mechanism that is not yet understood (Jaffe, 1976). The next known change is a rise in intracellular calcium which has been measured in sea urchin (Steinhardt *et al.*, 1977), fish (Ridgeway *et al.*, 1977) and starfish eggs (Moreau *et al.*, 1978) utilizing the calcium-dependent light emission of aequorin. These measurements show that the calcium rise is transient and begins approximately coincident with the cortical granule exocytosis or cortical reaction and ends shortly after the exocytosis is over. The free calcium level does not remain high but decreases to the pre-fertilization level.

Very shortly after the beginning of the cortical reaction there begins a massive efflux of hydrogen ions from the egg, referred to as the fertilization acid (Paul and Epel, 1975). The initial part of this appears to be independent of sodium whereas the second phase is sodium-dependent and apparently proceeds through a Na^{+}-H^{+} exchange mechanism (Johnson *et al.*, 1976). At approximately the same time the enzyme NAD kinase is activated, which results in a $2^{1}/_{2}$ fold increase in the NADP-NADPH levels of the cell (Epel, 1964). Shortly thereafter there occurs the transient activation of a lipid-oxidizing system, probably through a lipoxygenase (Perry, 1979). There is also a large increase in respiration associated with the production of hydrogen peroxide utilized for the intramolecular cross-linking and hardening of the fertilization membrane (Foerder *et al.*, 1978). During this period the continuing activity of the Na^{+}-H^{+} exchange system has resulted in a 2.5–5 mM loss of hydrogen ion from the cell with a resultant increase in intracellular pH (pH_i) from pH 6.8 to about 7.3 (Johnson *et al.*, 1976; Shen and Steinhardt, 1978). The activity of the Na^{+}-H^{+} exchange system ceases at this time, but the intracellular pH remains at this higher level (Shen and Steinhardt, 1978).

By about five minutes after fertilization the late changes have begun which include a 5 — 15-fold increase in the rate of protein synthesis (Epel, 1967). This increase does not depend on the synthesis of new mRNA but ensues from the recruitment of pre-existing mRNA from a yet uncharacterized cytoplasmic pool (see, e.g., Humphreys, 1971). This activation is therefore regulated at the translational level. Coincident with this increase in protein synthesis are large increases in the transport of amino acids (Epel, 1972) nucleosides (Piatigorsky and Whiteley, 1965) phosphate (Whiteley and Chambers, 1966) and potassium, the last change being responsible for a hyperpolarization of the membrane (Steinhardt *et al.*, 1971). During this period the sperm and egg nucleus are being propelled through the cytoplasm; they fuse about 20 minutes after insemination and at this time the first cycle of DNA synthesis is initiated (Hinegardner *et al.*, 1964).

II. EVIDENCE THAT TWO IONIC CHANGES ARE CRITICAL IN ACTIVATION

An increasing body of experimental evidence now indicates that the transient rise in calcium and the transient activation of the Na^+–H^+ exchange system can by themselves account for the activation of egg metabolism. The evidence for this comes from two lines of evidence which center around the findings that (1) one can induce part or all of the fertilization responses by either increasing calcium or pH_i or (2) that one can prevent these responses by inhibiting the rise in calcium or pH_i.

A. *Evidence that the Calcium Rise is Necessary*

We have noted above that the free calcium level increases after fertilization. One form of evidence that this rise is critical is that if one artificially increases the calcium level with the calcium-transporting ionophores (as A23187, or X537A), the normal sequence of events is initiated and is similar to that seen following normal fertilization (Chambers *et al.*, 1974; Steinhardt and Epel, 1974). The activation by either sperm, the ionophore and other parthenogenic agents can occur in the absence of extracellular calcium, suggesting that during normal or artificial activation, calcium is released from intracellular stores (Steinhardt *et al.*, 1977). Using ionophore A23187, this independence from extracellular calcium has been shown for eggs from numerous organisms, including tunicates, amphibians and mammals (Steinhardt *et al.*, 1974).

There is one interesting exception to this generalization, which however affirms the importance of calcium. The eggs of the bivalve mollusc, *Spisula solidissima,* cannot be activated by ionophore A23187 in the absence of extracellular calcium; however, if calcium is present they will become activated (Schuetz, 1975). This seeming exception therefore indicates that the ionophore is acting through its effects on calcium. A similar type of evidence is that the activation of amphibian eggs seen upon pricking with a pin also requires extracellular calcium. This has been interpreted as indicating that the influx of calcium that could occur during this injury can activate the egg (see Gilkey *et al.*, 1978).

The alternative line of evidence is that if one prevents the calcium rise, one also can prevent activation. The above examples of the Ca^{+2} requirements for artificial activation of *Spisula* and frog eggs is one instance. A more direct test is to inject the calcium-chelating agent EGTA into eggs; this prevents their subsequent activation by sperm (Zucker and Steinhardt, 1978).

B. *Evidence that the pH_i Rise is Necessary*

An increase in pH_i from about pH 6.8 in the unfertilized state to about 7.2 — 7.3 in the fertilized state appears to be the major consequence of the transient Na^+-H^+ exchange activity. This pH rise has been detected by measurements of the pH of cell lysates (Johnson *et al.*, 1976) pH microelectrodes (Shen and Steinhardt, 1978) and by measuring the partitioning of C^{14}-weak acids (Johnson and Epel, 1979).

One form of evidence that the pH_i change is important comes from experiments in which the pH_i is experimentally increased. This occurs during incubation in weak bases, such as NH_4Cl or procaine (Johnson *et al.*, 1976; Shen and Steinhardt, 1978; Winkler and Grainger, 1978). The changes that occur include many of those associated with the "late changes," suggesting that these changes emanate from the pH_i rise (Epel *et al.*, 1974; Vacquier and Brandriffe, 1975). As summarized in Fig. 2, the early changes that precede the pH_i rise do not take place in ammonia whereas those that are coincident with or follow the rise in pH_i do occur. An exception is the increased transport of amino acids, which appears to require the cortical reaction (Epel and Johnson, 1976). These types of studies therefore, correlate the rise in pH_i with the onset of a number of the "late changes" such as protein and DNA synthesis and chromosome condensation.

The other and even stronger line of evidence that the pH_i rise is critical comes from experiments in which the pH_i rise is prevented. Subsequent development is inhibited but these arrested eggs can be "rescued" if the pH_i is raised, as by incubating them in ammonia (Johnson *et al.*, 1976).

One important experimental approach to inhibiting the pH_i rise is based on the aforementioned Na^+-requirement for the Na^+-H^+ exchange and resultant pH_i rise; the pH change can be therefore inhibited by preventing the Na^+-influx (Johnson *et al.*, 1976). This is most easily accomplished by activating eggs with ionophore in Na^+-free seawater (Shen and Steinhardt, 1979) or by fertilizing eggs in the presence of sodium and then very quickly washing them and resuspending them in Na^+-free seawater (Chambers, 1976; Johnson *et al.*, 1976). Both approaches prevent the normal rise in pH_i.

An alternate approach, similar in principle, is to prevent the Na^+-influx with the sodium-channel blocker, amiloride. This diuretic drug acts on a channel that is apparently different than that seen in amphibian skin or bladder or in kidney epithelia. It blocks the sodium transport associated with H^+-exchange (Johnson *et al.*, 1976; Epel, 1978b) but at much higher concentration than seen in the aforementioned vertebrate systems (Cuthbert and Cuthbert, 1978). If eggs are incubated in high (2–5 x 10^{-4}M) amiloride concentrations in low sodium (25–50 mM Na^+, in choline-substituted seawater) one can inhibit both the Na^+-influx and H^+-efflux.

Under any of the above conditions which prevent the pH_i rise, the normal progression of development is inhibited. More importantly, one can "rescue" these eggs by simply adding millimolar concentrations of ammonium chloride, similar to that used to raise the pH_i of unfertilized eggs. This suggests that the effect of either Na^+-free conditions or high concentrations of amiloride is solely on the Na^+–H^+ exchange system and their action in raising pH_i.

It has been suggested that amiloride may be acting non-specifically (Cuthbert and Cuthbert 1978; Shen and Steinhardt, 1979). However, the finding that the effect can be reversed by ammonia would suggest that its action is through its effects on pH_i (Epel and Paul, 1979). More germane to the question of non-specificity is that if one adds similar concentrations of amiloride *after* the Na^+–H^+ exchange is over, development proceeds normally (Epel and Paul, 1979). This would also suggest that the action of this amiloride-sensitive system is only critical during this earliest period.

A third line of evidence has come from important studies of Grainger *et al.*, (1979) who have correlated pH_i with the rate of protein synthesis. When pH_i is raised (as by ammonia) there is a concomitant increase in the rate of protein synthesis; if the pH_i is dropped (as when eggs are incubated in acetate at low extracellular pH's) the rate of protein synthesis is reduced.

An interesting aspect of the increased protein synthesis is that although polysomal recruitment in normally fertilized and ammonia-activated eggs are similar (Epel *et al.*, 1974; Brandis and Raff, 1979) the

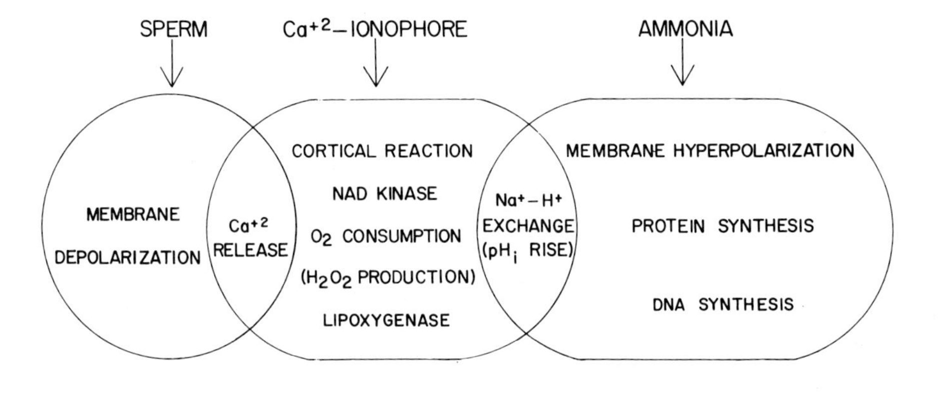

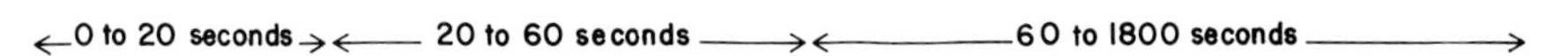

Fig. 2 The dissociation or dissection of the post-fertilization sequence of events by sperm, Ca^{+2}-ionophore or ammonia. The normal sequence, initiated by the *sperm*, results in the initial membrane depolarization, Ca^{+2}-release and all subsequent changes. *Ca^{+2}-ionophore* does not initiate the initial depolarization, but does trigger all subsequent events. Incubation in *ammonia* or other weak bases causes the pH_i rise. The earlier changes do not take place, but those changes following the pH_i rise now are initiated.

rate of protein synthesis is not as high in the ammonia-activated case. To explain this difference, it has been suggested that the augmentation of protein synthesis might occur by two separate mechanisms (Brandis and Raff, 1979; Grainger *et al.*, 1979). One is through recruitment of mRNA, which is pH-sensitive. The other is an increase in the transit time of mRNA along the polysome, which is not initiated during ammonia incubation and which is therefore presumed relatively insensitive to pH_i. One possibility is that the augmented transit time may be related to the early rise in calcium (Brandis and Raff, 1979; Grainger, *et al.*, 1979).

C. *Summary of Ionic Hypothesis*

The interpretation of these results points to two ionic changes which by themselves can result in all or part of egg activation. The first is the transient rise in calcium ion, which by itself can trigger the entire sequence of developmental events (such eggs do not necessarily divide, probably because they do not possess the centrosome necessary for the normal functioning of the mitotic apparatus, see Mazia, 1978). The second event, which I interpret as arising somehow from the earlier calcium increase, is the transient activation of Na^+–H^+ exchange with the resultant rise in pH_i. This change, as described above and indicated in Fig. 2, does not initiate any of the "early changes" but results in the turning on of the "late changes."

More comparative work is necessary to discern whether both of these changes are part of the activation process in other eggs. The rise in calcium is probably a universal concomitant of fertilization but the rise in pH_i may be restricted to eggs of only a few phyla. For example, if one assumes that the release of acid heralds a rise in pH_i, then pH_i rises may be restricted. For example, acid release is seen following fertilization of sea urchins, *Urechis caupo* (Paul, 1975) and *Spisula solidissima* (Ii and Rebhun, 1979), but is not seen following fertilization of tunicate (Lambert and Epel, 1979) or amphibian (Epel, 1979) eggs.

III. HOW MIGHT CALCIUM ACT TO TRIGGER DEVELOPMENT?

Given that calcium is the primary trigger of fertilization the major problem that I address in the remainder of this article relates to the sites of calcium action. Is calcium acting directly, for example, through its effects on calcium-activated proteases or protein kinase? Or is calcium acting indirectly through intermediary proteins which regulate or modulate enzyme activity when these intermediates have calcium bound to them? A currently popular example of the intermediary hypothesis centers on the regulatory protein, calmodulin. Calmodulin is a 17,000 MW heat-stable protein which binds calcium with high affinity and shares considerable homology with the calcium-binding and regulatory protein of muscle, troponin C (Kretsinger, 1976). When calmodulin is bound to calcium, the protein interacts with target enzymes or proteins and affects their structure and function. The best described examples are the activation of phosphodiesterase (Cheung, 1970) protein kinase of smooth muscle (Yagi *et al.*, 1978) and Ca^{+2}–ATPase of red blood cells (Jarrett and Penniston, 1977). Calmodulin can also effect structural proteins and somehow promotes the disaggregation of polymerized tubulin (Marcum *et al.*, 1978).

In the following sections I want to look at three early events, which occur during the time when the calcium level is elevated. I shall present evidence that these events/changes are regulated by calcium. I shall especially ask whether the changes are calmodulin-dependent, since this is the best described calcium-regulated protein. The criteria for deciding whether calmodulin is involved are summarized in Table I. Besides the requirement of calcium for activity and inhibition by EGTA, the major

TABLE I

Criteria for Calmodulin — Mediated Processes

(1) Sensitivity to low levels of Ca^{+2} and EGTA
(2) Separation of activator from protein in presence of EGTA
(3) Co-identity of activator as calmodulin
(4) Restoration of activity by calmodulin
(5) Effect of calmodulin inhibitors (chlorpromazine & trifluoperazine)

criterion is whether one can separate the effected target from its putative regulator and whether the separated activator is calmodulin or some other protein. The final criterion is the sensitivity of the process to perazine tranquilizers, such as chloropromazine and stelazine, which may be diagnostic inhibitors for calmodulin-mediated processes.

A. *The Cortical Exocytosis*

Exocytotic events are known to be profoundly affected by calcium (see review by Papahadjopoulous, 1978) and Vacquier has demonstrated a direct calcium involvement in the cortical exocytosis of sea urchin eggs. Vacquier (1975) isolated mats of cortical granules attached to their plasma membrane and showed that addition of calcium in micromolar amounts (Vacquier, 1976) resulted in the fusion of the granules to each other in a manner similar but not identical to that occurring during the cortical granule exocytosis seen *in vivo*. The calcium levels necessary to initiate exocytosis were originally established as being 12 — 20 μM (Vacquier, 1976; Steinhardt *et al.*, 1977) but recent work of Baker and Whitaker (1978) suggests that fresh preparations in the presence of ATP have a much higher affinity for calcium, in the range of 1 μM; this value is in dispute, however, and the actual value may be nearer to 12 μM (Zucker and Steinhardt, 1979; Baker and Whitaker, 1979). It is not clear whether calcium is acting directly or indirectly to induce exocytosis. The high affinity for calcium suggests that calmodulin might be involved as a mediator of the process, but there is as yet no direct evidence for or against its participation.

B. *Calcium-Stimulated Lipoxygenase Activity*

The respiratory burst at fertilization might also be related to the rise in calcium. Hultin (1950) first suggested such a correlation when he found that the oxygen consumption of homogenates was tremendously stimulated by the addition of this cation. George Perry, as part of his Ph.D. thesis, has re-examined this phenomenon and his work indicates that at least part of the calcium-induced respiratory activity of homogenates can be attributed to the activation of the lipid-oxidizing enzyme, lipoxygenase (Perry, 1979). The calcium-induced respiratory activity was found to be cyanide-insensitive, leading him to examine activities of non-mitochondrial oxygenases. A very active Ca^{+2}-dependent cyanide-insensitive system for the oxidation of unsaturated fatty acids such as arachidonic acid was found.

Arachidonic acid can be oxidized by two alternate pathways in cells. One, involving cyclo-oxygenase, produces the well-known family of the prostaglandin-like compounds; the other, termed lipoxygenase, results in the production of hydroxy-eicostetraenoic acid (HETE) and other products (Markus, 1978). Perry found that this latter enzyme is present and highly active in homogenates and accounts for most of the oxygen-consuming capacity when calcium is added to homogenates. The major product produced in homogenates when Ca^{+2} is added is HETE. As shown in Fig. 3 a similar product is also produced in intact eggs after fertilization.

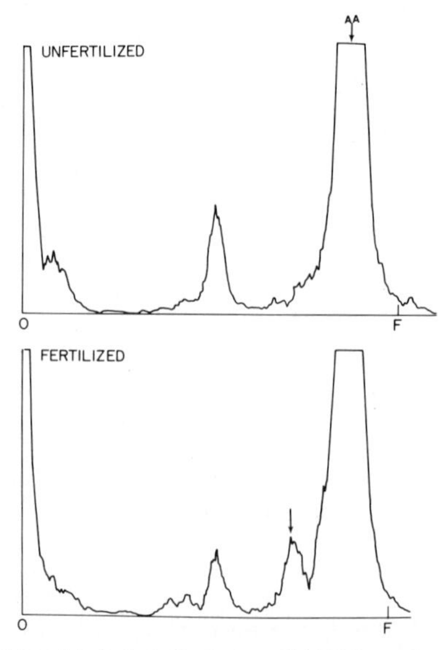

Fig. 3. The conversion of C^{14}-labelled arachadonic acid (AA) by unfertilized and fertilized eggs. *S. purpuratus* eggs or embryos were incubated in AA for 10 minutes, extracted, and then separated by thin-layer chromatography in benzene: dioxane: acetic acid (60:30:3). The radioactivity was then determined by radioscanning with a Packard 7201 Radiochromatogram Scanner. The position of HETE in fertilized eggs is indicated by the arrow.

Does Ca^{+2} act directly on lipoxygenase or through some intermediate such as calmodulin? Studies on the K_D of the enzyme for Ca^{+2} reveal that the affinity for Ca^{+2} is extremely high, in the area of 1×10^{-7}M. As the K_D for calmodulin-mediated processes is in the micromolar range, lipoxygenase would not appear to be calmodulin-mediated. Also consistent with the non-involvement of calmodulin is the finding that perazine drugs, such as chlorpromazine and stelazine have no effect on the Ca^{+2}-stimulated respiratory activity.

C. *NAD Kinase*

This enzyme is activated very shortly after fertilization, within seconds after the beginning of the cortical exocytosis (Epel, 1964). Its action, shown in equation 1 below, results in the phosphorylation of 30% of the cells' NAD into NADP; since total NADP is much lower than NAD, the net result is a 2–3 fold increase in the NADP (& NADPH) levels of the cell (see Table II).

$$(1)\ \text{NAD} + \text{ATP} \longrightarrow \text{NADP} + \text{ADP}$$

The timing of activation of this enzyme suggests that the increase in Ca^{+2} might be involved in triggering its activity; the conversion of NAD into NADP (i.e., enzyme activity) first becomes apparent at the time of the Ca^{+2} rise and the conversion ceases about the time that the Ca^{+2} level has decreased back to the pre-fertilization level. The later pH_i increase would not appear to be involved since there is no conversion of NAD into NADP when only the pH_i is raised, as by incubating eggs in ammonia (Wiley *et al.*, 1977; Epel *et al.*, 1979).

TABLE II

Changes in Pyridine Nucleotide Content at Fertilization of S. purpuratus Eggs

	Unfertilized	Fertilized
NAD	267 ± 4	166 ± 7
NADH	22 ± 10	46 ± 8
NADP	26 ± 2	56 ± 3
NADPH	26 ± 9	81 ± 1
TOTAL	341	348
% NADP + NADPH	16	40

Values in 10^{-9} moles/ml packed cells (from Epel, 1964)

Recent work in my laboratory in collaboration with C. Patton, R. Wallace and G. Cheung, has shown that this enzyme activity is extremely sensitive to low levels of calcium; its activity can be inhibited by addition of EGTA and re-activated by the subsequent provision of low amounts of calcium. The K_D is approximately 8 x 10^{-6}M which is in the affinity range for calmodulin-mediated reactions.

The definitive evidence that calmodulin is involved is that a calmodulin-like activator can be separated from the enzyme activity by DEAE-cellulose chromatography. A typical separation, shown in Fig. 4, depicts the elution profile of NAD kinase activity from a DEAE-cellulose column as the salt concentration is raised. The absorption and elution are done in the presence of EGTA so that if calmodulin is associated with the enzyme, it would be disassociated upon application to the column, thus allowing the separation of the enzyme from the putative activator.

The enzyme activity, shown as the filled bars on the histogram,

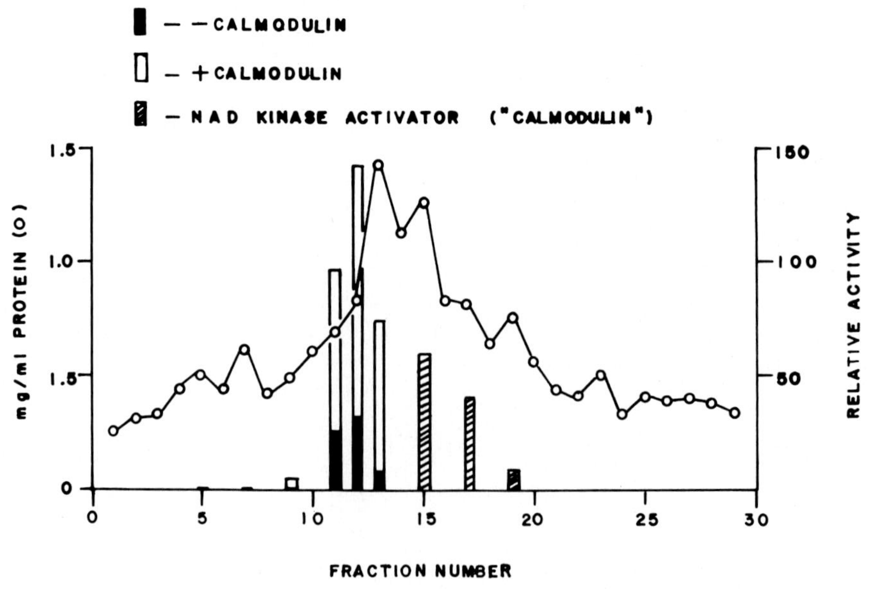

Fig. 4. Separation of NAD kinase from its activator in the presence of EGTA. *S. purpuratus* eggs were homogenized in a homogenization buffer containing 0.5M KCL-0.05 M MES buffer, pH 6.5-10mM EGTA — 5mM DTT. The 100,000g supernatant was dialyzed overnight against 0.02M MES-lmM EGTA — 5mM DTT, pH 6.5. The pH was adjusted to pH 8 and absorbed to DEAE-cellulose and eluted with a step gradient of NaCl (0.1M to 0.5M) containing 25mM Tris, pH 8.0 — 1mM EGTA and 5mM DTT.

appears in fractions 9–13. This activity is augmented 4–5 fold upon addition of bovine calmodulin (unfilled bars). The existence of activators was assessed by adding the other fractions not containing NAD kinase activity back to the enzyme-containing fractions. Activator was found in fractions 15–19 (shown as hatched bars). This activator activity is only seen in the presence of Ca^{+2} and the activator action is also heat-stable (5 minutes boiling had no effect); both properties are consistent with the activator being calmodulin-like. It is probable that this activator is identical to a calmodulin-like protein recently decribed in sea urchin gametes (Jones *et al.*, 1978; Head *et al.*, 1979).

A final feature of the activation by calmodulin is that there is no augmentation of NAD kinase activity if calmodulin is added to a crude homogenate. This is most easily interpreted as indicating that the enzyme is already maximally activated by calmodulin already bound to it from calmodulin present in excess in the homogenate. The augmentation can only be seen if the calmodulin is dissociated from the enzyme, as following DEAE-chromatography.

The final evidence that calmodulin is involved is that the enzyme activity is inhibited by low concentrations of the perazine drug, trifluoperazine. Consistent with the idea that these drugs are effecting calmodulin binding is that the kinase activity is not affected by the drug in the absence of calcium; the basal level is unaffected by the drug and only that portion stimulated by calcium is inhibited.

IV. CALCIUM LEVELS AND ENZYME ACTIVATION — SIGNIFICANCE FOR REGULATION OF CELL ACTIVITY

The above review and analysis, which identifies three targets for calcium at fertilization, also shows that the critical calcium levels for these three changes are quite different, being in the micromolar range for the cortical exocytosis and NAD kinase activity but an order of magnitude lower for lipoxygenase activity. This raises the possibility that there might be *hierarchies* of Ca^{+2} regulation, with low Ca^{+2} levels affecting some activities (lipoxygenase) and higher levels affecting other activities (NAD kinase and exocytosis).

There are alternative possibilities. One, specific to the lipoxygenase case, is based on the fact that this lipid-oxidizing system depends on its substrate, free unsaturated fatty acids, for activity. The level of such fatty acids is probably governed by a calcium-activated phospholipase present in sea urchin eggs and preliminary estimates suggest that the affinity for Ca^{+2} by these lipases is not as high as that of lipoxygenase. If so, although lipoxygenase activity might be active at 10^{-7}M Ca^{+2}, its actual activity would be substrate-limited and not be apparent until the calcium level achieved much higher concentrations and the phospholipase activated.

An alternate possibility is based on the premise that there actually are microdomains of calcium in the cell. Steinhardt and his associates (1977, 1979) have suggested that the increased calcium levels are restricted to the cortical region of the egg. This is based on calculations of the calcium level reached during aequorin light emission, which is much lower than the critical calcium levels required for the cortical exocytosis. They therefore suggested restriction of Ca^{+2} to the cortex (see, however, Baker and Whitaker, 1978, 1979).

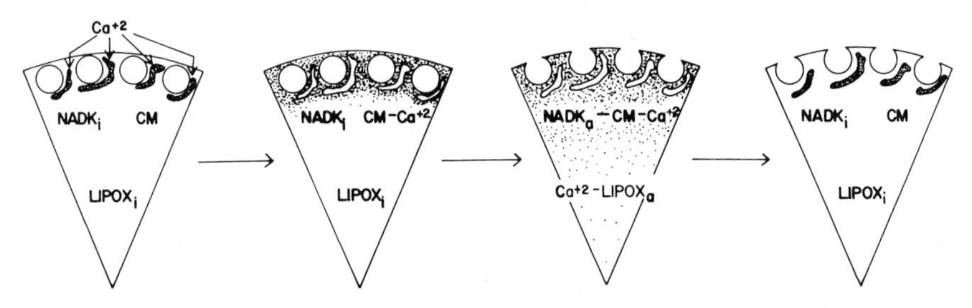

Fig. 5. A hypothetical representation of Ca^{+2} distribution and its effect on enzyme activity and the cortical granule exocytosis following fertilization. Calcium is shown sequestered in a subcortical vesicular system in the unfertilized egg. NAD kinase (NADK) and lipoxygenase (LIPOX) are indicated as inactive (i), and calmodulin (CM) is not associated with calcium (depicted as dots). Fertilization releases Ca^{+2} at micromolar levels in the cortex, which then associates with CM and also initiates the cortical exocytosis. The CM–Ca^{+2} complex is formed which also activates NADK ($NADK_a$). Calcium diffuses into other areas of the cytoplasm, and these lower Ca^{+2}– levels can activate enzymes, such as lipoxygenase ($LIPOX_a$). The Ca^{+2} is subsequently resequestered and the enzyme activities return to their pre-fertilization level.

If lipoxygenase activity is controlled by calcium at 10^{-7} levels *in vivo*, one scenario is that a Ca^{+2}–gradient is transiently produced at fertilization, with high levels of calcium in the cortical region of the egg grading to much lower concentrations in the internal cytoplasm. The high levels in the cortex would be adequate to initiate the cortical granule exocytosis and NAD kinase activity and the lower levels that diffuse into the cytoplasm would then be adequate to initiate lipoxygenase activity (Fig. 5). These ideas, of course, are highly speculative but bring to the fore the idea that calcium levels need not necessarily increase uniformly throughout the cell nor to very high levels within the cell to effect cell change. On other grounds one can imagine microdomains of calcium; otherwise the differential contraction of microfilaments and disaggregation of microtubles, which are supposedly linked to Ca^{+2} (Marcum *et al.*, 1978) could not occur.

V. SUMMARY

The data reviewed in this article are consistent with the idea that the triggering of development of the sea urchin egg at fertilization results from two ionic events. The first is a transient rise in calcium which most probably triggers all of the "early events." Some of these early events somehow result in the subsequent arousal or activation of metabolism. In the case of the sea urchin, the most important permanent change wrought in the cell during this period is the activation of the Na^+-H^+ exchange system. As noted, this exchange results in a prodigious extrusion of hydrogen ions from the cell with a resultant increase in pH_i from pH 6.8 to pH 7.2–7.3.

The question of how the rise in calcium results in activation has been approached by asking which events are calcium-mediated and whether calcium is acting directly on the process or through some intermediary molecule. Analysis of three reactions suggests that two to three different mechanisms may be involved. NAD kinase is activated by micromolar levels of calcium in combination with calmodulin. The cortical exocytosis is also initiated by micromolar levels of Ca^{+2}; the role, if any, of calmodulin is unknown. Lipoxygenase activity is activated by 10^{-7}M levels of calcium and activity is probably not calmodulin-mediated. It is possible that domains of differing calcium levels transiently occur in the cell during fertilization and that the activation resulting from the Ca^{+2} increase is through binding to different target molecules.

ACKNOWLEDGEMENTS

I thank the National Science Foundation for support of the research from my laboratory described in this article.

REFERENCES

Baker, P.F. and Whitaker, M.J. (1978). *Nature* **276**, 515-517.
Baker, P.F. and Whitaker, M.J. (1979). *Nature* **279**, 820-821.
Brandis, J.W. and Raff, R.A. (1979). *Nature* **278**, 467-469.

Chambers, E.L. (1976). *J. Exp. Zool.* **197**, 149-154.
Chambers, E.L., Pressman, B.C. and Rose, B. (1974). *Biochem. Biophys. Res. Commun.* **60**, 126.
Cheung, G. (1970). *Biochem. Biophys. Res. Commun.* **38**, 533-538.
Cuthbert, A. and Cuthbert, A.W. (1978). *Exp. Cell Res.* **114**, 409-415.
Epel. D. (1964). *Biochem. Biophys. Res. Commun.* **17**, 62-69.
Epel, D. (1967). *Proc. Nat. Acad. Sci., U.S.A.* **57**, 899-905.
Epel, D. (1972). *Exp. Cell Res.* **72**, 74-89.
Epel, D. (1978a). *In* "Current Topics in Developmental Biology" (A.A. Moscona and A. Monroy, eds.), Vol. 12, pp. 186-246, Academic Press, New York.
Epel, D. (1978b). *In* "Cell Reproduction," (E. Dirksen, D. Prescott and C.F. Fox, eds.), pp. 367-378. Acad. Press, New York.
Epel, D. (1979). unpublished results.
Epel, D. and Johnson, J.D. (1976). *In* "Biogeneses and Turnover of Membrane Macromolecules" (J.S. Cook, ed.), pp. 105-120. Raven Press, New York.
Epel, D. and Paul, M. (1979). unpublished results.
Epel, D., Steinhardt, R.A., Humphreys, T. and Mazia, D. (1974). *Develop. Biol.* **40**, 245-255.
Epel, D., Patton, C., Wallace, R. and Cheung, G. (1979). unpublished results.
Foerder, C., Klebanoff, S.J. and Shaprio, B.M. (1978). *Proc. Nat. Acad. Sci., U.S.A.* **75**, 3183-3187.
Gilkey, J.C., Jaffe, L.F., Ridgway, E.B. and Reynolds, G.T. (1978). *J. Cell. Biol.* **76**, 448-466.
Grainger, J.L., Winkler, M.M., Shen, S.S. and Steinhardt, R.A. (1979). *Develop. Biol.* **68**, 396-406.
Hagiwara, S. and Jaffe, L.A. (1979). *Ann. Rev. Biophys. Bioeng.* in press.
Head, J.F., Mader, S. and Kaminer, B. (1979). *J. Cell. Biol.* **80**, 211-218.
Hinegardner, R.T., Rao, B. and Tedlman, D.E. (1964). *Exp. Cell Res.* **36**, 53-61.
Hultin, T. (1950). *Exp. Cell Res.* **1**, 159-168.
Humphreys, T. (1971). *Develop. Biol.* **26**, 201-208.
Ii, I. and Rebhun, L. (1979). *Develop. Biol.* **72**, 195-200.
Jaffe, L.A. (1976). *Nature* **261**, 68-71.
Jarrett, H.W. and Penniston, J.T. (1977). *Biochem. Biophys. Res. Commun.* **77**, 1210-1216.
Johnson, C. and Epel, D. (1979). unpublished results.
Johnson, J.D., Epel, D. and Paul, M. (1976). *Nature* **262**, 661-664.
Jones, H.P., Bradford, M.M., McRorier, R.A. and Cormier, M.J. (1978). *Biochem. Biophys. Res. Commun.* **82**, 1264-1272.
Kretsinger, R.H. (1976). *Ann. Review Biochem.* **45**, 239-266.
Lambert, C.C. and Epel, D. (1979). *Develop. Biol.* **69**, 296-304.
Marcum, S.M., Dedman, J.R., Brinkley, B. and Means, A.R. (1978). *Proc. Nat. Acad. Sci. U.S.A.*, **75**, 3771-3775.
Markus, A.J. (1978). *J. Lipid Research* **19**, 793-826.
Mazia, D. (1978). *In* "Cell Reproduction: In Honor of Daniel Mazia" (E.R. Dirksen, D.M. Prescott and C.E. Fox, eds.), pp. 1-14, Acad. Press, N.Y.
Moreau, M., Geurrier, P., Doree, M. and Ashley, C.C. (1978). *Nature* **272**, 251-253.
Papahadjopoulous, D. (1978). *In*"Membrane Fusion" (G. Poste and G. Nicolson, eds.), pp. 766-791. North-Holland Publishing Co., Amsterdam.
Paul, M. (1975) *Develop. Biol.* **43**, 299-312.
Paul, M. and Epel, D. (1975). *Exp. Cell. Res.* **94**, 1-6.
Perry, G. (1979). Ph.D. Thesis., University of California, San Diego.
Piatigorsky, J. and Whiteley, A.H. (1965). *Biochem. Biophys. Acta,* **108**, 404-418.
Ridgway, E.B., Gilkey, J.C. and Jaffe, L.F. (1977). *Proc. Nat. Acad. Sci. U.S.A.* **74**, 623-627.

Schuetz, A.W. (1975). *J. Exp. Zool.* **191**, 443-446.
Shen, S.S. and Steinhardt, R.A. (1978). *Nature* **272**, 253-254.
Shen, S.S. and Steinhardt, R.A. (1979). *Nature,* in press.
Steinhardt, R.A. and Epel, D. (1974). *Proc. Nat. Acad. Sci., U.S.A.,* **71**, 1915-1919.
Steinhardt, R.A., Lunden, L. and Mazia, D. (1971). *Proc. Nat. Acad. Sci. U.S.A.* **68**, 2426-2430.
Steinhardt, R.A., Epel, D., Carroll, E.J. and Yanigamachi, R. (1974). *Nature* **252**, 41-43.
Steinhardt, R., Zucker, R. and Schatten, G. (1977). *Develop. Biol.* **58**, 185-196.
Vacquier, V.D. (1975). *Develop. Biol.* **43**, 62-74.
Vacquier, V.D. (1976). *J. Supramolec. Structure* **5**, 27-35.
Vacquier, V.D. and Brandriffe, B. (1975). *Develop. Biol.* **47**, 12-31.
Whiteley, A.H. and Chambers E.L. (1966). *J. Cell Physiol.* **68**, 309-324.
Wiley, H.S., Johnston, R.N., Beach, D. and Epel, D. (1977). *Biol. Bull.* **153**, 449.
Winkler, M.M. and Grainger, J.L. (1978). *Nature* **273**, 236-238.
Yagi, K., Yazawa, M., Kakiuchi, S., Oshima, M. and Uenishi, K. (1978). *J. Biol. Chem.* **253**, 1338-1340.
Zucker, R.S. and Steinhardt, R.A. (1978). *Biochem. Biophys. Acta* **541**, 459-466.
Zucker, R.S. and Steinhardt, R.A. (1979). *Nature* **279**, 820.

Initiation of the Cell Cycle and Establishment of Bilateral Symmetry in *Xenopus* Eggs

Marc Kirschner

Department of Biochemistry and Biophysics
School of Medicine
University of California, San Francisco
San Francisco, California 94143

John C. Gerhart

Department of Molecular Biology
University of California, Berkeley
Berkeley, California 94720

Koki Hara

and

Geertje A. Ubbels

Hubrecht Laboratory
Uppsalalaan 8
Utrecht, Netherlands

Abbreviations: MMR, modified Ringer (for formulation see legend to figure 2); SEP or SES, sperm entrance point or spot.

ISBN 0-12-612984-3

I. INTRODUCTION

The unfertilized egg is radially symmetric and arrested in a nondividing state. After fertilization it becomes bilaterally symmetric with every position on its surface corresponding to a specific location in the embryo and adult. It also becomes biochemically more active, with rapid cycles of DNA synthesis and mitosis. Both of these effects are produced by fusion of a single spermatozoan with the egg surface. It is the purpose of this paper to ask two questions: 1) which events at fertilization are involved in establishing the cell cycle, and 2) which events are involved in producing bilateral symmetry in the embryo.

As for the question of the establishment of the cell cycle, in unfertilized *Xenopus laevis* eggs the cell cycle is arrested at second meiotic metaphase where no DNA synthesis is taking place. After fertilization, there are a number of well timed cortical and cytoplasmic events, followed by first cleavage, 90 minutes after fertilization. After first cleavage, 10 rapid nearly synchronous cleavages (every 30 minutes) ensue until the mid-blastula period when the cells change their cell cycle and become asynchronous. The cell cycle in the early blastula period is a rudimentary one, having only M and S phases (Graham and Morgan, 1966). Thus, fertilization elicits from the egg a highly regulated but simple cell cycle. Recent experiments by Hara (1977) have demonstrated that in *Xenopus* and in axolotls, blastomeres dissociated from the egg continue to cleave synchronously suggesting strongly that any endogenous mechanism regulating the division behavior of the egg is subdivided among the individual blastomeres during cleavage.

As for the question of establishing the dorsal ventral asymmetry in the egg, fertilization also produces a number of cortical reactions, one of which, grey crescent formation, is spatially organized along the future dorsal-ventral axis. Many experiments over the past 90 years have correlated the position of the grey crescent with the future dorsal side of the embryo. Some have suggested that the cortical material itself has morphogenetic potential in organizing either directly or indirectly the nervous system, spine, musculature, etc., which comprise the dorsal axial structures.

In the first section of this paper we will consider how fertilization initiates the cell cycle and what components are necessary to drive the cell cycle in early blastula stages. In the second section we will consider the role of the sperm and the response of the egg cortex, the role of the grey crescent, and cytoplasmic components in the establishment of the prospective dorsal-ventral axis.

II. INITIATION OF THE CELL CYCLE AFTER FERTILIZATION

A. *External and Internal Events of the First Cleavage Period*

The period after fertilization in amphibian eggs is concerned with the completion of meiosis, fusion of the pronuclei, restarting the cell cycle, and the establishment of the dorsal-ventral asymmetry of the future embryo. The last event is only manifest on a cellular level at gastrulation. Directly after fertilization there are a number of cortical and internal

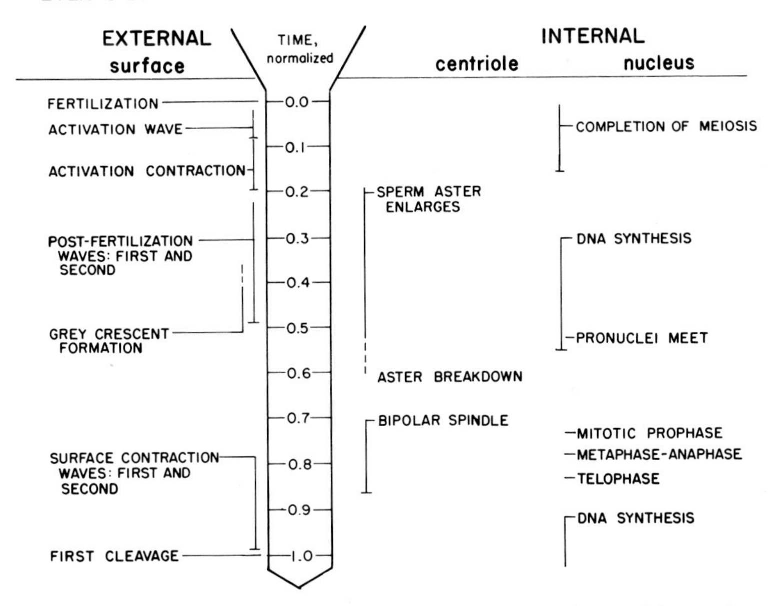

Fig. 1. Internal and surface events during the first cleavage period in amphibians. This table is constructed from a number of studies on various amphibians. The first cleavage period is normalized to 1.0. In *Xenopus laevis* it lasts 90 minutes at 21°C, in *Rana pipiens* 3 hours, in *Ambystoma mexicanum* approximatley 5 hours, reviewed by Gerhart (1979, in press.)

movements which are related to these events. They are diagrammed in Fig. 1 on a relative time scale from 0 (fertilization) to 1 (first cleavage). In *Xenopus* this period lasts about 90 minutes.

Some of the internal events listed in Fig. 1 such as DNA synthesis and mitosis have been well studied in a number of species. The surface events have been more recently studied with the help of time lapse cinematography. Directly after fertilization (during the time period 0–0.1) the cortex undergoes a rapid contraction in the animal or pigmented hemisphere. This wave of contraction, which has been termed the activation wave (Hara and Tydeman, 1979) is probably closely related to release of the cortical granules, perhaps mediated by a propagated wave of calcium release (Grey *et al.*, 1974; Gilkey *et al.*, 1978). It is set off by both normal fertilization and artificial activation. A second kind of wave which proceeds from the point of sperm penetration to the opposite side of the embryo at 0.3 to 0.5 has been called the post fertilization wave (Hara *et al.*, 1977) and might be a reponse of the cortex to the growth of the sperm aster, which is occurring about the same time. This wave does not occur in locally activated eggs, which lack a sperm aster. A third kind of wavelike surface activity composed of two closely spaced contraction waves occurs just prior to first cleavage. They are a genuine contraction of the cortex, starting near the animal pole, and propagating as circular waves over the equator to the vegetal pole. They have been termed the surface contraction wave (Hara, 1971). They are initiated at about the time of metaphase of the first division (0.8) and are followed immediately by first cleavage. By their points of initiation, their dependence on the aster, and by their distinctive morphology, these waves are distinguishable from each other. These waves are easily visualized by time lapse cinematography, but not by direct observation.

In the course of studies on the relationship of surface activities to internal movements during the first cleavage period, we attempted to interfere with some of these events by inhibiting them with specific drugs. We started by injecting colchicine and vinblastine to block those events which depend on an intact aster or microtubule system. As expected, eggs treated with colchicine or vinblastine fail to cleave. In addition, there is no post fertilization wave, and some internal movements are inhibited (Ubbels, Hara, Gerhart, and Kirschner, unpublished observations). To our surprise, the surface contraction waves occurred in the arrested eggs at precisely the normal time, indicating that this event is not tied to aster formation as is the general cleavage process. In addition, as the time of subsequent cleavage approached, the eggs underwent periodic surface contraction waves timed very precisely to the cleavage cycle in untreated eggs. Thus we think these waves reflect a basic timing mechanism of the cell cycle. The experiments reported here probe what cellular components are required for this timing mechanism. They suggest that there is an intrinsic oscillator in the cytoplasm or cortex of the amphibian egg which times the cell cycle independently of the centriole or nucleus. These experiments suggest further that the amphibian egg may be a useful system for studying the regulation of the cell cycle.

B. *Periodic Surface Contraction Waves*

Eggs activated by pricking undergo the activation contraction in a manner similar to fertilized eggs. After 80 minutes, a surface contraction wave begins near the animal pole. As in fertilized eggs, two waves, one following quickly behind the first wave are produced. However, unlike fertilized eggs, there is no subsequent cleavage. Instead the egg relaxes its cortex and later undergoes a second pair of surface contraction waves at about the time a fertilized egg would have undergone a second cleavage. Usually this is repeated a total of five or six times, each set of waves occurring at the time of cleavage in the fertilized egg.

The visibility of the waves is exaggerated if the vitelline membrane is first removed from the egg. Under these conditions the cortex is more

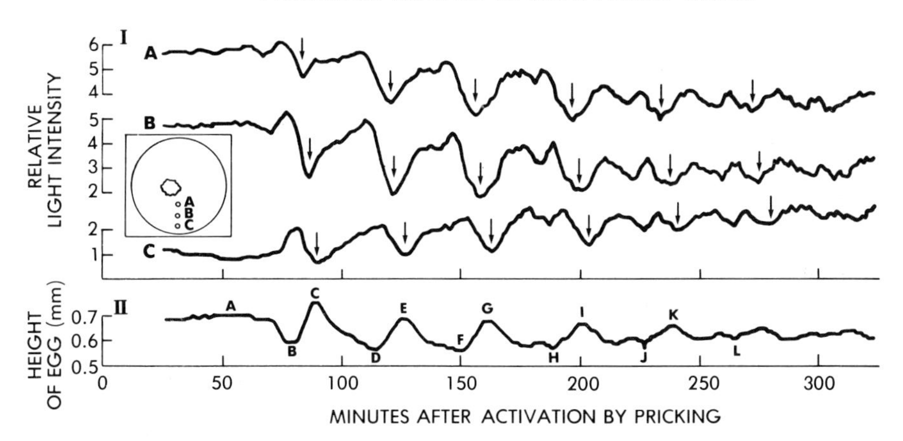

Fig. 2. Periodic waves in activated eggs. Unfertilized *Xenopus* eggs were collected by squeezing from the female, dejellied with a solution of 10 mM dithiothreitol in modified Amphibian Ringers, MMR (0.1 M NaCl, 2 mM KCl, 1 mM $MgSO_4$, 2 mM $CaCl_2$, 5 mM Hepes, 0.1 mM EDTA, pH 7.8) which was adjusted to pH 8.3, and washed extensively in 20% MMR. They were demembranated by placing in 5% Ficoll (400,000 MW, Pharmacia) in MMR and the vitelline membrane removed with watchmaker forceps. The eggs in MMR were then activated by pricking at a position approximately half way between the equator and the animal pole and the surface waves recorded by time lapse cinematography simultaneously from the top and side (Hara, 1970). (I) The propagation of the surface contraction waves was recorded by measuring the light intensity of the projected image at three spots A, B, and C located between the animal pole and the equator. The graphs represent relative light intensity versus the time after activation (pricking). The vertical arrows indicate the position of the darkest zone of the waves. (II) Periodic changes in the actual height of the same egg.

easily deformable. Especially when viewed from the side, it can be seen that during the time the waves are propagated, the egg also undergoes a periodic "rounding up." When light intensity measurements are made from the film at various positions on the egg surface, the periodic nature of the waves, as well as their speed of propagation can be determined as shown in Fig. 2. The waves in activated eggs are propagated at a speed of 60 μ/min in agreement with earlier measurements of Hara *et al.* (1977) in fertilized eggs. When the activated egg is filmed simultaneously from the top and side, one can easily record the rounding up and relaxation of the egg at the same time one is recording the propagation of the waves from the animal pole. The egg first flattens and then elevates and then relaxes to the original height. This rounding and flattening has the same period as the waves (Fig. 2). The amplitude of both measurements diminishes with subsequent cycles. In normal cleavage, the start of furrow formation coincides with the maximum height of the egg. In activated eggs, the egg cannot cleave presumably because it is lacking a functional centriole (Fraser, 1971; Heidemann and Kirschner, 1975; Maller *et al.*, 1976). After rounding up it enters directly into a relaxation phase, without first cleaving. When similar experiments were performed with fertilized eggs treated with colchicine or vinblastine, the results were the same; the eggs underwent periodic waves and periodic "rounding up" without cleavage but timed very precisely with the cleavage cycle in control eggs.

Activated eggs and fertilized eggs blocked with vinblastine or colchicine contain a haploid and a diploid nucleus respectively. It seemed quite possible that even though the normal mitotic cycle was arrested, the nuclei were undergoing cyclic changes and were still contributing to the periodic phenomena. To determine whether nuclear expression was required for producing periodic surface contraction waves, we examined the behavior of non-nucleated fragments.

Non-nucleate fragments of *Xenopus* eggs were made by constricting the egg with a loop of newborn baby hair by the method utilized by Spemann in 1901. Eggs were demembranated 30 minutes post fertilization and as shown in Fig. 3 were constricted at 40 minutes slowly with a loop of baby hair, with the expectation that the fusing pronuclei would be trapped in one half of the egg. After constriction both halves were filmed, side by side. The side containing the nuclei continued to cleave at the expected times, whereas the other side remained as a single cell (see Fig. 3E and 3F). Cytological examination of fixed egg fragments from ten experiments revealed that the noncleaving fragment never contained asters or nuclei while the cleaving fragment always showed both.

When time lapse films were examined, however, the noncleaving

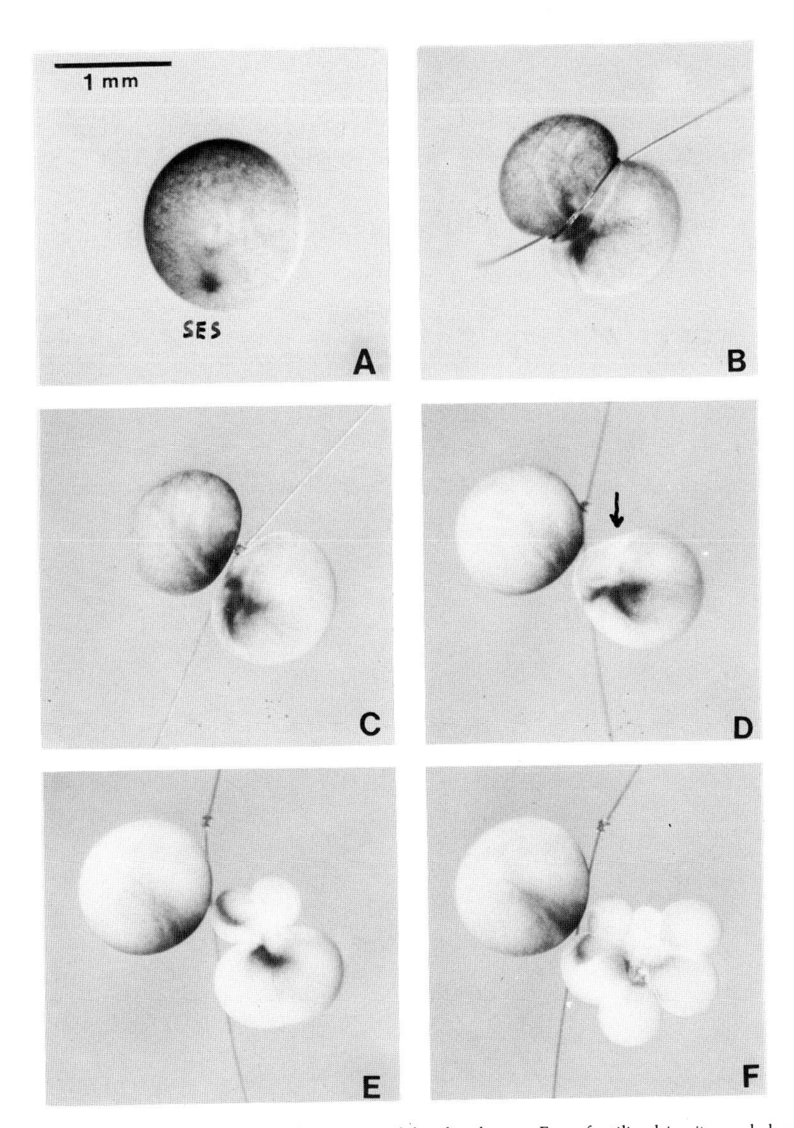

Fig. 3. Production of non-nucleate fragments of fertilized eggs. Eggs fertilized *in vitro* and clearly showing a sperm entry point were dejellied and demembranated as stated in the legend of Fig. 2. Fig. 3A shows an unconstricted egg with vitelline membrane intact at 25 minutes post pertilization (pf.) at 25°C. In Fig. 3B at 38 min. pf. the egg is demembranated and a loop of newborn human hair is placed around the egg in the plane defined by the animal pole, the sperm entrance point, and the center of the egg, i.e. a medial constriction (along the plane of bilateral symmetry). In 3C the hair loop is pulled tight 48 min. pf. In 3D the egg fragments are separated and the right fragment is cleaving, 64 min. pf. In 3E 90 min. pf. the nucleated side is starting second cleavage. In 3F 116 min. pf., third cleavage has begun.

fragment clearly showed periodic surface contraction waves of the same form as activated eggs. The periodic waves and the periodic rounding up occurred with the same period as the cleavage cycle in the nucleated fragments. The results of height measurements in the non-nucleated fragment from one series are shown in Fig. 4, along with the time of

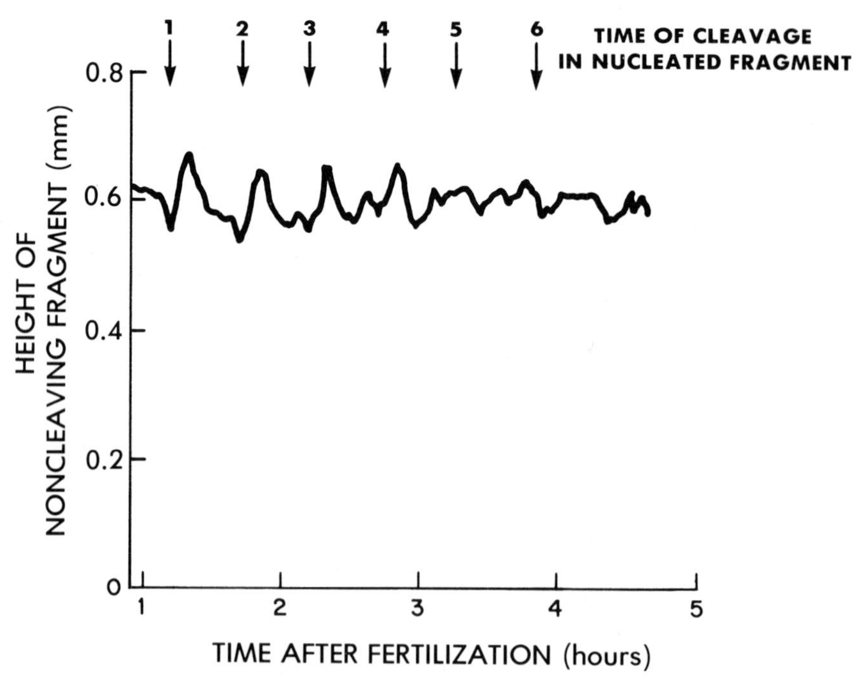

Fig. 4. Periodic surface contraction waves in a non-nucleated fragment obtained by constriction of a fertilized *Xenopus* egg. The height of the egg fragment, measured from the projected image, is plotted versus the time after fertilization. The arrows (1-6) indicate the time of onset of cleavage in the cleaving (nucleated) fragment (21°C.)

onset of cleavage observed in the cleaving fragment. The average period of cleavage and of changes in height are nearly identical, even though the maxima in the height of the non-nucleated half are 6 to 8 minutes retarded as compared to the cleavage time in the nucleated fragment. The average period measured for the surface contraction waves in the non-nucleated half is 30.4 ± 1.3 minutes, while the average period of cleavage in the nucleated fragment is 32.0 ± 1.5 minutes.

C. *Regulation of the Cell Cycle*

Several discrete and accurately timed cortical events take place between fertilization and first cleavage in amphibian eggs. One such event, the surface contraction wave, occurs just before first cleavage and can only be seen by speeding up the natural rate of development by time lapse cinematography. We have found that when cell division is inhibited by antimitotic drugs, the surface contraction waves take place at the expected time, and continue periodically. The onset of these waves is spaced at about 30 minute intervals after first cleavage, times corresponding closely to the cleavage period in normal eggs. Thus at some level the cell cycle continues despite the interruption of the normal events of karyokinesis and cytokinesis. Associated with the waves is a periodic "rounding up" of the egg with the maximum in egg height at about the time cleavage would normally occur. This periodic rounding up of blastomeres prior to each cleavage has been previously observed during the normal cleavage cycle of salamander eggs by Selman and Waddington (1955) and Harvey and Fankhauser (1933), although then it seemed that this rounding up might be obligatorily tied to the cleavage process. Recently Sawai (1979) independently also observed a very similar set of periodic changes in the tension and height of the cortex in non-nucleated egg fragments of the newt, *Cynops pyrrhogaster*. These periodic changes were also timed closely to the cleavage cycle in fertilized eggs. Thus in nondividing eggs there seems to be some operative remnant of the normal cell cycle which is manifest as periodic contractions of the cortex.

In normal development fertilization initiates cortical events in the quiescent unfertilized egg. We wished to know what role the sperm, its centriole and nucleus, and the female pronucleus play in timing the events of the cell cycle. We found that eggs activated by pricking, as is well known, fail to cleave, presumably due to the lack of a functioning centriole (Heidemann and Kirschner, 1975; Maller *et al.*, 1976). However, they do undergo periodic surface contraction waves timed with the cleavage cycle. Finally, an egg fragment devoid of either the maternal or paternal nucleus and having no evidence of an aster, also fails to cleave but undergoes periodic surface contraction waves with the same period as the cleaving egg fragment from which it was derived.

All of these observations point to an autonomous oscillatory regulator of the cell cycle, residing in the cytoplasm (or cortex) of the egg. Neither the cycle of DNA synthesis nor the cycle of centriole duplication, mitosis, and cytokinesis is required for the expression of these periodic cortical waves. As depicted in Fig. 5, both DNA synthesis and mitosis, as well as

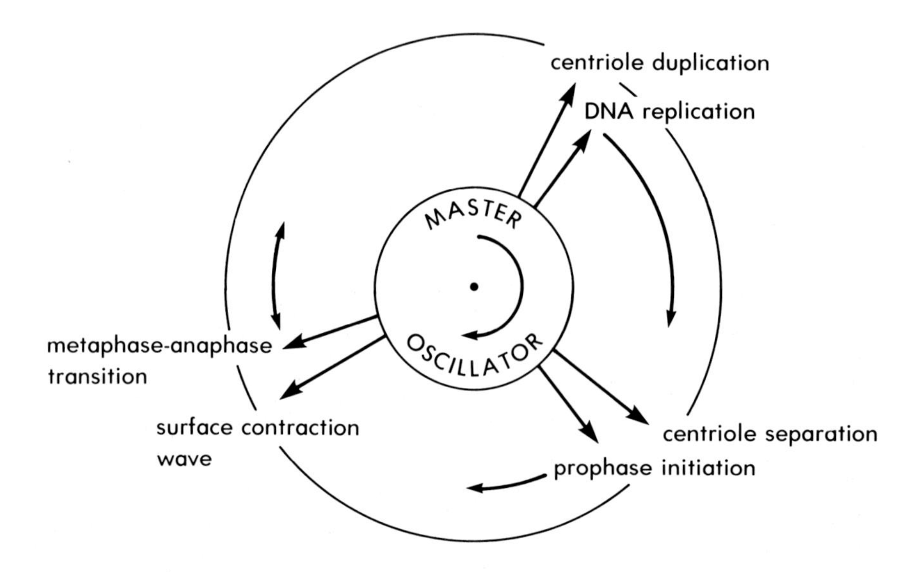

Fig. 5. Model of cell cycle regulation in *Xenopus* eggs.

possibly the cortical contractions themselves, can be seen as effects of some more basic timing mechanism. This timing mechanism, which must be some oscillatory biochemical reaction, can act independently of the things it affects. The notion of a master oscillator timing the cell cycle in Physarum has been suggested by Kauffman and Willie (1975).

These observations in a vertebrate egg can be added to a long series of related observations in invertebrate eggs. Bell, in 1962, reported periodic contractions of anucleate fragments of ascidian eggs, timed with the cleavage cycle. Recently, Yoneda *et al.* (1978) extended the experiment of Kojima (1960 a,b) in demonstrating periodic changes in tension of the egg surface timed with the cell cycle in activated sea urchin eggs, and in anucleate fragments. These experiments all point to a cell cycle timing mechanism acting independently of the nucleus and the centriole. In addition, there have been reports in activated eggs of periodic changes in chromosome condensation (Mazia, 1974), protein synthesis (Mano, 1970), enzyme activities (Petzelt, 1976), and sulfhydryl reactivity (Sakai, 1960), suggesting that these cycles may reflect broad biochemical changes in the cytoplasm. In addition, Nishioka and Mazia (1977) have recently discussed the notion that the cell can be activated periodically by a switch which has pleiotropic effects on a variety of synthetic activities associated with distinct phases of the cell cycle.

The amphibian egg offers some unique advantages for studying the cell cycle and for understanding the master oscillator. In addition to the easy visibility of the cyclic cortical changes, it has a simple cycle and is relatively independent of biosynthetic requirements during the early cleavage period. Future experiments should focus on the biochemical nature of the oscillator and its interaction with the nucleus and centrioles.

III. ESTABLISHMENT OF DORSAL-VENTRAL ASYMMETRY IN THE EGG AFTER FERTILIZATION

A. *Historical Background*

Whereas the fertilized egg of most anurans is radially symmetric about its animal-vegetal axis, the definitive embryo is clearly bilaterally symmetric with an anterior-posterior axis and a dorsal-ventral axis. Almost a century ago, Roux, Morgan and Boring, and others observed 1) that shortly after fertilization the uncleaved egg acquires bilateral symmetry when it forms the grey crescent, a less pigmented region on one side of the animal hemisphere opposite the side of sperm entry, and 2) that the asymmetry of the egg in general has a precise relationship to the asymmetry of the embryo. In particular, the lower limit of the crescent appeared to coincide with the site at which the blastopore first appeared a day or so after fertilization, and the crescent as a whole corresponded to the position of eventual dorsal embryonic structures such as the notochord, somites, and neural tube. Since local insemination experiments demonstrated that the sperm can penetrate anywhere in the animal hemisphere, and that the grey crescent and dorsal structures invariably arise at a position opposite the point of sperm entry, it seemed clear that the radially symmetric unfertilized egg has around its entire circumference the potential to become asymmetric under the influence of the sperm. Furthermore, it seemed clear that the early asymmetry committed the egg to a definitive pattern of regional developmental assignments expressed at later stages. The spatial relationships in the development of the establishment of the dorsal-ventral axis in *Xenopus* are diagrammed in Fig. 6.

Later, Spemann and Mangold (1924; reviewed by Spemann, 1938) discovered that the early gastrula of the urodele contained in the region above the blastopore a small group of cells capable of organizing the development of dorsal structures from surrounding cells; for example, organizing an entire secondary embryonic axis when transplanted into a

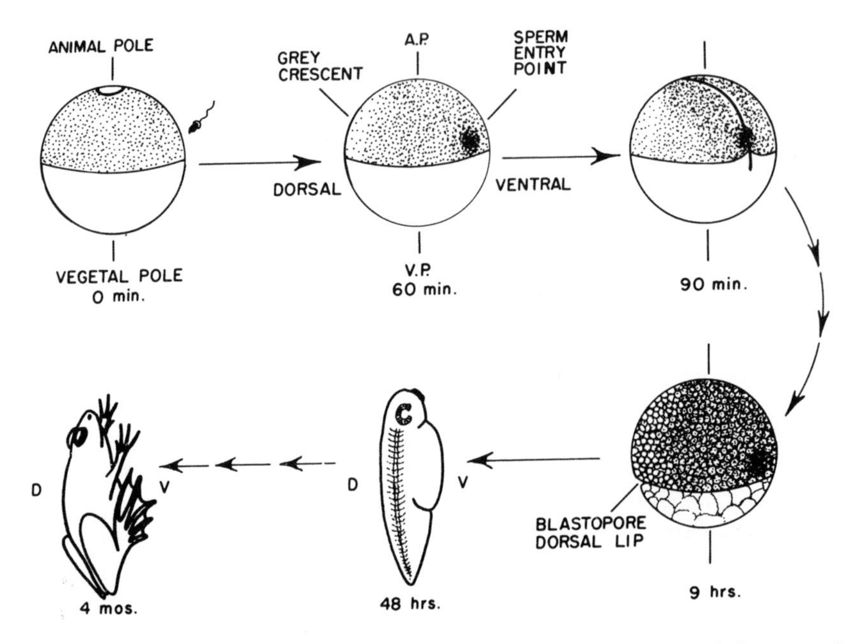

Fig. 6. Dorsal-ventral axis formation in *Xenopus*. The first drawing depicts a radially symmetric unfertilized egg with the animal (pigmented) hemisphere and vegetal hemisphere, showing an approaching sperm. In the second picture, one hour after fertilization, the egg is bilaterally symmetrical and a fate map can be estabished. The ventral side is distinguished by a pigment accumulation at the point of sperm entry. Opposite to that, in the prospective dorsal region is the grey crescent. At 90 minutes the first cleavage plane usually forms along the medial line (third picture). It bisects the grey crescent and sperm entrance point, separating the egg into left and right bilaterally symmetrc halves. At approximately 9 hours the first structural dorsal-ventral asymmetry is produced. The blastopore lip first forms on the dorsal side, in the area of the grey crescent, which is at the boundary of the animal and vegetal hemispheres. Two days later a definitive embryo is produced having gut on the ventral side, and musculature, notochord, and spinal cord on the dorsal side. During gastrulation, various dorsal and ventral cortical regions have contributed to internal structures as well (for a recent discussion on *Xenopus* see Keller, 1975). Several months later the tadpole has metamorphosed into an adult form preserving the dorsal-ventral arrangements in the early embryo as derived from the fertilized egg.

ventral equatorial position in a recipient early gastrula. Clearly, the small region above the lip of the blastopore was a source of inductors which would "dorsalize" the embryo. The prospective dorsal-ventral axis of the early gastrula was defined by the position of this region. Since the organizer region occupied the equivalent position of the grey crescent, it seemed plausible that the organizer gained its special dorsalizing powers directly from the crescent itself or from subjacent cytoplasm. Thus, the definitive dorsal-ventral axis of the embryo appeared to originate by the action of the grey crescent, the position of which was fixed shortly after fertilization by the entering sperm.

The most elaborate proposal of the crescent's role in dorsalization is found in the cortical field hypothesis of Dalcq and Pasteels (1937), in which the developmental fate of each region of the egg is said to derive from local values of two kinds of quantitative material gradients; one, a cortical gradient centered in the grey crescent at the egg surface and declining in intensity at increasing distance; and two, an internal cytoplasmic "vitelline" gradient related to the amount of yolk platelets, which abound as a coherent mass in the vegetal hemisphere and decrease in size and number in the animal hemisphere. The gradients were said to be distinguishable experimentally by their response to gravity or centrifugation, insofar as the dense yolk mass of the vegetal hemisphere could be easily rearranged to affect its gradient whereas the cortical gradient of the egg surface could not. Since the importance of the yolk mass for determination of the fate of egg regions already had been stressed by other authors, such as Penners and Schleip (1928 a,b), the novel contribution of Dalcq and Pastells was the idea of the cortical gradient and the application of a double gradient to the amphibian egg. The cortical gradient hypothesis of Dalcq and Pasteels suggested the following properties for the grey crescent:

1. it originates only once in early development, at a defined time before first cleavage,
2. it is a unique source of dorsalizing activity for the egg and cannot be replaced or circumvented by cytoplasmic determinants, and
3. it either acts over a long time period or acts at a much later time than the period of its formation. Perhaps its activity is passed on to or expressed in the organizer, its topographic successor.

Pasteels (1938, 1946, 1948, reviewed in 1964) devised three tests of the hypothesis, involving rotation and inversion of eggs at times after crescent formation or first cleavage. These experiments were extensions and re-evaluations of older experiments by Born and Schultze, as summarized in Table I. Pasteel's results, while interesting and informative, are not altogether unambiguous and have been rather neglected compared to the direct and dramatic proof by Curtis (1960, 1962a) of the dorsalizing activity of the grey crescent cortex.

As shown in Fig. 7, Curtis excised a piece of grey crescent cortex 150 μm x 150 μm x 3 μm from a host egg shortly before first cleavage or at the 8-cell stage, and grafted that piece of grey crescent in the ventral marginal zone of a recipient egg at the 1-cell stage, shortly before or during first cleavage. Eggs receiving cortex from the grey crescent region developed into twin embryos in 100% of the cases, whether the graft came from a one cell or an 8-cell donor. The two equal sized axes in the

TABLE I

Shifting of the Dorsal-Ventral Axis by Rotation of the Amphibian Egg: Historical Summary

Author	SPECIES	TIME OF ROTATION	RESULT
Ancel and Vintemberger (1948)	*Rana fusca*	before grey crescent formation	plane and direction of rotation determines position of grey crescent and of blastopore lip
Born (1885)	*Rana temporaria*	after grey crescent formation and before first cleavage	blastopore lip and final dorsal-ventral axis shifted toward or into plane of rotation
Pasteels (1946, 1948)	*Rana temporaria* Axolotl		
Schultze (1894)	*Rana fusca*	inversion at first cleavage, or slightly before	double axes formed in as many as 25-50% of the cases
Penners and Schleip (1928)	*Rana fusca*		
Pasteels (1938)	*Rana temporaria* *Rana esculenta*		

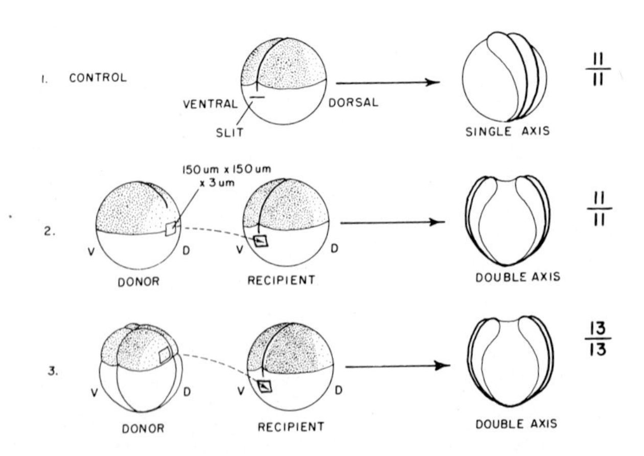

Fig. 7. Schematic outline of the cortical transplantation experiments of Curtis (1960, 1962a). The figure presents only those experiments concerning the transplantation of a piece of grey crescent cortex (150 μm x 150 μm x 3 μm) from a donor egg at 1 to 2 cell stage (line 2) or at the 8 cell stage (line 3), into the prospective ventral marginal zone of a recipient egg at the 1 cell stage or just starting first cleavage. The control (line 1) for transplantation is provided by recipient eggs cut in the ventral marginal zone, but not receiving a graft. The ventrally grafted eggs develop to twin embryos in 100% of the cases, whereas controls give only normal single axis embryos. The number of cases are indicated to the right of the figure.

resulting embryo were presumed to derive, one from the original grey crescent of the recipient, and the second from the grafted grey crescent cortex placed on the original ventral side. This is the most widely accepted and strongest validation of the cortical field hypothesis. It demonstrated that a small piece of the grey crescent cortex carries sufficient morphogenetic information to establish a full sized secondary dorsal axis on the original ventral side of the egg. However, the results did not fit theory in all respects, for additional experiments by Curtis showed on the one hand that the 8-cell recipient embryo no longer responded to crescent grafts by forming a second axis. Curtis suggested that the crescent is, therefore, only transiently essential for dorsalization, acting somewhere in the period from 1 to 8 cells, and then transfers its dorsalizing powers to the cytoplasm or to the remainder of the cortex.

There were other results, those of Schultze and Born (Table I) which more strongly taxed the resilience of the cortical field idea, and Dalcq and Pasteels tried to incorporate these effects into their theory in terms of the quantitative interactions of the two gradients. In the case of Born's observations, eggs held in horizontal positions after crescent formation underwent internal rearrangements of their contents and established their dorsal-ventral axis in the plane of slippage of the vegetal yolk mass. Dorsal structures arose from the side where movements of the yolk mass had left a trail along the cortex of the egg, the so-called "Born's crescent." Born's crescent as a dorsalizing agent presented a problem for the uniqueness of the cortical gradient centered in the old crescent. Furthermore, the cortical field hypothesis did not provide explicit accounts for the observations of Schultze as extended by Penners and Schleip (1928 a,b), in which eggs completely inverted at the 2 cell stage can form full sized secondary axes which Penners (1936) showed could derive on occasion from the expected ventral side, as if dorsalization had occurred a second time, far from the locus of the grey crescent and long after its formation. The frequency of twins was variable but in some experiments, reached 25% or 33% (Penners and Schleip, 1928 b) or 50% (Schultze, 1894).

Since these results in the literature contained a number of paradoxes for the cortical field idea, and because the egg of X. *laevis* had not been tested for the effectiveness of gravity in determining the positions of the dorsal-ventral axis or in causing twin axes, we chose to repeat the entire series of classical rotation and inversion experiments to obtain a single consistent set of data on one species and to provide a complete background for the Curtis experiment which was performed on X. *laevis*.

B. *Role of the Sperm in Establishment of the Dorsal-Ventral Axis of the X. laevis Egg.*

Approximately 15 minutes after fertilization the sperm entry point (SEP) is easily recognized in X. *laevis* eggs as a dark spot in the animal hemisphere (Palaček *et al.*, 1978). Midway into the period between fertilization and first cleavage, the grey crescent becomes visible opposite the SEP. The crescent is somewhat difficult to see in X. *laevis* eggs compared to its counterpart in eggs of *Discoglossus* or R. *fusca*, and therefore extensive documentation of the cresent position vis a vis the SEP has not been reported. Nonetheless, Palaček *et al.* (1978) note its orientation opposite the SEP. Instead, in our experiments the sperm entry point itself rather than the grey crescent has been used as a reference for the prospective dorsal-ventral axis. The equator of the egg was marked with Nile Blue at a point below the equator but nearest the SEP and the egg was allowed to cleave and to develop 9 hours to gastrulation, when the dorsal lip of the blastopore appeared. The lip was then scored for its position vis a vis the position of the SEP, as revealed by the mark. As shown in Fig. 8, 70% of the eggs originated the blastopore lip in the sector 180-160° from the SEP, that is, an opposite position. Furthermore, 27% of the eggs showed lips in the next sector from 160 to 120°, and the remaining 3% fell within 120-100°. Thus the blastopore lip in general arises opposite the sperm entrance point, a position in general occupied by the grey crescent. The SEP is a good but not perfect indicator of the position of the lip. This level of dispersion has been observed for other species.

The classical rotation experiments of Ancel and Vintemberger (1948) are an important test of the role of the sperm itself in axis determination. These authors used activated eggs of R. *fusca* and showed they could control with perfect predictability the position of the grey crescent by brief rotation of the egg in the period *prior* to crescent formation. They tipped the egg 135° off the vertical axis and allowed it to right itself slowly within the vitelline membrane. Rotation occurs because the egg is surrounded by the perivitelline fluid in a space between the plasma membrane and the extracellular fertilization or vitelline membrane. This space acts as a lubricated bearing allowing the egg to attain a stable position determined by its center of gravity in the dense yolk mass of the vegetal hemisphere. For activated eggs, a single rotation of 135° was sufficient to fix the future position of the grey crescent in 100% of the cases. The rotation could be done any time until the grey crescent actually formed.

It should be recalled that artifically activated eggs lack a sperm

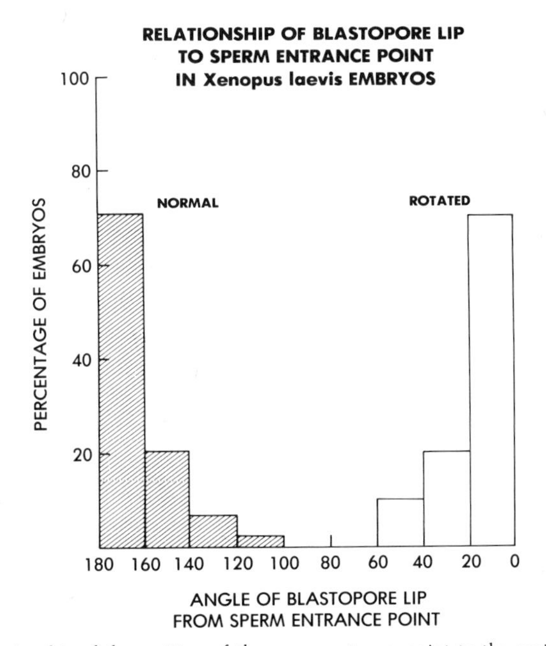

Fig. 8. The relationship of the position of the sperm entrance point to the position of the dorsal blastopore lip in *Xenopus* embryos which developed from normal and from rotated eggs. Dejellied fertilized eggs were obtained as in the legend of Fig. 2. They were marked at the margin beneath the sperm entrance point by application of a small crystal of Nile Red, a sparingly soluble form of Nile Blue, as described by Kirschner and Hara (1979). The eggs were either allowed to develop or at 50 minutes post fertilization were rotated 90° ventral side up and immobilized by placing them in a solution of 5% Ficoll in 20% MMR. After 30 min they were placed in 20% MMR and allowed to rotate back to the animal pole-up position. At the time of the appearance of the blastopore lip (9-15 hours depending on temperature) they were examined for the relative position of the blue mark, the initial blastopore lip, and the vegetal pole. With the vegetal pole as the center, the angle between two radii, to the blue mark and to the dorsal blastopore lip, was measured. The number of eggs within a certain range of angle is displayed as a histogram. Normal eggs group around 180°, blastopore opposite mark (sperm entrance side), rotated eggs group around 0°, blastopore, on same side as mark.

centriole and aster, as well as a sperm entry point, and yet they form a crescent; thus, crescent formation does not require these sperm agents even though they may serve to orient the position of crescent formation in fertilized eggs. The experiments show, however, that the sperm aster is not needed to orient the grey crescent under conditions of rotation, where slight shifting of the internal contents is provoked by gravity in the short period while the egg is off axis. Ancel and Vintemberger also tested fertilized eggs and found rotation could overcome the effect of the sperm in orienting the crescent, but only if the egg was given 4 successive rotations of 90° off the vertical axis each time the egg righted itself. In the case of fertilized eggs, however, a single 135° rotation was not sufficient to overwhelm the sperm influences. This competition between the sperm and the gravitational shift may suggest that the sperm itself, like gravity, is accomplishing a small shift of internal contents important for the orientation of the crescent. With the fertilized eggs, Ancel and Vintemberger established that the blastopore lip appeared in the plane of rotation, on the crescent side. Thus, the gravity-oriented crescent was a true indicator of the dorsal-ventral axis.

We have repeated these experiments with X. *laevis* eggs, using a modified procedure of significant advantage over the original one. The eggs marked near the SEP are placed in a solution of 5% Ficoll, a hydrophilic sucrose polymer which is too large to penetrate the fertilization membrane. The Ficoll counteracts the hydrophilic glycoproteins retained in the perivitelline space; the space loses water and the vitelline membrane collapses within five minutes onto the egg surface, preventing the egg from re-orienting itself when placed off-axis. The Ficoll treatment is completely reversible. Eggs in Ficoll can be rotated onto the side with the SEP up or down, left for a chosen period, and later returned to the upright position. In a fixed position at such an angle, the internal contents continue to rearrange until the treatment is intentionally ended, instead of when the egg rights itself. In Fig. 9 is shown the perivitelline space before and after its collapse by Ficoll. With this procedure, two kinds of rotation experiments were performed as diagrammed in Fig. 10. In the first, eggs were rotated and left for thirty minutes with the sperm entrance point down (prospective ventral side down). These eggs formed blastopore lips opposite to the SEP, as did

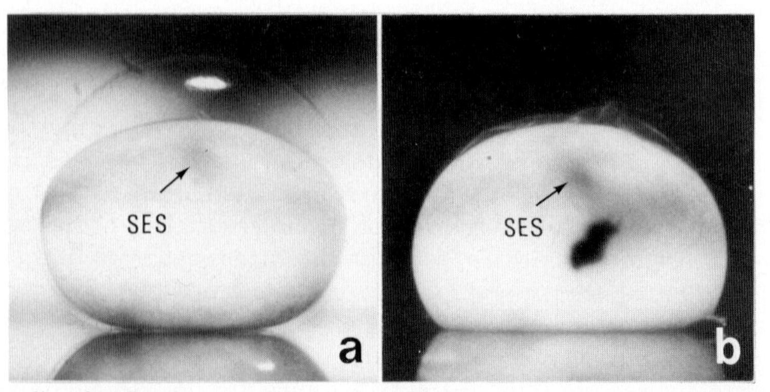

Fig. 9. Effect of Ficoll on the perivitelline space. Fig. 9a shows a side view of a dejellied egg 40 minutes after fertilization. Note the vitelline membrane and the perivitelline space, particularly prominent over the egg. The sperm entrance spot is denoted by SES. Fig. 9b shows the same egg after placement for 5 minutes in a solution of 5% Ficoll in MMR. It has a Nile Blue mark below the SES. The vitelline membrane has collasped over the egg and the perivitelline space has shrunk appreciably.

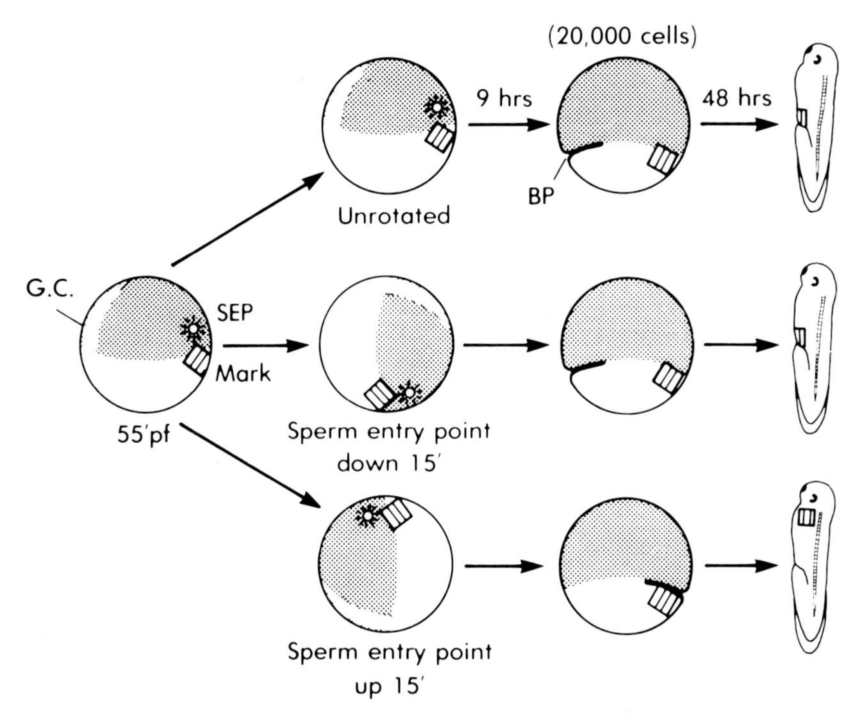

Fig. 10. Schematic diagram of the 90° rotation experiments in *Xenopus*. As discussed in the text, under the appropriate conditions a 90° rotation of a *Xenopus* egg can lead to complete reversal of the prospective dorsal-ventral axis. In the experiments diagrammed here, fertilized eggs are dejellied and marked below the sperm entrance point. The sperm entrance mark is denoted by a small circle with rays, the dye mark by a square striped patch, the animal hemisphere by stippling, and the grey crescent by lighter stippling. In the control series (top line) the unrotated egg initially forms its blastopore lip, 180° opposite the mark, which denotes the side of the point of sperm entrance. In eggs rotated on their side sperm entrance side down (grey crescent up), at 50 minutes post fertilization for 10 to 60 minutes as shown in the second line, the same result is obtained: the mark is found 180° opposite the sperm entrance point. In eggs rotated sperm entrance side up (grey crescent down) for the same period, the opposite result is obtained: the blastopore forms on the same side as the mark (sperm entrance side). Confirmation that the mark corresponds to ventral structures in the first two series, and dorsal structure in the latter series was obtained by locating the mark in a few randomly chosen late tailbud stage embryos. The mark was found in ventral posterior endoderm in cases where the blastopore first formed opposite the mark and in anterior endoderm in cases where it formed on the same side of the mark, as predicted by the fate maps of Keller (1975).

normal unrotated eggs. On the other hand, when eggs were left for thirty minutes with the SEP directly up, that is, with the prospective ventral side up, the blastopore lip appears directly at the side of sperm entry, instead of 180° away, as did the controls. Thus, the rotation treatment effectively reversed the dorsal-ventral axis relative to the sperm entry point. In the early period before crescent formation, as little as 10 minutes off axis with the SEP up was sufficient to reverse the axis. Thus, in X. *laevis* it is clear that gravitational rearrangements can control the position of the axis, with the dorsal side derived from the equatorial position uppermost during the rotation period. The results in histogram form are given in Fig. 8.

When the eggs were submitted to a 90° rotation of 60 min duration beginning at various times after fertilization, we found that the egg becomes increasingly resistant to alignment of the dorsal-ventral axis by gravity, but that the time of lability extended well into the period of grey crescent formation. This is shown in Fig. 11. Controls of unrotated eggs and eggs rotated to a position of SEP down are shown to the right of the figure; they have blastopore lips opposite the SEP. In this figure individual eggs are shown as lines with the average response given as a circle. In fact, rotation with the SEP down accentuates the normal relationship of the SEP and lip, and sharpens the distribution (compare the right-hand entries in Fig. 11), as if gravity reinforces the sperm. At times between 0.25 and 0.70 (22 minutes to 63 minutes) the axis of all eggs can be established nearly 180° opposed to its normal sperm-controlled orientation. At times approaching first cleavage, it becomes more difficult to achieve complete reversal. In addition, a few examples of double blastopores and twin embryos were noted when rotations were performed near the time of first cleavage. However, as late as 0.8 and 0.9 of the interval to first cleavage a few eggs still form axes at a position determined by the rotation rather than by the sperm. Thus, near the time of first cleavage, the egg population is heterogeneous in its response. In contrast to these results, Ancel and Vintemberger failed to achieve gravitational reorientation in fertilized eggs after the start of crescent formation (0.45), but they, of course, could not apply comparably severe gravitational treatment since their eggs always re-rotated within the perivitelline space to an upright position within a short time. Furthermore, in their experiments with unfertilized activated eggs, the period of study automatically ended when the crescent appeared since development arrested long before the blastopore appears.

C. *Reservations about the Curtis Experiments.*

These results with X. *laevis* led us to reconsider the results and procedures of the Curtis experiment. The exact experimental protocol used by Curtis raises several questions in view of the above rotation experiments in X. *laevis.* To begin with, in the Curtis experiment the piece

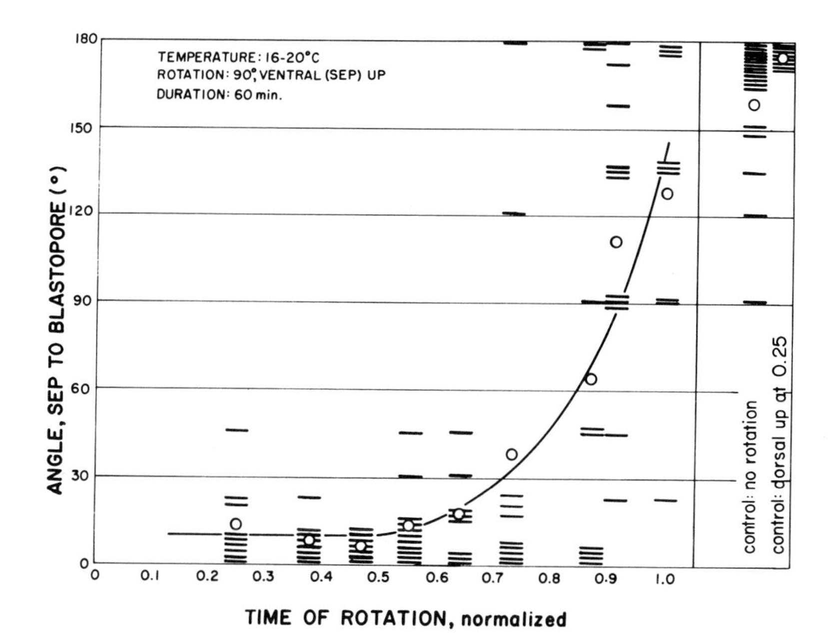

Fig. 11. Position of the dorsal-ventral axis in X. *laevis* eggs rotated at various times after fertilization. Eggs were fertilized and dejellied as described in the legends of Fig. 2 and Fig. 3 and placed in 20% MMR containing 5% Ficoll. The eggs were marked at the prospective ventral equator as revealed by the dark sperm entry point (SEP), either by an injection of 1% Nile Blue sulfate in 20 mM sodium phosphate, pH 7.2, or by application of a small crystal of "Nile Red" to the egg surface (Kirschner and Hara, 1979). Eggs were left in a vertical position with the animal pole up, until the time of rotation when they were turned on the side to orient the dye mark directly up (90° rotation with SEP up; line 3 Fig. 10) or directly down (90° rotation with SEP down; line 2 Fig. 10) Eggs remained in the rotated position for sixty minutes and were then returned to an upward position and allowed to develop in 20% MMR. The position of the first appearance of the blastopore lip, approximately 9 to 15 hours after fertilization (16-21°), was scored relative to the position of the dye mark. Zero degrees indicates superposition of the dye mark and lip, whereas an angle of separation of 180° indicates a position of the mark opposite the lip, as in Fig. 8. The angle of separation is given on the vertical axis, and the time at which the egg was rotated is given on the horizontal axis. The time is normalized to set fertilization as 0.0, the start of the first cleavage as 1.0. This period ranged from 90 to 130 minutes depending on the temperature. So that the variation in the egg response can be appreciated, data for individual eggs at each time are given as horizontal dashes, while the average for that time is given as an empty circle. Control results are shown at the right hand margin: eggs were either left without rotation, or were turned dye mark (SEP side) down. Notice that the latter controls give an even sharper population response than do unrotated controls, as far as the opposition of the lip and mark. Also notice in the experimental series the heterogeneity of the egg response in the time range of 0.7 to 0.9, when a few eggs are totally affected by gravity, a few are not at all affected, and others give intermediate angles of separation.

of cortex was transplanted into an egg from which the vitelline membrane had been necessarily removed. These recipient eggs, lacking a perivitelline space, could not rotate to a gravitationally stable position. They were immobilized in a position ventral side up, in order to receive the graft in their ventral margin. This is the position which in our experiments can affect the position of the dorsal-ventral axis, even in eggs approaching first cleavage. Thus, we were skeptical about the assumption that all recipients had a well-established dorsal side opposite the side of the grafting. Curtis essentially found that every egg put through this treatment, with the ventral side up, gave rise to twins no matter what the source of the graft. His control involved slitting eggs in the ventral margin without implanting cortex. The "perfect" and necessary control would have been to implant a piece of "inactive" ventral cortex which according to theory would have no dorsalizing activity and would not give rise to a secondary axis. Instead, the transplantation controls were done by putting ventral cortex into the dorsal margin in eggs immobilized with the prospective ventral side *down*, conditions which in our experiments just reinforce the sperm-controlled effects. As a separate reservation, the recipient eggs in the Curtis experiment were used at a time before or during first cleavage. This was the very time that mere inversion of the eggs of other species was found by several authors to provoke an appreciable fraction of twins (See Table I).

Because of these reservations, we "repeated" the Curtis experiment in part, by removing the vitelline membrane and turning the demembranated eggs ventral margin upward, that is, SEP up, in the period of 0.8 and after, to see the effect on axis formation. To our surprise, twin embryos developed at rather high frequencies without resort to cortical grafting. Fig. 12A shows two marked eggs just after the time of rotation, ventral side up. Fig. 12B shows an embryo having two distinct and separated blastopores, produced under the experimental conditions of Curtis, but without any cortical grafting. Fig. 12C and 12D show embryos with definitive equal sized double axes. As shown in Table II, the frequency of twins ranges from 20 to 70% depending on the batch of eggs and on factors we are still in the process of clarifying. The act of flattening the egg and poking it in the ventral margin, as part of giving a Nile Blue injection at the time the furrow first appears at the animal pole, may enhance twin formation. Lower temperature, which may prolong the period for internal rearrangement before successive cleavage furrows impede gravitational effects, may also raise the frequency. But at present it can be concluded that a large number of twins do occur without the grafting of the grey crescent cortex. These twins represent a

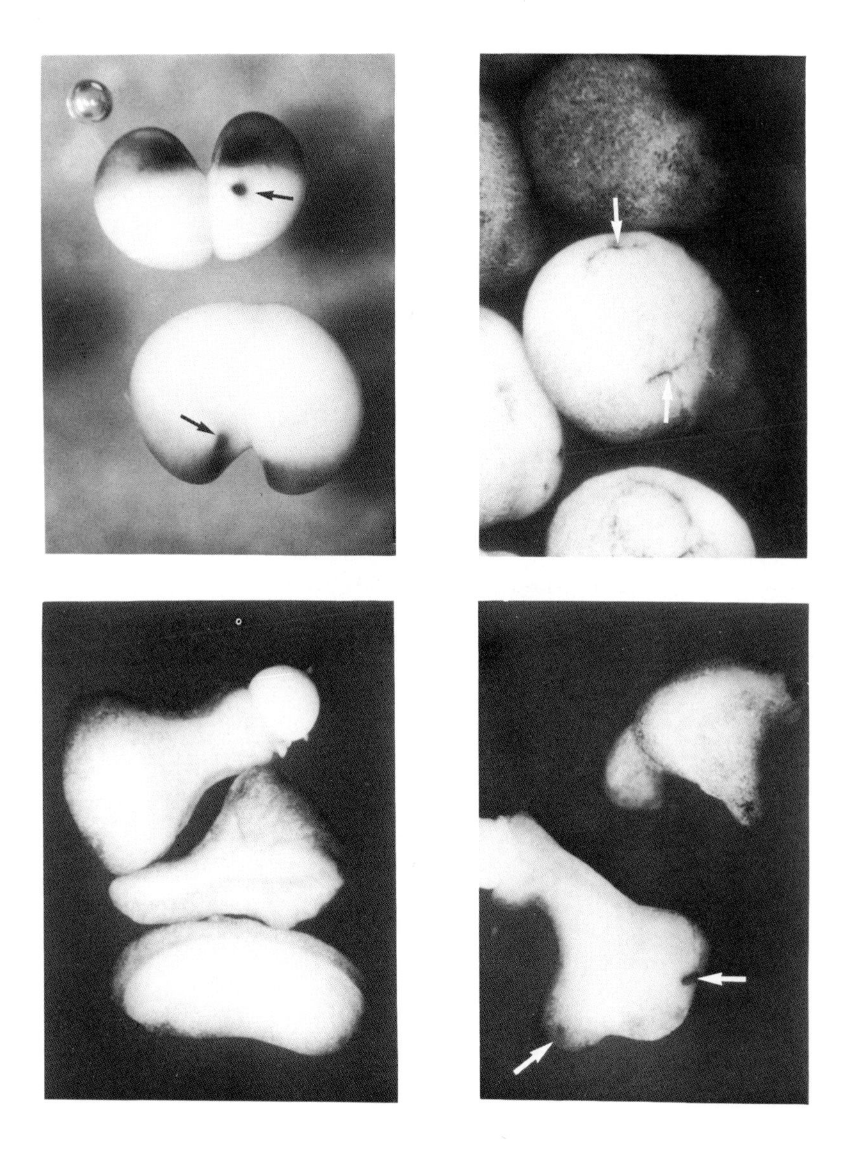

Fig. 12. Twin embryos produced by rotation of demembranated eggs. Fig. 12A shows two demembranated eggs turned ventral side up, just after first cleavage. The dark spot below the equator denoted by the arrows are Nile Blue dye marks indicating the ventral side. Fig. 12B (center) shows an embryo with a double blastopore lip (noted by arrows) obtained from a demembranated egg turned ventral side up at the time of first cleavage, and allowed to develop. Fig. 12C shows two neurula stage embryos with double axes (bottom two embryos) and one normal embryo with a single axis (top embryo) obtained as in 12B. Fig. 12D shows two late tailbud stage embryos with double axes obtained as in 12C (arrows denote cement glands in the lower embryo).

TABLE II

Twinning by Demembranated Eggs (X. laevis) Turned Ventral Side Up Near First Cleavage

Experiment	Time of Rotation normalized	Margin Wounded	Total Number	DEVELOPMENT OF EMBRYOS No Axis	Single Axis (S)	Double Axes (D)	$\% \frac{D}{D+S}$
1.	0.8	—	22	13	7	2	22%
2.	0.8	+	20	6	4	10	71%
3.	0.9	—	20	8	8	4	33%
4.	1.0	—	21	10	8	3	27%
5.	1.0	+	27	15	6	6	50%
TOTALS					33	25	43%

Dejellied fertilized eggs were transferred to 5% Ficoll in full strength MMR and in some experiments were marked on the equator of the prospective ventral side by Nile Blue injection. The eggs remained in a vertical position. Just prior to the time of rotation at 0.8 or 1.0, eggs were demembranated with watchmaker's forceps and transferred to an agar surface (0.5% agarose) under 100% MMR. Demembranated eggs were turned ventral side up and flattened onto the surface with the aid of a glass capillary. In some experiments, eggs previously unmarked were injected with Nile Blue at the ventral equator in the path of the oncoming furrow. After 6 to 8 cleavages, eggs were transferred to hemispheric wells in agar in 20% MMR. Gentamycin, 50 μg/ml, was included to reduce bacterial infection. At the late neurula stage, embryos were scored for the number of axes, as indicated by closed neural folds. Most embryos failed to develop beyond the tailbud stages.

wide range of double axiation, from extreme cases with completely separate and equal sized opposite axes, to others with equal sized axes joined at increasingly anterior levels, and finally to ones with small headless secondary "trunks" which we are analyzing histologically for full characterization.

It is interesting that Curtis found that the 8-cell egg was no longer responsive to crescent transplantation, in that this stage no longer formed twins. Penners and Schleip (1928 b) had shown that eggs of *R. fusca* inverted at the 8 to 16 cell stage no longer give inversion twins, presumably because furrows impede internal rearrangements. Thus, we think it feasible that the cortical grafting results are invalid for the lack of sufficient controls. Conservatively stated, the experimental conditions used by Curtis are highly effective in inducing twinning without the need for cortical grafting. More likely the twinning is provoked by rotation and flattening, conditions related to the compression and inversion conditions first used by Schultze in 1894 for twinning. We have not yet obtained 100% twinning in our experiments, and so we cannot rule out firmly that the cortical graft has some stimulatory effect on twinning. However, the *specific* action of the grey crescent cortex in provoking secondary axes cannot be assumed until other cortical regions are tested by implantation into the *ventral* cortex. As mentioned before, these controls were not reported, even though Curtis did many control grafts to the *dorsal* margin.

D. *Summary and Speculation*

Our results do not of course address the role of the grey crescent in normal development but just cast doubt on conclusions about the transplantability of the crescent's activity. As far as the role of the grey crescent under normal conditions, there is little evidence to indicate the crescent has continued activity beyond the 8-cell stage, since numerous authors including Curtis have removed the crescent or the animal hemisphere gradient carrying the crescent at various cleavage and blastula stages, and have found the egg to establish its dorsal-ventral axis normally (Curtis, 1962b). Also, in X. *laevis*, the recent careful fate maps of Keller (1975, 1976) show that the "organizer" region of dorsal mesoderm originates from internal cells and not from cells containing the crescent cortex. And furthermore, Nieuwkoop and his colleagues have built a strong experimental case for the conclusion that at least by the mid-blastula stage it is the dorsal vegetal hemisphere which carries the cytoplasmic pattern for the dorsal-ventral axis, and not the animal hemisphere. For simplicity, we follow Nieuwkoop (1969, 1973) in suggesting that by the 8-cell stage the vegetal hemisphere rather than the grey crescent and animal hemisphere carries the prospective axis.

As regards cytoplasmic asymmetries which might correlate with the prospective dorsal-ventral axis of the vegetal hemisphere, Pasteels (1946, 1964) and Ancel and Vintemberger (1948) called attention to a tail of vegetal yolk pulled up to the prospective dorsal side of the cortex in fertilized eggs. This yolk extension was known as the vitelline wall. As these authors pointed out, the vitelline wall also formed under the force of gravity in rotation experiments on fertilized and activated eggs of *R. fusca*. We have recently found that fertilized *Xenopus* eggs when rotated SEP down acquire the same yolk configuration as do unrotated eggs, i.e. vitelline wall on the dorsal side, whereas *Xenopus* eggs rotated with the SEP up, acquire a mirror image configuration of vegetal yolk. Thus, this rearrangement of yolk correlates well with the reversal of the axis orientation by gravity, perhaps indicating a role for the yolk. Alternately, the yolk rearrangement may be a marker of other important cytoplasmic rearrangements which occur reciprocally as the yolk is displaced by gravity.

If the prospective axis is carried in the vegetal hemisphere soon after fertilization, and if the grey crescent has no role in dorsalization after the 8-cell stage, the question remains of the crescent's role in the brief period of its formation midway to the first cleavage. While the mechanism of crescent formation is unknown, it has been proposed by Løvtrup (1965) to reflect an asymmetric contraction of the egg cortex as a whole toward the side of sperm entry, with the pigment granules displaced ventrally leaving a region of reduced pigmentation. It is possible such an asymmetric cortical contraction might also pull the adherent vegetal yolk mass up the dorsal side of the egg, forming the vitelline wall. Thus, the crescent would reflect a contraction of the animal hemisphere, a contraction important for achieving an asymmetric arrangement of vegetal materials. By this view, the crescent would reflect a transient "dorsalizing" process, but would not itself be a dorsal determinant.

What, finally, is the egg asymmetry which precedes the asymmetric movement of the cortex and which controls the orientation in which the cortical and internal movements will occur? Our experiments address this question only insofar as the rotation of the egg can orient the dorsal-ventral axis of the embryo. We suggest that the sperm by way of its centriole and large aster can act on some target in the egg, inducing an asymmetry in it, and that its target is the direct determinant of the direction of cortical movement. We refrain from suggesting that the centriole and aster directly orient the cortical contraction, because activated *R. fusca* eggs lack a centriole and yet form a grey crescent at a position controlled by gravitational rearrangements (Ancel and Vintemberger, 1948). That the aster can affect the internal arrangements of cytoplasmic materials in fertilized eggs is documented in the studies of Ubbels (1977) on the movement of a yolk-free cytoplasmic region from a central position to the prospective dorsal side by the time of crescent formation in *X. laevis* and also by the experiments of Manes and Barbieri (1976), 1977; Manes *et al.*, 1978) on orientation of the grey crescent, specifically by injections of centriole preparations into activated eggs. In *Discoglossus* the dorsal shift of yolk-free cytoplasm rich in DNA and glycogen is also well established (Klag and Ubbels, 1975). In the latter case, the yolk-free cytoplasm finally abuts the vitelline wall on the dorsal side. In the pre-crescent period these rearrangements seem aster-driven in fertilized eggs and fail to occur in artifically activated *X. laevis* eggs, or in eggs treated soon after fertilization with colchicine or vinblastine which block aster growth (Ubbels, Gerhart, Hara and Kirschner, unpublished observations). Thus, we think the aster is the normal device by which materials are moved asymmetrically in the egg cytoplasm in the pre-crescent period, and that on the basis of these arrangements the cortex orients its large scale contracting movement which produces further internal rearrangements of the animal as well as vegetal hemisphere.

In summary, a plausible mechanism to account for the establishment of the dorsal-ventral axis is as follows:

1. Fertilization introduces the sperm at an asymmetric position in the egg. The sperm entry point is a record of this position.
2. The sperm centriole through its large aster causes the first slight displacement of cytoplasmic materials along the prospective dorsal-ventral axis. Materials of both the animal hemisphere cytoplasm and vegetal yolk mass may be perturbed.
3. The animal hemisphere cortex responds to the slight cytoplasmic and yolk asymmetry by contracting strongly to one side. This asymmetric contraction produces the grey crescent area and large redistributions of internal materials, such as the characteristic dorsal elevation of vegetal yolk known as the vitelline wall.
4. The asymmetric redistribution of vegetal materials establishes the prospective dorsal-ventral axis in this hemisphere. This distribution is achieved in the early cleavage stages. Later the dorsal vegetal blastomeres induce the neighboring animal hemisphere blastomeres to become dorsal mesoderm precursors and to take on the inductive and dorsalizing activities of the Spemann organizer in the late gastrula, as demonstrated by Nieuwkoop and his colleagues.
5. The rotation experiments serve to show that the first 3 steps outlined above, can be bypassed if gravity provides the motive force for internal rearrangements, and that the crucial cytoplasmic distributions of step 4 can be achieved without the sperm aster or cortical contraction normally associated with grey crescent formation.

ACKNOWLEDGEMENTS

The first two authors thank Professor Pieter Nieuwkoop for the opportunity in the Spring of 1978 to visit the Hubrecht Laboratory. We thank also Drs. Pieter Nieuwkoop and Job Faber for discussions and reading the manuscript, and Peter Tydeman for technical assistance. This research was supported in part by grants from the USPHS to M.K., #GM-26875-01, to J.G., GM-19363, by a grant to M.K. from the American Cancer Society, #VC213, and in part by support to the Hubrecht Laboratory from the Royal Netherlands Academy of Arts and Sciences.

REFERENCES

Ancel, P. and Vintemberger, P. (1948). *Bull. Biol. Fr. Belg.* (Suppl.) **31**: 1-182.
Bell, L.G.E. (1962). *Nature* **193**:190-191.
Born, G. (1885). *Arch. Mikrosk. Anat.* **24**:275-545.
Curtis, A.S.G. (1960). *J. Embryol. Exp. Morphol.* **8**:163-173.
Curtis, A.S.G. (1962a). *J. Embryol. Exp. Morphol.* **10**:410-422.
Curtis, A.S.G. (1962b). *J. Embryol. Exp. Morphol.* **10**451-463.
Dalcq, A., and Pasteels, J. (1937). *Arch. Biol.* **48**:669-710.
Fraser, L.R. (1971). *J. Exp. Zool.* **177**:153-172.
Gerhart, J. (1979). Biological Regulation and Development, Vol. II. (R. Goldberger, ed.), Plenem Press, in press.
Gilkey, J.C., Jaffe, L.F., Ridgway, E.B. and Reynolds, G.T. (1978). *J. Cell Biol.* **76**:448-466.
Graham, C.F. and Morgan, R.W. (1966). *Develop. Biol.* **14**:439-460.
Grey, R.D., Wolf, D.P. and Hedrick, J.L. (1974). *Develop. Biol.* **36**:44-61.
Hara, K. (1970). *Mikroscopie* **26**:181-184.
Hara, K. (1971). *Wilhelm Roux' Archiv.* **167**:183-186.
Hara, K. (1977). *Wilhelm Roux' Archiv.* **181**:73-87.
Hara, K. and Tydeman, P. (1979). *Wilhelm Roux' Archiv.* **186**:91-94.
Hara, K., Tydeman, P. and Hengst, R.T.M. (1977). *Wilhelm Roux' Archiv.* **181**:189-192.
Harvey, E.N. and Fankhauser, G. (1933). *J. Cell Comp. Physiol.* **3**:463-475.
Heidemann, S.R. and Kirschner, M.W. (1975). *J. Cell Biol.* **67**:105-117.
Kauffman, S.A. and Willie, J.J. (1975). *J. Theoret. Biol.* **55**:47-93.
Keller, R.E. (1975). *Develop. Biol.* **42**:222-241.
Keller, R.E. (1976). *Develop. Biol.* **51**:118-137.
Kirschner, M.W. and Hara, K. (1979). *Microscopie,* in press.
Klag, J.J. and Ubbels, G.A. (1975). *Differentiation* **3**:15-20.
Kojima, M. (1960a). *Embryologia* **5**:1-7.
Kojima, M. (1960b). *Embryologia* **5**:178-185.
Løvtrup, S. (1965). *Wilhelm Roux' Archiv.* **156**:204-248.
Maller, J., Poccia, D., Nishioka, D., Kidd, P., Gerhart, J. and Hartman, H. (1976). *Exp. Cell Res.* **99**:285-294.
Manes, M.E. and Barbieri, F.D. (1976). *Develop. Biol.* **53**:138-141.
Manes, M.E. and Barbieri, F.D. (1977). *J. Embryol. Exp. Morphol.* **40**:187-197.
Manes, M.E., Elinson, R.P. and Barbieri, F.D. (1978). *Wilhelm Roux' Archiv.* **188**:99-104.
Mano, Y. (1970). *Develop. Biol.* **22**: 433-460.
Mazia, D. (1974). *Proc. Nat. Acad. Sci.* U.S.A. **71**:690-693.
Nieuwkoop, P.D. (1969). *Wilhelm Roux' Archiv.* **163**:298-315.
Nieuwkoop, P.D. (1973). *Adv. Morphogen.* **10**:1-39.
Nishioka, D. and Mazia, D. (1977). *Cell Biol. Int'l. Rep.* **I**:23-30.
Palaček, J., Ubbels, G.A. and Rzehak, (1978). *J. Embryol. Exp. Morphol.* **45**:203-215.
Pasteels, J. (1938). *Arch. Biol.* **49**:629-667.
Pasteels, J. (1946). *Acta. Anat.* **2**:1-16.
Pasteels, J. (1948). *Folia Biotheor.* (Leiden) **3**:83-108.
Pasteels, J. (1964). *Adv. Morphogen.* **3**:363-388.
Penners, A. (1936). *Z. Wiss. Zool.* **148**:189-220.
Penners, A. and Schleip, W. (1928a). *Z. Wiss. Zool.* **130**:305-454.
Penners, A. and Schleip, W. (1928b). *Z. Wiss. Zool.* **131**:1-156.
Petzelt, C. (1976). *Exp. Cell Res. 102*:200-204.

Sakai, H. (1960). *J. Biophys. Biochem. Cytol.* **8**:609-615.
Sawai, T. (1979). *J. Embryol. Exp. Morphol.* **51**: 183-193.
Schultze, O. (1894). *Wilhelm Roux' Archiv.* **I**:160-204.
Selman, G.G. and Waddington, C.H. (1955). *J. Exp. Biol.* **32**:700.
Spemann, H. (1938). "Embryonic Development and Induction." New Haven, Conn., Yale Univ. Press.
Spemann, H. and Mangold, H. (1924). *Wilhelm Roux' Archiv.* **100**:599-638.
Ubbels, G.A. (1977). *Mem. Soc. Zool. France* **41**:103-116.
Yoneda, M., Ikeda, M. and Washitani, S. (1978). *Develop. Growth and Differen.* **20**:329-336.

The Amphibian Egg Cortex in Fertilization and Early Development

Richard P. Elinson

Department of Zoology
University of Toronto
Toronto M5S 1A1 Ontario
Canada

I. INTRODUCTION

The unfertilized frog egg has a pigmented animal half and a non-pigmented vegetal half. The sperm enters the animal half, and before the egg cleaves, a distinctly pigmented surface, the grey crescent, appears between the animal and vegetal regions on the side opposite to sperm entry. Later in development, gastrulation begins at the junction between the grey crescent and the vegetal region. The side of the egg which the sperm entered becomes the ventral side of the embryo and the grey crescent side becomes the dorsal side of the embryo (Fig. 1).

ISBN 0-12-612984-3

The grey crescent has sparked the imagination of developmental biologists over the years. No fewer than six reviews have considered the grey crescent and dorsal–ventral polarization in the past fifteen years (Pasteels, 1964; Løvtrup, 1965, 1978; Nieuwkoop, 1973, 1977; Brachet, 1977). Despite its role in prompting speculation, very few developmental biologists actually work on this problem. In the same fifteen year period, only about twenty–five research papers have been published which relate to the dorsal–ventral polarization of the egg, and only four have included the words "grey crescent" in the title.

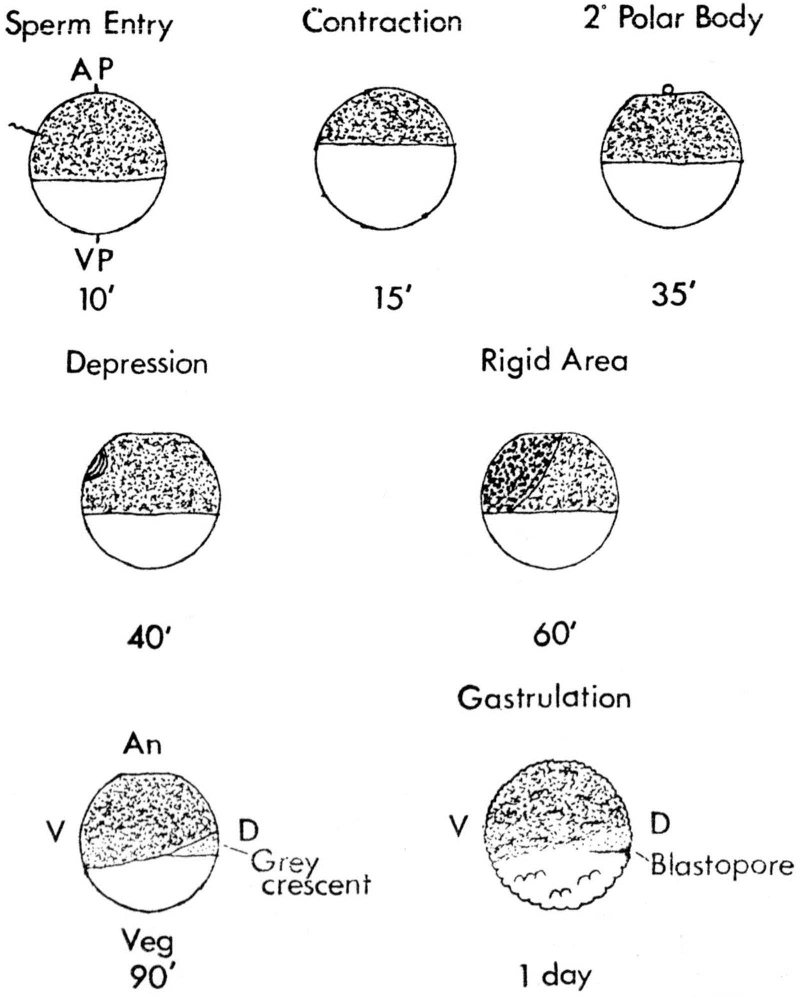

Fig. 1. Time course of events in *Rana pipiens* at about 22°C. Abbreviations: AP-animal pole; VP-vegetal pole; An-Animal; Veg-Vegetal; D-dorsal; V-ventral.

Not wishing to add to greatly to the theoretical burden being carried by the grey crescent, I will first review the properties of the amphibian egg cortex and the changes which occur at fertilization. I will then consider various cortical features and changes including the formation and the properties of the grey crescent cortex.

In this review, the cortex is defined as the plasma membrane plus a thin layer of underlying cytoplasm which can be distinguished either structurally or functionally from the rest of the cytoplasm. Hopefully, these two characterizations are describing the same entity. The amphibians discussed are principally anurans, the tailless amphibians. They include the clawed toad *Xenopus laevis,* the painted frog *Discoglossus pictus,* the toad *Bufo arenarum,* and the frogs *Rana pipiens* and *Rana temporaria (=Rana fusca).* Some of the reports are on eggs of urodeles, the tailed amphibians, which are a different order.

II. CORTEX OF THE UNFERTILIZED EGG

The mature egg ready for fertilization is arrested at metaphase II of meiosis, and the small metaphase II spindle lies at the animal pole, the center of the animal half. The surface of the unfertilized egg is covered with microvilli, a feature best illustrated by scanning electron microscopy (Grey *et al.,* 1974; Monroy and Baccetti, 1975; Elinson, present results). On eggs of *Rana pipiens,* the microvilli near the animal pole are long and sometimes ridgelike while those near the vegetal pole are short and stubby (Fig. 2-Unf). The transition between the two areas is gradual. Several attempts have been made to examine the egg surface with lectins, but except for a few special sites, the surface does not bind lectins (O'Dell *et. al.,* 1974; Denis-Donini *et al.,* 1974; Denis-Donini and Campanella, 1977; Stenuit *et al.,* 1977).

As in the eggs of many other animals, there is a layer of membrane bound granules known as cortical granules underlying the plasma membrane. These have been described many times in anuran eggs (see Kemp and Istock, 1967; Grey *et al.,* 1974 for references), but usually have not been reported in urodele eggs (Wartenberg and Schmidt, 1961; Wartenberg, 1962; Picheral, 1977b). Since cortical granules are involved in the block to polyspermy and since urodele eggs are normally polyspermic, it is expected that they would lack cortical granules. The one exception is a report by Ginsberg (1971) of cortical granules in the axolotl.

In general, anuran cortical granules are 1.5-2.5 μ in diameter and contain acid and neutral polysaccharides. In *Xenopus,* the size and

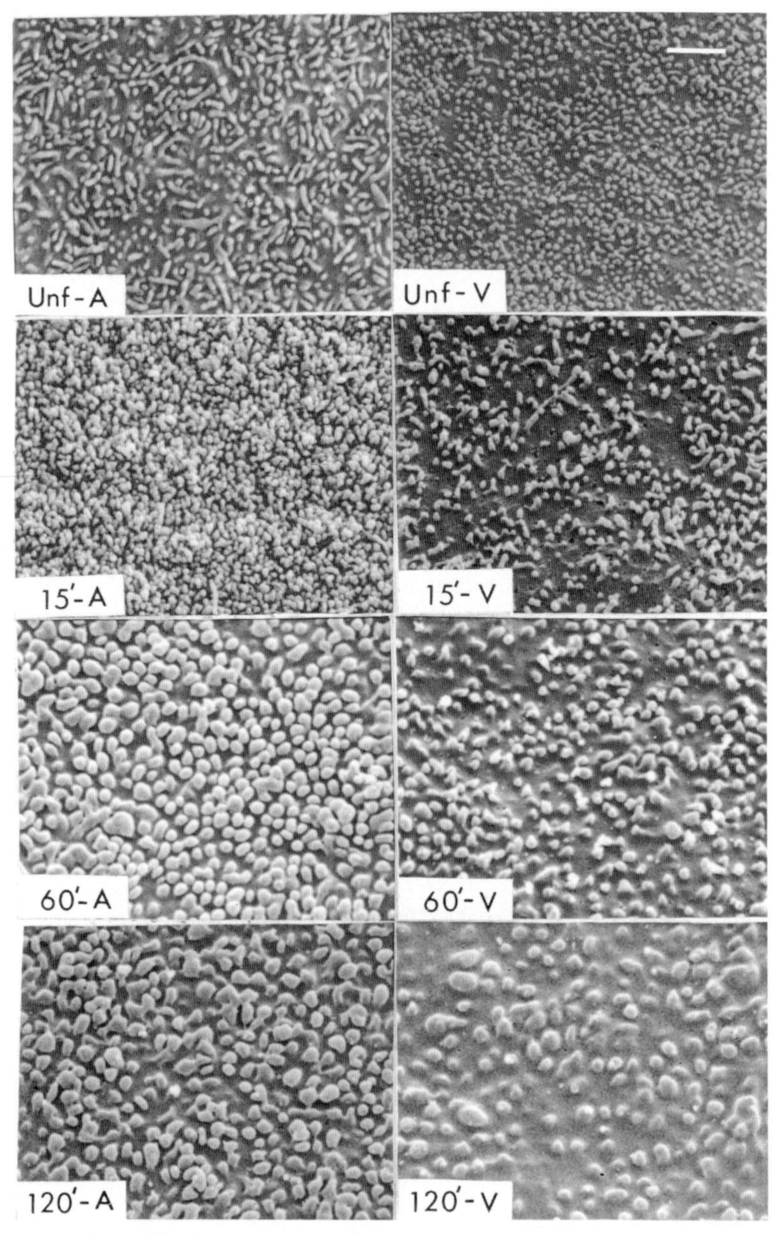

Fig. 2. Surface of the frog egg. These scanning electron micrographs show areas near the animal pole (A) and vegetal pole (V) on unfertilized eggs (Unf) and on fertilized eggs at 15, 60, and 120 minutes post-insemination. Scale line = 3 μm.

ultrastructural appearance of the granules differ between the animal and vegetal halves (Grey *et al.*, 1974; Campanella and Andreuccetti, 1977), but the functional significance, if any, of this difference is unknown. The only biochemical analysis of the granules has been done by Hedrick and co-workers who have isolated a cortical granule lectin (Wyrick *et al.*, 1974; Greve and Hedrick, 1978). The lectin participates in the formation of the fertilization envelope after sperm entry. As in sea urchin (Vacquier, 1975), frog cortical granules remain attached to the cortex when the egg is broken open (Goldenberg, 1979) and thus are an integral part of the cortex.

Besides the cortical granules, other components of the cortex are membrane vesicles or cisternae (Grey *et al.*, 1974; Campanella and Andreuccetti, 1977). Campanella and Andreuccetti (1977) reported that the cisternae are connected with the endoplasmic reticulum, and they proposed that this membrane system functions during propagation of the wave of cortical granule breakdown after fertilization. In addition, actin and microfilamentous structures were detected in the cortex of growing oocytes (Franke *et al.*, 1976), and it is likely that actin is present in some form in the cortex of the ovulated unfertilized egg.

III. CORTICAL CHANGES AT FERTILIZATION

A. *Site of Sperm Entry*

In some respects, our knowledge of sperm entry in amphibians is deficient compared to other animals. For instance, although amphibian sperm have acrosomes (Burgos and Fawcett, 1956; Buongiorno-Nardelli and Bertolini; 1967; James, 1970; Poirier and Spink, 1971), the acrosome reaction, a prerequisite for fertilization in many marine invertebrates and in mammals, has been observed only recently in sperm of *Leptodactylus chaquensis* (Raisman and Cabada, 1977) and *Pleurodeles waltlii* (Picheral, 1977a). Similarly, membrane fusion between the sperm and the egg as well as many other stages of sperm penetration have not yet been reported.

Despite this lack of basic knowledge, there are several aspects of sperm entry described in amphibians which await comparative study in other animals. For instance, effective sperm entry is restricted to part of the egg surface in many anurans (see Elinson, 1975, for references). In *Rana pipiens,* sperm are able to penetrate the jelly and vitelline envelope and approach the plasma membrane anywhere; yet, the fertilizing sperm enters only the animal half (Elinson, 1975). Sperm do not appear to enter the vegetal half, although the unlikely possibility of an entry which triggers no egg response has not been completely excluded. In *Discoglossus*, sperm enter only at the animal dimple, but whether this restriction is due to the plasma membrane or to the arrangement of the extracellular coats has not been determined (Campanella, 1975). Unlike anurans, sperm enter urodele eggs in both the animal and vegetal halves (see Elinson, 1975, for references). This lack of restriction on position of entry is correlated with the fact that fertilization in urodeles is normally polyspermic while fertilization in anurans is monospermic.

In many other animals such as insects or fish, the site of sperm entry is restricted to a single point due to the extracellular coats rather than due to the plasma membrane. There appears, however, to be a situation analogous to *Rana pipiens* in the mouse. Sperm entry is restricted to the 80% of the surface covered with microvilli (Nicosia *et al.*, 1977).

The entry of the amphibian sperm produces a small, microvillus-free structure at the site (Fig. 3a) which is analogous to the fertilization cone in eggs of other animals (Picheral, 1977b; Elinson and Manes, 1978). This structure disappears and is replaced by a small clump of elongated microvilli (Fig. 3b). Picheral (1977b) presented evidence that the microvillus-free structure was pinched off like a polar body leaving the clump of microvilli. If this is so, then it is likely that sperm membranes and other structures are eliminated from the egg. In contrast, sperm surface components persist following fertilization in rabbit (O'Rand, 1977) and in mouse and sea urchin (Shapiro *et al.*, 1979). It would be interesting to label amphibian sperm and to see whether membrane components persist or are sloughed off as suggested by the morphological observations.

Whether or not the sperm membrane persists, the site of sperm entry remains detectable as a clump of microvilli (Fig. 3c) for at least two hours (Elinson and Manes, 1978). The function, if any, of this scar is unknown.

B. *Fertilization Potential*

The initial response of the anuran egg to sperm entry or to activation by pricking is a membrane depolarization due to an efflux of chloride ions (Maeno, 1959; Ito, 1972; Cross and Elinson, 1978). The membrane potential changes from −28 to +6 mv in less than one second upon fertilization in *Rana pipiens* and the depolarization constitutes a fast block to polyspermy. When the potential of an unfertilized egg is raised to a positive potential by passing current into it, fertilization is prevented (Cross and Elinson, 1978). Induction of polyspermy by ion-substituted media in *Bufo americanus* (Cross and Elinson, 1978) and in *Xenopus laevis*

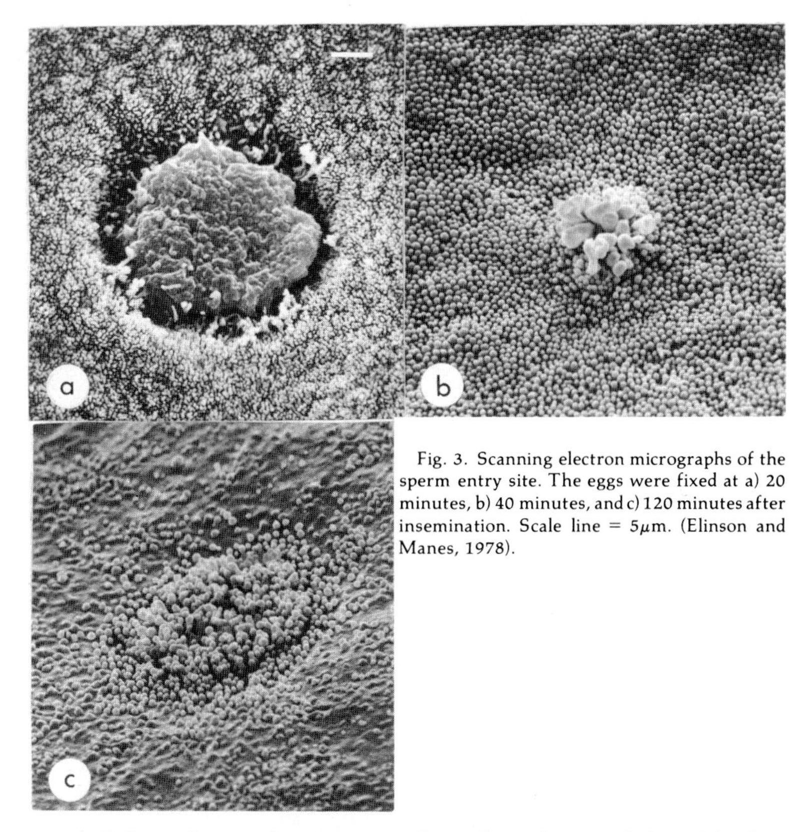

Fig. 3. Scanning electron micrographs of the sperm entry site. The eggs were fixed at a) 20 minutes, b) 40 minutes, and c) 120 minutes after insemination. Scale line = 5μm. (Elinson and Manes, 1978).

(Grey and Schertel, 1978) suggests that this electrical fast block to polyspermy is widespread among anurans. Anuran eggs are thus similar to eggs of sea urchins and *Urechis* which exhibit a fast block to polyspermy based on a membrane depolarization (Jaffe, 1976; Gould-Somero *et al.*, 1979). The depolarization in these two salt-water animals is due to a sodium influx rather than a chloride efflux as in the freshwater anurans.

C. *Cortical Reaction*

Following entry of the sperm or activation by pricking, the cortex undergoes several morphological changes. These include exocytosis of the cortical granules, elongation of the microvilli, and formation of a layer of dense cytoplasm. The cortical granules fuse with the plasma membrane and release their contents to the outside. As in eggs of other animals, cortical granule breakdown appears to be due to an increase in cytoplasmic free calcium (Kas'yanov *et al.*, 1971; Steinhardt *et al.*, 1974; Wolf, 1974; Hollinger and Schuetz, 1976; Lohka, 1978; Goldenberg, 1979). The contents of the cortical granules interact with the vitelline envelope and the jelly leading to the formation of the fertilization envelope which serves as a block to polyspermy (Wyrick *et al.*, 1974; Grey *et al.*, 1976).

The breakdown of the cortical granules propagates as a wave from the point of pricking in an activated egg (Fig. 4), and presumably the wave is similar with respect to the site of sperm entry in a fertilized egg. The rate of propagation is about 20-30 μ/sec in *Rana pipiens* (Kemp and Istock, 1967), 12-24 μ/sec at 15° in *Rana temporaria* (Kas'yanov *et al.*, 1971) and 10 μ/sec at 21° in *Xenopus* (Hara and Tydeman, 1979). At these rates, the wave takes 1-3 minutes to reach the opposite side of the egg, and this timing points out the necessity of a fast electrical block to polyspermy prior to the cortical granule mediated one.

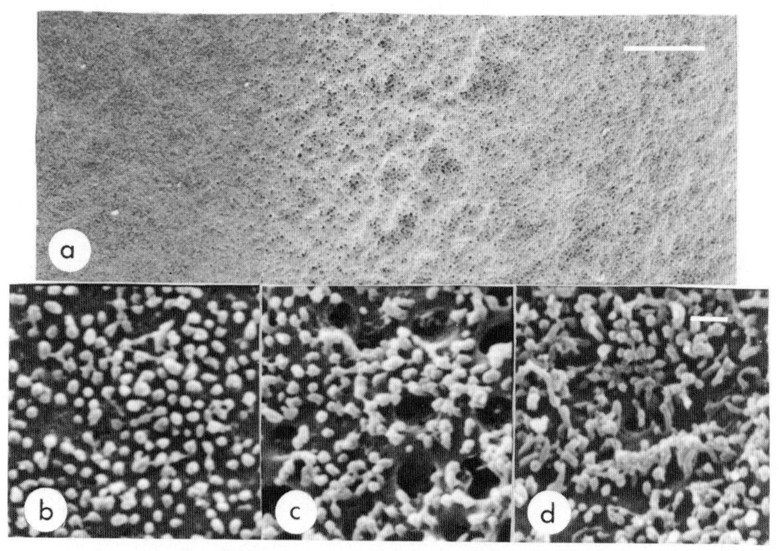

Fig. 4. Scanning electron micrograph of the wave of cortical granule breakdown. a) The left-hand side of the picture is unactivated egg surface while the right hand side is activated. The wave of cortical granule breakdown is in the center. Scale line: 30 μm. b) Unactivated surface. c) Cortical granule breakdown. d) Activated surface. Scale line for b, c, d: 1 μm.

Scanning electron microscopy of *Rana pipiens* eggs undergoing activation reveals that just following the wave of cortical granule breakdown, the microvilli become greatly elongated (Fig. 4). The presence of elongated microvilli following the wave of cortical granule breakdown has also been observed on sea urchin eggs (Eddy and Shapiro, 1976; Schroeder, 1979). Inside the egg, a layer of dense cytoplasm which excludes all organelles forms under the plasma membrane (Grey *et al.*, 1974; Campanella and Andreuccetti, 1977). The formation of this dense layer may provide the egg with structural support which is required due to the lifting of the vitelline envelope from the egg surface.

D. *Cortical Contraction*

After the morphological changes associated with the wave of cortical granule breakdown, the cortex contracts symmetrically towards the animal pole. The contraction has been described in *Rana temporaria* (Ancel and Vintemberger, 1936; Løvtrup, 1962), *Xenopus laevis* (Gurdon, 1960; Ortolani and Vanderhaeghe, 1965; Rzehak, 1972; Palaček *et al.*, 1978) and *Rana pipiens* (Elinson, 1975). The contraction begins between three and four minutes after pricking or sperm entry in *Rana pipiens,* reaches a maximum within five minutes, and then relaxes slowly (Elinson, 1975). As a result of the contraction, the area of the pigmented animal cortex is greatly reduced (Fig. 1). The microvilli in the animal half become tightly packed together while those of the vegetal half are spread apart (Fig. 2-15'). The cortical contraction probably depends on calcium since contractions can be produced experimentally by injection of calcium into the egg (Gingell, 1970) and by treatment with the ionophore A23187 (Schroeder and Strickland, 1974).

One of the functions of the cortical contraction at activation is to serve as an elevator which assists in moving the male pronucleus closer to the animal pole where the female pronucleus is. This is particularly important when the sperm enters at the border between the animal and vegetal halves, about 100° from the animal pole. The male pronucleus enters the cortex; the cortex contracts carrying the pronucleus polewards; the pronucleus migrates into the cytoplam, and the cortex relaxes (Elinson, 1975).

Another function of the cortical contraction may be to pull the egg surface from the forming fertilization envelope which together with the uptake of water into the perivitelline space (Nishihara and Hedrick, 1977), allows the elevation of the fertilization membrane. As a result, the egg is free to rotate with respect to gravity, and it assumes the characteristic orientation of an activated egg with the animal pole uppermost. These and other functions of the cortical contraction could be better defined if the contraction could be inhibited, but attempts to stop the contraction with cytochalasin B have thus far failed (Manes *et al.*, 1978).

IV. CORTEX OF THE FERTILIZED EGG

The grey crescent appears between 75 and 90 minutes after insemination in *Discoglossus pictus* (first cleavage at 155 minutes) (Klag and Ubbels, 1975) and between 70 and 130 minutes in *Rana temporaria* (first cleavage at 150 minutes) (Ancel and Vintemberger, 1948). A similar time course is found in *Rana pipiens* (Elinson, unpublished) with *Xenopus laevis* eggs forming a grey crescent by 50 minutes with first cleavage at 95 minutes (Palaček *et al.*, 1978). Before discussing the formation of the grey crescent, I would like to describe the general characteristics of the cortex of the fertilized egg.

The egg surface in *Rana pipiens* remains covered with microvilli which become larger and thicker with time. Microvilli at the vegetal pole remain shorter and sparser than those at the animal pole (Fig. 2). As with unfertilized eggs, fertilized eggs prior to cleavage do not bind lectins (O'Dell *et al.*, 1974; Denis-Donini *et al.*, 1976; Denis-Donini and Campanella, 1977). Bluemink and Tertoolen (1978) in a thorough description of the intramembranous particles of the *Xenopus laevis* egg, found little difference between the membranes of the animal and vegetal halves.

The limits of the cortex of the fertilized egg are not clearly defined. The two types of data related to this problem are the thickness of cortical pieces which can be dissected from the egg and the thickness of a morphologically distinct layer in intact eggs. All of the following observations have been made on *Xenopus laevis* eggs. Cortical pieces dissected free by Curtis (1960) were 0.5 to 3.0 μ thick and included mitochondria and pigment granules. Similar pieces isolated by Bluemink (1972) were about 5μ thick and included mitochondria, pigment granules, lipid droplets, and small yolk granules. As mentioned previously, electron micrographs of intact eggs show a dense layer of cytoplasm under the plasma membrane which is 0.1 to 1μ thick, and is described as finely textured or feltlike (Hebard and Herold, 1967; Bluemink, 1972; Grey *et al.*, 1974). Hebard and Herold (1967) also delineated a zone of yolk-free cytoplasm which includes the dense layer and varies in thickness between 0.5—5 μ. Taken together, these studies define a cortex consisting of a dense layer plus some underlaying cytoplasm to a depth of less than 5 μ. Some of the variability would disappear if the time after fertilization and the region of the cortex examined were carefully controlled.

The fact that the cortex can be isolated as intact pieces demonstrates that it has considerable structural coherence. The cortex is elastic

(Selman and Waddington, 1955; Sawai and Yoneda, 1974), and it is capable of active contractions upon wounding (Holtfreter, 1943; Gingell, 1970; Luckenbill, 1971; Bluemink, 1972) and in response to calcium (Gingell, 1970; Schroeder and Strickland, 1974). These observations argue for the presence of a contractile, coherent structure of which a prime candidate is a network of actin-containing microfilaments. Indeed, microfilaments were seen at wounds (Bluemink, 1972) and following an ionophore-induced contraction (Schroeder, 1974), and actin-containing filaments have been identified in the cleavage furrow (Perry *et al.*, 1971). In general, however, microfilaments have been rarely found in the cortex of the uncleaved fertilized egg of a number of species (Perry *et al.*, 1971; Schroeder, 1974; Perry, 1975). The contradiction between the nature of the cortex of the living egg and the lack of structural components in an assembled form may be resolved by saying that actin stored in the cortex can rapidly assemble into a functional form with appropriate stimuli. Alternately, a better understanding of the organization of structural proteins in the cortex and their relation to the dense layer may settle the question.

V. FORMATION OF THE GREY CRESCENT

A. *Role of the Spermaster*

The earliest experimental observations on the frog embryo demonstrated that the entering sperm determined the plane of bilateral symmetry (Newport, 1854; Roux, 1887). Ancel and Vintemberger (1948) found that although a number of experimental conditions could override the influence of the sperm, the grey crescent forms on the side of the egg opposite to sperm entry during normal development. What activity of the sperm is involved in grey crescent formation?

Kubota (1967) reported that a small area of the cortex becomes more rigid shortly after the formation of the second polar body in eggs of *Rana nigromaculata.* He measured rigidity by the decreased ability of centrifugation to drive the pigment granules from the cortex. The rigid area expands with time until the whole animal half cortex is rigid, and at all times, the rigid area is underlain by the sperm aster. By the end of the spread of the rigidity, bilateral symmetry becomes fixed. Kubota proposed that the rigid cortex causes an asymmetrical load in the cortex which leads to its rotation to produce the grey crescent on the opposite side.

The patterns seen by Kubota can be seen on living eggs of *Rana pipiens* without centrifugation (Elinson and Manes, 1978). At first, a depression appears on the egg surface (Fig. 1). The membrane site of sperm entry is found in the depression and the sperm aster underlies the depression. With time, the depression expands to a rigid area. In contrast to Kubota, we argued that the rigidity is cytoplasmic rather than cortical, and that it is probably due to the consistency of the astral region. These observations indicate that due to the aster, there is an asymmetric increase in cytoplasmic rigidity which just precedes the appearance of the grey cresent (Fig. 1).

In *Xenopus laevis*, Hara *et al.* (1977) have detected two post-fertilization waves which emanate from the area of sperm entry. Whether these waves and the astral area have any relationship remains to be seen.

A direct test of the aster's role in determining the position of the grey crescent was provided by Manes and Barbieri (1976, 1977). When they

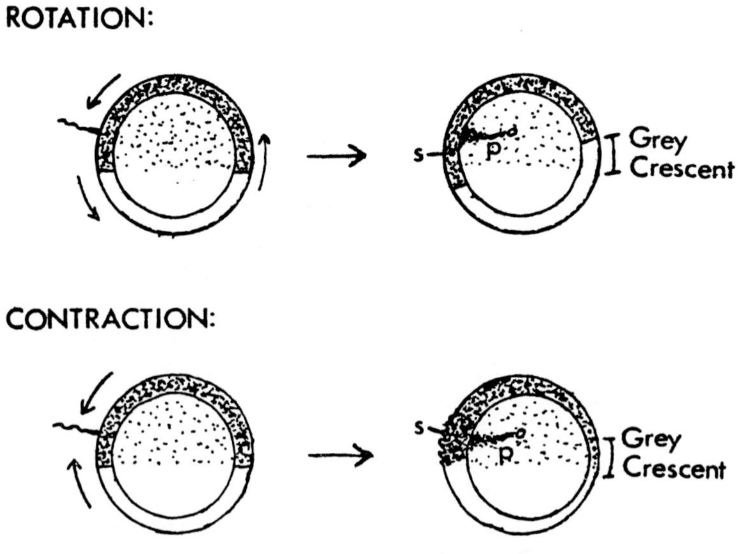

Fig. 5. Hypotheses of grey crescent formation. According to the rotation hypothesis, the entire cortex rotates relative to the cytoplasm producing a displacement vegetally of the membrane site of sperm entry (s) from the cytoplasmic pronuclear trail (p). According to the contraction hypothesis, the cortex contracts around the site of sperm entry causing no movement of the sperm entry site (s). The effect of the cortical contraction at 15 minutes on the displacement between the membrane site of sperm entry and the pronuclear trail is ignored for simplicity's sake in this illustration. Also, the thickness of the cortex is greatly exaggerated.

injected a sperm homogenate into *Bufo arenarum* eggs, the grey crescent appeared on the side opposite to the injection. The activity of the homogenate in mimicking the normal action of the sperm was correlated with the activity of the homogenate in causing aster formation. The growing sperm aster then is somehow involved in the reorganization of the egg to produce the grey crescent.

B. *Cortical Rotation or Cortical Contraction?*

Two hypotheses have been advanced to explain how the grey crescent arises. The first is that the whole cortex rotates relative to the underlying cytoplasm (Ancel and Vintemberger, 1948). The rotation is directed by the sperm aster and leads to vegetal half cortex moving over animal half cytoplasm on the side of the egg opposite sperm entry. The grey crescent, then represents a juxtaposition of vegetal cortex and animal cytoplasm (Fig. 5). The second hypothesis is that a local contraction occurs around the site of sperm entry (Løvtrup, 1958, 1965). This leads to a stretching of the cortex on the opposite side of the egg, and this cortex becomes the grey crescent (Fig. 5).

By marking the cortex and the cytoplasm before grey crescent formation, Ancel and Vintemberger (1948) tested these hypotheses in *Rana temporaria* and found that the grey crescent forms by a cortical rotation. Our observations in *Rana pipiens* on the membrane site of sperm entry and the pigmented cytoplasmic trail of the male pronucleus provide further confirmation of the rotation hypothesis (Elinson, 1975; Elinson and Manes, 1978). We found that after grey crescent formation, the sperm entry site was displaced vegetally with respect to the cytoplasmic trail as predicted by the rotation hypothesis (Fig. 4). For *Rana temporaria* and *Rana pipiens* as well as for *Discoglossus pictus* (Klag and Ubbels, 1975), the grey crescent arises by rotation.

As a result of the rotation, the vegetal cortex covers the margin between the animal and vegetal cytoplasms at the dorsal (grey crescent) side, and the animal cortex covers this margin on the opposite, ventral side. The reason that the grey crescent appears grey is because we see pigmented animal cytoplasm through the transparent vegetal cortex which has rotated over it. The difficulty in finding grey crescents on eggs of some species or between eggs of the same species is probably due to the relative pigmentation of the cytoplasm and the cortex.

Despite the evidence in favour of the rotation hypothesis, the idea of a contraction has remained an attractive possibility because it provides a mechanism as well as an explanation for grey crescent formation. Observations of pigment patterns on the egg of *Xenopus laevis* suggested to

Rzehak (1972) and to Palaček *et al.* (1978) that in this species, the grey crescent forms by a contraction around the sperm entry site. The difference in results may reflect a species difference since *Xenopus* and *Rana* are about as distantly related phylogenetically as two anurans can be. Before concluding that the eggs of these animals differ, however, it would be of interest to see whether the membrane site of sperm entry persists on *Xenopus* eggs and to see where it is found relative to the pigment patterns observed.

C. *Mechanism of the Cortical Rotation*

It is difficult to envisage a mechanism for a movement of the cortex relative to the cytoplasmic mass. This is especially so given that the cytoplasmic mass is fixed in orientation by gravity, and the cortex should be fixed at least at the vegetal pole by friction since it is squeezed between cytoplasm on one side and the fertilization envelope on the other. Furthermore, the rotation does not involve a gross rearrangement or flow of different regions of cytoplasm since the pigment trail marking the pronuclear path remains intact (Ancel and Vintemberger, 1948; Manes and Barbieri, 1977).

One attempt has been made to inhibit grey crescent formation with drugs. Cytochalasin B does not block grey crescent formation while colchicine does (Manes *et al.*, 1978). The inhibition by colchicine could be due to a general collapse of cytoplasmic structure.

A lead on the mechanism of rotation may come from experiments using ultraviolet (UV) light. When the vegetal half of the fertilized egg of *Rana pipiens* or *Xenopus laevis* is UV-irradiated, the resulting embryos exhibit delays of gastrulation and gross abnormalities of neural development. The eggs are most sensitive to UV before grey crescent formation and are insensitive after (Grant, 1969; Grant and Wacaster, 1972; Malacinski *et al.*, 1975, 1977; Manes and Elinson, unpublished). We have found that UV irradiation of the vegetal half prevents the formation of the grey crescent itself (Manes and Elinson, unpublished). How much of the UV- induced syndrome can be attributed to the inhibition of grey crescent formation is presently unclear. The delay in dorsal lip formation as well as the irregular, broader appearance of the dorsal lip (Grant and Wacaster, 1972; Malacinski, *et al.*, (1977) may reflect the embryo's indecisiveness in knowing where to begin invaginating. Since UV permits continued development and its effects are reversible (Malacinski, *et al.*, 1974), UV should prove useful in examining the mechanism of cortical rotation.

VI. IS THE GREY CRESCENT CORTEX A DISTINCT CORTICAL AREA?

It is clear from a variety of experimental manipulations that the egg cortex is important in the dorsal–ventral polarization of the embryo (Pasteels, 1964). This has led to the idea that the grey crescent cortex has special morphogenetic properties, and several recent investigations have been directed towards finding cortical differences between the grey crescent and elsewhere. An alternative view based on the formation of the grey crescent is that the grey crescent cortex is displaced vegetal cortex and that the activity in the grey crescent region is due to the normal expression of differences between animal and vegetal cortices, albeit in a reoriented position. I shall close by reviewing papers on the grey crescent cortex stressing the second view.

In a series of papers, Dollander examined the permeability of eggs of the urodele *Triturus* to Nile blue (Dollander and Melnotte, 1952; Dollander, 1953, 1956, 1957). The dye penetrates little or not at all into the animal and ventral sides (see Fig. 1 for orientation) but penetrates into the egg at the vegetal and dorsal sides. The dorsal side, marked by a clear crescent in these species, is clearly different from the ventral side in dye permeability. The permeability difference, however, reflects the original animal–vegetal permeability difference found before the formation of the clear crescent (Dollander, 1957).

One attempt has been made to look at the grey crescent cortex ultrastructurally (Hebard and Herold, 1967). The cortex from different regions of the *Xenopus* egg is slightly different in the thickness and in the continuity of the dense layer. In general, the dorsal (grey crescent) cortex resembles the vegetal cortex in thickness but is more discontinuous. One attempt has been made to look at the grey crescent biochemically (Tomkins and Rodman, 1971). The protein composition of the animal cortex and the grey crescent cortex appeared to differ, but the vegetal cortex was not examined.

This brings us to the experiments of Curtis (1960, 1962) which provide the best-known evidence for the biological distinctness of the grey crescent cortex. Curtis (1960) grafted small pieces of cortex from the grey crescent, animal pole, or mid-ventral margin into different regions of the cortex of a donor egg in *Xenopus laevis.* He found that the cortex from the grey crescent behaves differently compared to cortex from either the animal pole or the mid-ventral margin. Unfortunately, Curtis (1960) was unable to remove cortical pieces from the vegetal pole and so, did not compare the activity of vegetal cortex. The cortex at the vegetal pole must differ from cortex in the grey crescent at least in ease of isolation.

In summary, neither the lone biochemical study nor the cortical transplantation experiments have examined the vegetal cortex. The ultrastructural and dye permeability studies hint that the grey crescent cortex and the vegetal cortex resemble each other. There are many differences between the animal and vegetal cortices prior to grey crescent formation. These are apparent from diverse observations on sperm entry, cortical contractility, pigmentation, microvillar structure, dye permeability, thickness of the dense layer, and so on. Some difference already present in the animal and vegetal cortex when reoriented may be sufficient to explain the biological activity of the grey crescent cortex.

ACKNOWLEDGEMENT

This work was supported by a grant from the National Research Council, Canada.

REFERENCES

Ancel, P. and Vintemberger, P. (1936). *C.R. Soc. Biol.* **122**, 934-936.
Ancel, P. and Vintemberger, P. (1948). *Bul. Biol. Fr. Belg.* Suppl. **31**, 1-182.
Bluemink, J. G. (1972). *J. Ultrastruct. Res.* **41**, 95-114.
Bluemink, J. G. and Tertoolen, L.G.J. (1978). *Develop. Biol.* **62**, 334-343.
Brachet, J. (1977). *Curr. Top. Develop. Biol.* **11**, 133-186.
Buongiorno-Nardelli, M. and Bertolini, B. (1967). *Histochemie* **8**, 34-44.
Burgos, M.H. and Fawcett, D.W. (1956). *J. Biophys. Biochem. Cytol.* **2**, 223-239.
Campanella, C. (1975). *Biol. Reprod.* **12**, 439-447.
Campanella, C. and Andreuccetti, P. (1977). *Develop. Biol.* **56**, 1-10.
Cross, N.L. and Elinson, R.P. (1978). *Amer. Zool.* **18**, 642.
Curtis, A.S.G. (1960). *J. Embryol. Exp. Morphol.* **8**, 163-173.
Curtis, A.S.G. (1962). *J. Embryol. Exp. Morphol.* **10**, 410-422.
Denis-Donini, S. and Campanella, C. (1977). *Develop. Biol.* **61**, 140-152.
Denis-Donini, S., Baccetti, B. and Monroy, A. (1976). *J. Ultrastruct. Res.* **57**, 104-112.
Dollander, A. (1953). *Arch. Anat. Microsc. Morphol. Exp.* **42**, 185-193.
Dollander, A. (1956). *C.R. Soc. Biol.* **150**, 1416-1418.
Dollander, A. (1957). *C.R. Soc. Biol.* **151**, 977-979.
Dollander, A. and Melnotte, J.P. (1952). *C.R. Soc. Biol.* **146**, 1614-1616.
Eddy, E.M. and Shapiro, B.M. (1976). *J. Cell. Biol.* **71**, 35-48.
Elinson, R.P. (1975). *Develop. Biol.* **47**, 257-268.
Elinson, R.P. and Manes, M.E. (1978). *Develop. Biol.* **63**, 67-75.
Franke, W.W., Rathke, P.C., Seib, E., Trendelenburg, M.F., Osborn, M. and Weber, K. (1976). *Cytobiologie* **14**, 111-130.
Gingell, D. (1970). *J. Embryol. Exp. Morphol.* **23**, 583-609.

Ginsberg, A.S. (1971). Translated from *Ontogenez* **2**, 645-648 by Consultants Bureau (1972).
Goldenberg, M. (1979). Master's thesis, University of Toronto
Gould-Somero, M., Jaffe, L.A. and Holland, L.Z. (1979). *J. Cell Biol.* **82**, 426-440.
Grant, P. (1969). In "Biology of Amphibian Tumors" (M. Mizell, ed.) pp. 43-51. Springer-Verlag, New York.
Grant, P. and Wacaster, J.F. (1972). *Develop. Biol.* **28**, 454-471.
Greve, L.C. and Hedrick, J.L. (1978). *Gamete Res.* **1**, 13-18.
Grey, R.D. and Schertel, E.R. (1978). *J. Cell. Biol.* **79**, 164a.
Grey, R.D., Wolf, D.P., and Hedrick, J.L. (1974). *Develop. Biol.* **36**, 44-61.
Grey, R.D., Working, P.K. and Hedrick, J.L. (1976). *Develop Biol.* **54**, 52-60.
Gurdon, J.B. (1960). *J. Embryol. Exp. Morphol.* **8**, 327-340.
Hara, K. and Tydeman, P., (1979). *Roux's Arch. Develop. Biol.* **186**, 91-94.
Hara, K., Tydeman, P. and Hengst, R.T.M. (1977). *Roux's Arch. Develop. Biol.* **181**, 189-192.
Hebard, C.N. and Herold, R.C. (1967). *Exp. Cell Res.* **46**, 553-570.
Hollinger, T.G. and Schuetz, A.W. (1976). *J. Cell Biol.* **71**, 395-401.
Holtfreter, J. (1943). *J. Exp. Zool.* **93**, 251-323.
Ito, S. (1972). *Develop. Growth Differ.* **14**, 217-227.
Jaffe, L.A. (1976). *Nature* **261**, 68-71.
James, W.S. (1970). Ph.D. Thesis, University of Tennessee.
Kas'yanov, V.L., Svyatogor, G.P. and Drozdov, A.L. (1971). Translated from *Ontogenez* **2**, 507-511 by Consultants Bureau (1972).
Kemp, N.E. and Istock, N.L. (1967). *J. Cell Biol.* **34**, 111-122.
Klag, J.J. and Ubbels, G.A. (1975). *Differentiation* **3**, 15-20.
Kubota, T. (1967). *J. Embryol. Exp. Morphol.* **17**, 331-340.
Luckenbill, L.M. (1971). *Exp. Cell Res.* **66**, 263-267.
Lohka, M.J. (1978). Master's thesis, University of Toronto.
Løvtrup, S. (1958). *J. Embryol. Exp. Morphol.* **6**, 15-27.
Løvtrup, S. (1962). *J. Exp. Zool.* **151**, 79-84.
Løvtrup, S. (1965). *Roux's Arch. Entwick. Org.* **156**, 204-248.
Løvtrup, S. (1978). *Biol. Rev.* **53**, 1-41.
Maeno, T. (1959). *J. Gen. Physiol.* **43**, 139-157.
Malacinski, G.M., Allis, C.D. and Chung, H.-M. (1974). *J. Exp. Zool.* **189**, 249-254.
Malacinski, G.M., Benford, H. and Chung, H.-M. (1975). *J. Exp. Zool.* **191**, 97-110.
Malacinski, G.M., Brothers, A.J. and Chung, H.-M. (1977). *Develop. Biol.* **56**, 24-39.
Manes, M.E. and Barbieri, F.D. (1976). *Develop. Biol.* **53**, 138-141.
Manes, M.E. and Barbieri, F.D. (1977). *J. Embryol. Exp. Morphol.* **40**, 187-197.
Manes, M.E., Elinson, R.P. and Barbieri, F.D. (1978). *Roux's Arch. Develop. Biol.* **185**, 99-104.
Monroy, A. and Baccetti, B. (1975). *J. Ultrastruct. Res.* **50**, 131-142.
Newport, G. (1854). *Philos. Trans. Roy. Soc. London* **144**, 229-244.
Nicosia, S.V., Wolf, D.P. and Inoue, M. (1977). *Develop. Biol.* **57**, 56-71.
Nieuwkoop, P.D. (1973). *Adv. Morph.* **10**, 1-39.
Nieuwkoop, P.D. (1977). *Curr. Top. Develop. Biol.* **11**, 115-132.
Nishihara, T. and Hedrick, J.L. (1977). *Fed. Proc.* **36**, 811.
O'Dell, D.S., Tencer, R., Monroy, A. and Brachet, J. (1974). *Cell Different.* **3**, 193-198.
O'Rand, M.G. (1977). *J. Exp. Zool.* **202**, 267-273.
Ortolani, G. and Vanderhaeghe, F. (1965). *Rev. Suisse Zool.***72**, 652-658.
Palaček, J., Ubbels, G.A. and Rzehak, K. (1978). *J. Embryol. Exp. Morphol.* **45**, 203-214.
Pasteels, J.J. (1964). *Adv. Morphogen* **3**, 363-388.
Perry, M.M. (1975). *J. Embryol. Exp. Morphol.* **33**, 127-146.

Perry, M.M., John, H.A. and Thomas, N.S.T. (1971). *Exp. Cell. Res.* **65**, 249-253.
Picheral, B. (1977a). *J. Ultrastruct. Res.* **60**, 106-120.
Picheral, B. (1977b). *J. Ultrastruct. Res.* **60**, 181-202.
Poirier, G.R. and Spink, G.C. (1971). *J. Ultrastruct. Res.* **36**, 455-465.
Raisman, J.S. and Cabada, M.O. (1977). *Develop. Growth Differ.* **19**, 227-232.
Roux, W. (1887). *Arch. Mikrosk. Anat.* **29**, 157-212.
Rzehak, K. (1972). *Folia Biol.* (Krakow) **20**, 409-416.
Sawai, T. and Yoneda, M. (1974). *J. Cell Biol.* **60**, 1-7.
Schroeder, T.E. (1974). *J. Cell Biol.* **63**, 305a.
Schroeder, T.E. (1979). *Develop. Biol.*, **70**, 306-326.
Schroeder, T.E. and Strickland, D.L. (1974). *Exp. Cell Res.* **83**, 139-142.
Selman, G.G. and Waddington, C.H. (1955). *J. Exp. Biol.* **32**, 700-733.
Shapiro, B.M., Gabel, G.A. Foerder, C.A., Eddy, E.M. and Klebanoff, S.J. (1979) *Fed. Proc.* **38**, 465.
Steinhardt, R.A., Epel, D., Carroll, Jr., E.J. and Yanagimachi, R. (1974). *Nature* 252, 41-43.
Stenuit, K., Gueskens, M. Steinert, G. and Tencer, R. (1977). *Exp. Cell Res.* **105**, 159-168.
Tomkins, R. and Rodman, W.P. (1971). *Proc. Nat. Acad. Sci. U.S.A.* **68**, 2921-2923.
Vacquier, V.D. (1975). *Develop. Biol.* **43**, 62-74.
Wartenberg, H. (1962). *Z. Zellforsch. Mikrosk. Anat.* **58**, 427-486.
Wartenberg, H. and Schmidt, W. (1961). *Z. Zellforsch. Mikrosk. Ant.* **54**, 118-146.
Wolf, D.P. (1974). *Develop. Biol.* **40**, 102-115.
Wyrick, R.E., Nishihara, T. and Hedrick, J.L. (1974). *Proc. Nat. Acad. Sci. U.S.A.* **71**, 2067-2071.

Cytostatic Factor and Chromosome Behavior in Early Development

Yoshio Masui

Department of Zoology
Univerity of Toronto
Toronto, Ontario M5S 1A1
Canada

Peter G. Meyerhof

Department of Zoology
University of California, Berkeley
Berkeley, California 94720
U.S.A.

Margaret A. Miller

Department of Zoology
University of Toronto
Toronto, Ontario M5S 1A1
Canada

ISBN 0-12-612984-3

I. THE CONCEPT OF CYTOSTATIC FACTOR (CSF)

A. *Fertilization and Activation of Oocytes*

The fact that an egg needs to be fertilized to initiate development may be interpreted as an indication that the egg is equipped with an inhibitory mechanism which will be removed by fertilization. From this viewpoint the aim of studying fertilization is to elucidate this inhibitory mechanism and its removal. Studies of the time course of physiological parameters in an egg following insemination are therefore undertaken in order to uncover the sequence of reactions that ultimately lead to release of the egg from inhibition.

Measurements of physiological parameters of inseminated eggs show that it is the surface membrane properties that first undergo a change. Parameters such as the membrane potential and ultrastructure of the cortex are changed almost instantaneously following insemination, while activities of internal cytoplasm such as respiration and protein synthesis remain initially unchanged. Nuclear changes such as resumption of arrested meiosis and DNA synthesis are even further delayed (for reviews see Epel, 1975, 1978; Schuel, 1978). This sequence of events may reflect the pathway through which the activation signal is transmitted. The signal starts at the surface, passes through the cytoplasm and eventually reaches the nucleus or chromosomes.

The changes of surface membrane caused by fertilization are remarkably similar among eggs of different species. Depolarization of the membrane and cortical granule breakdown are found in a variety of animal eggs (for reviews see Epel, 1978; Schuel, 1978). The release of Ca^{2+} across the membrane barrier (Mazia, 1937; Ridgeway *et al.*, 1978; Zucker *et al.*, 1978) appears to be an ubiquitous signal for egg activation, since the ionophore A23187, which abolishes membrane permeability barriers to divalent cations, acts as an agent to activate eggs in various species (Steinhardt *et al.*, 1974; Chambers, 1974).

However, it should be noted that the activation signal occurring in the surface membrane, despite its common nature, has different effects on behavior of chromosomes in eggs of different animals. In eggs having chromosomes arrested at prophase of the first meiosis such as *Urechis* eggs, the activation signal induces meiotic division as well as chromosome condensation, whereas in those with chromosomes arrested at the first or second meiotic metaphase, such as insects or vertebrates respectively, it causes chromosome movement and subsequent decondensation. On the other hand, sea urchin eggs, having completed meiosis before activation, commence DNA synthesis when given the activation signal. These examples suggest that a rather nonspecific change in cell membrane properties induced by an activation signal may have a specific effect on chromosomes through its interaction with specific cytoplasmic factors that govern chromosome behavior at specific stages of meiosis.

B. *Inhibitory Factors in Unfertilized Eggs*

Several investigators have searched for the inhibitory substance underlying the specific cytoplasmic activity that can hold oocyte chromosomes from continuing meiosis or from entering mitosis until fertilization takes place. In amphibian eggs CO_2 has been suspected as such an inhibitor since eggs are reversibly inhibited in a high CO_2 milieu (Bataillon and Tchou-Su, 1930). Also it was reported that a heparin-like substance with the ability to inhibit cleavage when applied to fertilized eggs existed in ovaries as well as in spawned eggs of various animals (Heilbrunn *et al.*, 1954). Similarly, sea urchin oocytes were found to contain a polynucleotide which significantly retarded cell division of fertilized eggs (Menkin, 1959). On the other hand, comparison of the metabolism in unfertilized and fertilized eggs has suggested that the presence of enzyme inhibitors as well as excess of metabolites in unfertilized eggs may play a role in the inhibition, preventing the egg from the initiation of development (Monroy, 1965). Recent discoveries of a rise in intracellular pH following fertilization (Johnson *et al.*, 1976; Shen and Steinhardt, 1978) and the initiation of chromosome cycles following ammonia treatment, which causes a rise in intracellular pH without affecting the surface membrane (Mazia *et al.*, 1975) in sea urchin eggs have suggested that low intracellular pH is a factor responsible for the inhibition of unfertilized eggs.

As seen above the principal postulate adopted for studying the inertness of unfertilized eggs and their activation by fertilization has been the existence of inhibitory substances or factors in the unfertilized egg cytoplasm. Consequently the process evoked by activation must counteract these inhibitory factors. However, it is important to note that any cytoplasmic factor assumed to be responsible for the inhibition of the unfertilized egg must meet the folowing criteria. First, when the factor is applied to fertilized eggs, the eggs should exhibit characteristics that have been present before fertilization. Second, the factor should be found only in the unfertilized egg, and not in the fertilized egg. Third, the inhibitory factor should be inactivated under the conditions that cause activation of unfertilized eggs.

Adopting the above criteria we have investigated a cytoplasmic factor obtained from unfertilized amphibian eggs that inhibits mitosis when introduced into fertilized eggs. We have called this inhibitory factor "Cytostatic Factor" (CSF).

II. CYTOSTATIC FACTOR IN *RANA PIPIENS* OOCYTES

A. *Methodology*

To obtain eggs, the donor females, which had been kept at 4°C, were exposed to 18°C for one to two days, before they were given an injection of one to two pituitaries together with 1 mg progesterone to induce ovulation. After injection they were kept at 18°C for 40 to 72 h until eggs became mature. Some of the eggs were inseminated in a petri dish containing one macerated testis suspended in 5 ml of 10% Ringer solution. The rest were used as the donors of CSF.

Success of experiments appeared to depend primarily on the quality of the eggs used. This could be judged from fertilization of aliquot eggs, i.e. eggs from the same female. If the aliquot showed a relatively high rate of activation, but a low rate of cleavage, the rest of eggs were not used. These eggs, if used as the donor of CSF, frequently caused degeneration of recipient embryos, or if used as the recipient, they frequently also underwent degeneration. Therefore, eggs were used for experiments only if 80% or more of aliquot eggs cleaved normally following insemination.

Before subjecting eggs to the extraction procedures, they were dejellied by shaking gently in 1% Na_2HPO_4 solution containing 0.5% crude papain and 0.3% cystein HCl at room temperature. The eggs were washed with phosphate buffer followed by an extraction medium. The extraction medium routinely used consisted of 0.25M sucrose, 0.2M NaCl, 0.01M tris-maleate or phosphate buffer, pH 6.4 (Meyerhof and Masui, 1977). This medium has been used as standard extraction medium, to which substances to be tested were added.

In preparation of CSF samples it was essential to avoid homogenization of eggs in order to preserve CSF activity in the samples. To this end, eggs were packed in a centrifuge tube (13 x 50mm) containing extraction medium and crushed by applying a centrifugal force (25,000xg) for 15 min. under refrigeration. As illustrated in Fig. 1, the aqueous portion which resulted from the first centrifugation could be separated into three zones: fluid zone and upper and lower gelatinous zones. These portions were immediately collected either separately or together, and subjected to a second centrifugation at 105,000xg for 2h under refrigeration. The resultant supernatant called "extract" contained soluble egg cytoplasm and the extraction medium in variable proportions depending on the initial packing conditions of eggs. The extract used in most parts of this study contained 0.17 ml of the extraction medium and 30 mg protein per ml.(Meyerhof and Masui, 1977). It was also noted that CSF activity in extracts varied with the size of centrifuge tubes used for the second centrifugation. Therefore, in our study only small tubes (5x43mm) were used for this centrifugation.

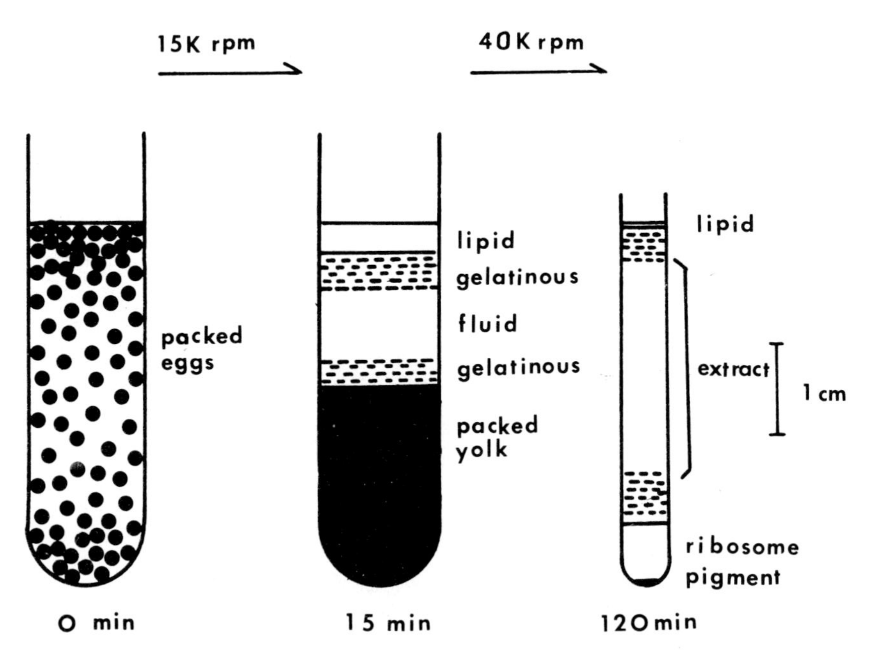

Fig. 1. CSF extraction procedure. The centrifuge used here was Beckman Model L3-4 with rotor SW-50.1.

For the assay of CSF activity a volume of extract or cytoplasm directly withdrawn from intact eggs was injected into a blastomere of 2-cell embryos using a graduated micropipette attached to a micromanipulator (Masui and Markert, 1971). When cytoplasm was assayed, donor eggs were immersed in 0.025M NaH_2PO_4 to prevent their activation (Ziegler and Masui, 1973). The injection caused cleavage arrest in the recipient blastomeres in a dose-dependent manner (Masui and Markert, 1971; Meyerhof and Masui, 1977). The larger the volume of injected cytoplasm or the higher the concentration of injected extracts, the higher the frequency with which the arrest occurred and the larger the blastomere remaining uncleaved. Therefore, the CSF activity in different samples could be compared semiquantitatively from percentages as well as from sizes of arrested recipient blastomeres.

B. *Characteristics of CSF-arrested Blastomeres*

When blastomeres were injected with cytoplasm or fresh extracts from unfertilized eggs in a sufficient amount, say 60nl, many of them were found to undergo no subsequent cleavage, but some ceased cleavage after cleaving once or twice following injection. The arrested blastomeres usually exhibited no sign of deterioration (Fig. 2) except for some showing a surface change similar to that of artificially activated eggs. Such cases were discarded from the results.

Almost all CSF-arrested blastomeres were found to contain a mitotic apparatus with metaphase chromosomes (Fig. 3). When nuclei of adult frog brains, which were isolated in a medium consisting of 0.2M sucrose, 10mM $MgCl_2$, 4mM EGTA, 0.4% bovine serum albumin and 10mM Tris-maleate buffer, pH 6.8, were injected together with the medium, they were induced to form metaphase chromosomes (Fig. 4, Meyerhof and Masui, 1979a). This ability of cytoplasm to induce chromosome condensation to the metaphase state has been described in mature oocytes in amphibians (Gurdon, 1968; Ziegler and Masui, 1973, 1976). Furthermore, the surface morphology of blastomeres was found to be altered during cleavage arrest in such a way that microvilli which were originally short were elongated (Figs. 5-7). These observations suggest a striking similarity in internal as well as external cytoplasmic characters between CSF-arrested blastomeres and unfertilized eggs.

However, a dissimilarity also existed between these two. For example, cytoplasm of CSF-arrested blastomeres sometimes contained vacuoles of various sizes filled with a lightgreen-stained colloidal substance, although these were rarely found in embryos which had formed a blastocoel or large intercellular space where the colloidal substance with similar stainability had been accumulated. These vacuoles were never found in unfertilized eggs. A difference was also found in cytoplasmic activities between the CSF-arrested blastomere and the unfertilized egg. While cytoplasm from unfertilized eggs, when injected into ovarian oocytes, could induce maturation in the oocytes, showing a high activity

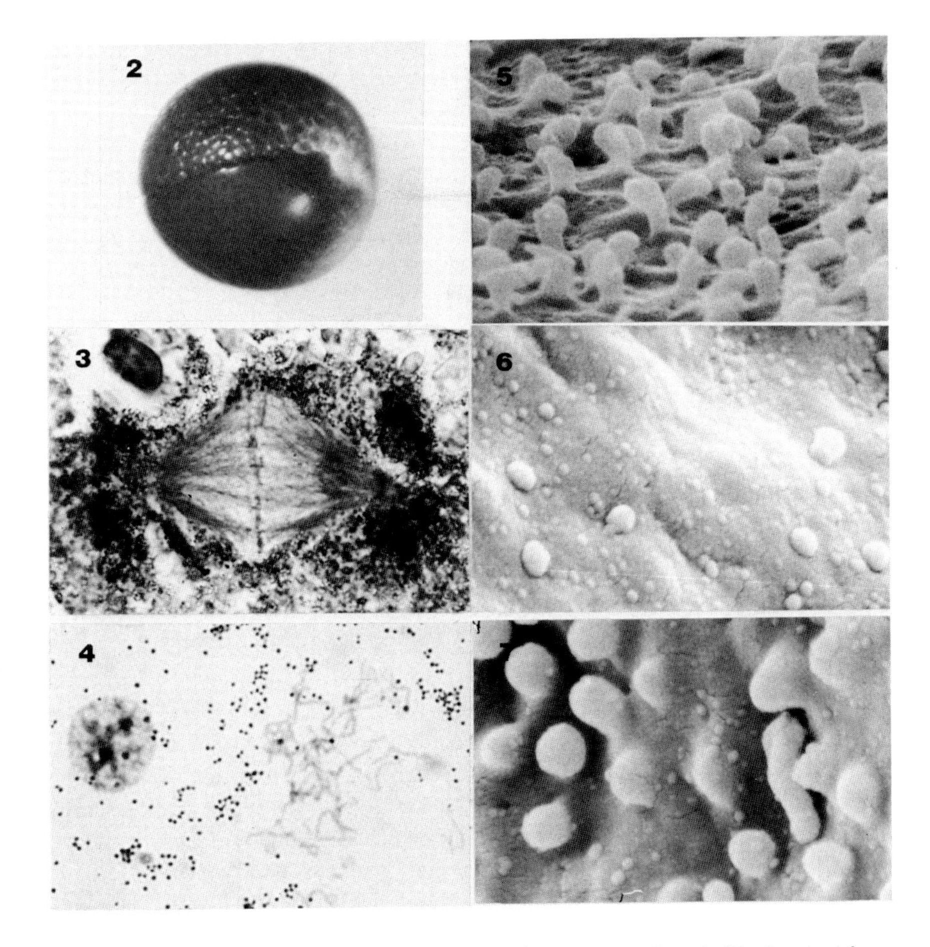

Fig. 2. An 18 h-old *Rana pipiens* embryo arrested after the injection with 60nl of fresh extract from unfertilized *Rana pipiens* eggs into a blastomere at 2-cell stage. 25x

Fig. 3. Metaphase spindle in a *Rana pipiens* embryo arrested by the injection of fresh extract from unfertilized eggs which took place 60 min after insemination. The embryo was fixed with Smith's formalin-acetic acid-bichromate, Feulgen-stained and counterstained with light green. 1000x

Fig. 4. A squash preparation of condensed chromosomes (right) and interphase nuclei (left) recovered from a CSF-arrested *Rana pipiens* blastomere. Nuclei of adult *Rana pipiens* brains were injected into the blastomere 20 h after its arrest by 1°CSF. 1200x

Fig. 5. Scanning electron micrograph of the surface of an unfertilized *Rana pipiens* egg. 30,000x

Fig. 6. Scanning electron micrograph of the surface of a cleaving blastomere of *Rana pipiens* at 2-cell stage (Meyerhof, unpublished). 30,000x

Fig. 7. Scanning electron micrograph of the surface of a CSF-arrested *Rana pipiens* blastomere at 2-cell stage (Meyerhof, unpublished) 30,000x

of maturation promoting factor (MPF), as well as CSF activity (Masui and Markert, 1971), neither CSF (Meyerhof and Masui, 1979a) nor MPF (Masui, unpublished) could be detected in CSF-arrested blastomeres. These observations suggest that CSF-arrested blastomeres are not entirely similar to unfertilized eggs with respect to their cytoplasmic activities. Therefore, apparently CSF-arrested blastomere cytoplasm has not completely reverted to the conditions previously present in unfertilized eggs.

C. *Inactivation of CSF During Egg Activation*

CSF appears in oocytes in the course of maturation after germinal vesicle breakdown (GVBD) has occurred (Masui and Markert, 1971; Meyerhof and Masui, 1979b). Oocytes having an intact germinal vesicle (GV) showed little CSF activity when their cytoplasm is injected into blastomeres. It was found that CSF activity also appeared in oocytes from which the GV had been removed if the enucleated oocytes were treated with progesterone and kept in culture for the length of time required for GVBD to take place in the normal course of maturation (Masui and Markert, 1971). Therefore, CSF is totally a product from cytoplasmic activities of maturing oocytes.

When mature oocytes were inseminated, pricked with a glass needle or stimulated by an electric shock (Masui and Markert, 1971; Masui, 1974), they lost CSF activity in the cytoplasm within 2 h following activation at 18°C. Again the loss of CSF activity was found to occur in oocytes from which the GV had been removed prior to progesterone treatment, indicating that the inactivation of CSF is a process totally controlled by cytoplasmic activities (Masui and Markert, 1971).

The signal to inactivate CSF may be propagated through the cytoplasm, since the activation stimulus given to the oocyte by pricking with a glass needle was limited to the area near the surface, and yet the loss of CSF activity occurred inside the oocyte. Therefore, it would be logical to expect that CSF, if introduced into an egg, can be inactivated when the activation signal reaches it. Indeed, unfertilized egg cytoplasm or its fresh extracts were found to lose the CSF activity to stop cleavage when introduced into an egg within the first 45 min. of insemination (Meyerhof and Masui, 1977). Furthermore, eggs at various stages of development appear to be endowed with varying capacities to inactivate CSF. As seen in the results of experiments shown in Fig. 8 (Meyerhof, 1978), in which various doses of CSF were injected into eggs at different times after insemination, CSF appears to have peaks of effect in the

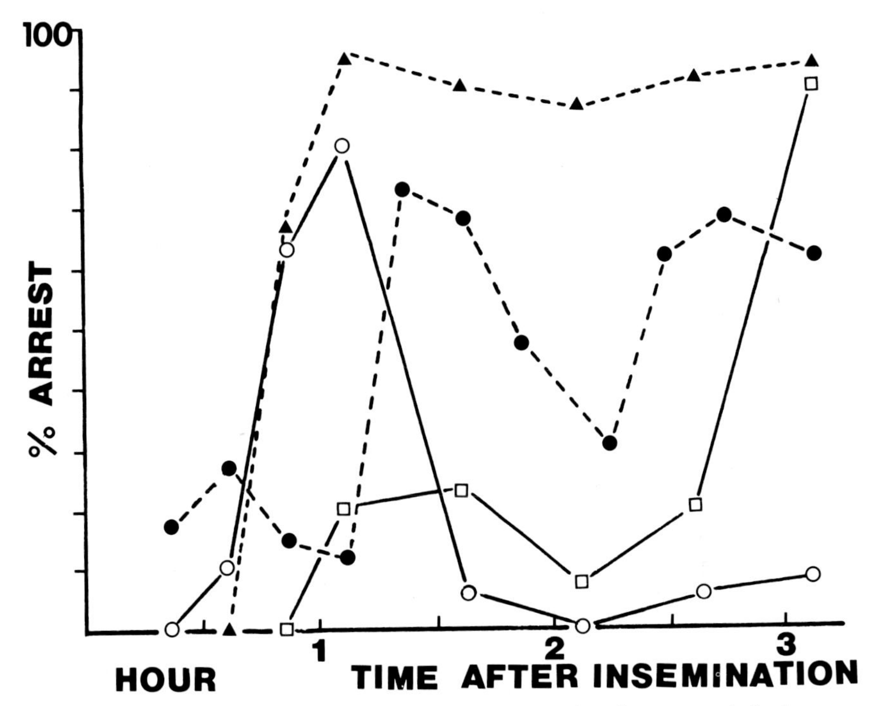

Fig. 8. Sensitivity of the *Rana pipiens* zygote during the first cell cycle to the injection of a fresh extract from unfertilized *Rana pipiens* eggs: Solid triangles — injection with 60nl extract; open circle — injection with 60nl extract; open square — injection with 30nl extract; and solid circle — injection with 60nl cytoplasm of unfertilized eggs. Extracts were made with a phosphate-buffered medium (0.25M sucrose, 0.2M NaCl, 0.01M $MgSO_4$, 0.01M NaH_2PO_4-Na_2HPO_4, pH 6.4) For different series of injections, extracts were prepared using eggs from different females.

period from 60 to 90 min and in that from 150 to 200 min. From this observation it could be conjectured that fertilized eggs have a cytoplasmic mechanism to regulate CSF activity, at least, in early periods of cleavage.

It is known that activation of *Rana pipiens* eggs can be triggered by Ca^{2+} ions (Hollinger and Schuetz, 1976), and that the level of Ca^{2+} ions in the cell is regulated during the course of the cell cycle (for review, Rebhun, 1977). Therefore, it is possible that Ca^{2+} ions play a role in regulation of CSF. When we tested the ability of Ca^{2+} ions to inactivate CSF in fresh extracts from unfertilized eggs, it was found that CSF could be inactivated within 30 min at 18°C following addition of Ca^{2+} to the extracts at concentrations as low as 10^{-5}M (Meyerhof and Masui, 1977). From these data we may assume that the inactivation of CSF in fertilized eggs is caused by an increased Ca^{2+} ion level in the cell during

TABLE I

*Effects of Divalent Cations on CSF Activity in Extracts**

Test time after extraction	Day 0				Day 1			
		Stage of Recipient at Arrest (%)				Stage of Recipient at Arrest (%)		
		1-3	4-8		1-2	4-8		
Addition	Recipient	cell	cell	no arrest	Recipient	cell	cell	no arrest
none	282	51	10	39	319	0	2	98
			A. Addition to Extraction Medium**					
$CaCl_2$	349	0	7	93	231	0	8	92
EGTA	266	93	5	2	235	75	6	19
$MgSO_4$	197	80	12	8	181	2	8	90
EDTA	102	0	1	99	73	9	0	100
			B. Addition to Final Extract***					
$CaCl_2$	259	0	10	90	303	2	12	86
EGTA	106	45	35	20	131	3	13	84
$MgSO_4$	250	47	12	41	238	1	7	92
EDTA	126	0	2	98	73	0	5	95

*Unfertilized eggs were extracted according to the procedure illustrated in Fig. 1. Extracts were injected into a blastomere of 2-cell embryos (60 nl/blastomere). Recipients were examined 18 to 24 h after injection.

**Chemicals were added to the extraction medium at a concentration of 5 mM which gave rise to a final concentration of about 1 mM in extracts.

***Chemicals were added to extracts immediately after preparation at a concentration of 2.5 mM.

fertilization. Conversely, the appearance of CSF activity in maturing oocytes as well as its maintenance in unfertilized eggs may be dependent on low levels of Ca^{2+} ions in the cytoplasm.

D. *Stability of CSF*

To test the effects of Ca^{2+} on stability of CSF *in vitro* the CSF activity was examined at various times after extraction under different ionic conditions (Table I). When EGTA was added to the extraction medium, the resulting extracts always retained CSF activity for at least 48 h after extraction, while extracts made with the standard extraction medium had lost the activity within 24 h at 0°C. However, the ability of EGTA to prolong CSF activity could not be seen if the chemical was added to extracts following the second centrifugation. Therefore, it appears that CSF is destablized almost immediately if eggs are crushed in the absence of EGTA and this initial destabilization is irreversible.

Similarly addition of Mg^{2+} to the extraction medium before crushing eggs was effective in enhancing CSF activity. However, this effect was also not observed when the ion was added to extracts after the second centrifugation. Therefore, Mg^{2+} ions, as well as EGTA, interfere with the destabilization of CSF that occurs in the first step of extraction. Conversely, removal of all endogenous divalent cations from extracts by adding EDTA either to the extraction medium or to extracts after the final centrifugation resulted in a loss of CSF activity, suggesting that continuous presence of Mg^{2+} ions is necessary for sustaining CSF activity. Thus, the destabilizing effect of Ca^{2+} on CSF may be a result of its effect to counteract Mg^{2+} action.

Although EGTA added to the extraction medium caused a delay in CSF inactivation, this delay was no longer than one day in duration even in cold storage. Therefore, mechanisms other than Ca^{2+} action must be involved in CSF inactivation. For instance, interaction of CSF with other cellular components as well as spontaneous degradation of CSF may be considered to be such mechanisms. The following experiment suggests the involvement of cytoplasmic molecules larger than CSF in its inactivation. In this experiment we separated the fluid portion of cytoplasm from the gelatinous portion immediately after eggs were crushed in the presence of ETGA, and then examined stability of CSF in both portions. It was found that CSF activity in samples from the gelatinous portion of cytoplasm was lost within 24 h of cold storage if the second centrifugation was eliminated, while samples from the fluid

TABLE II

Stability of CSF in Gelatinous and Fluid Portions of Egg Cytoplasm

Time of Injection		2 — 4 h				20 — 24 h			
			Stage at Arrest				Stage at Arrest		
Injected Material	Injected volume nl	No. of cases	1-2 cell %	4-8 cell %	No Arrest %	No. of cases	1-2 cell %	4-8 cell %	No Arrest %
gelatinous* 15K rpm	60	28	71	7	21	27	0	0	100
	120	18	72	6	22	20	10	15	75
gelatinous* 40K rpm	60	21	81	5	14	27	100	0	0
	120	25	88	0	12	17	88	6	6
fluid* 15K rpm	60	24	88	4	8	21	81	0	19
	120	14	86	7	7	21	86	0	14
fluid* 40K rpm	60	23	43	4	52	25	96	0	4
	120	29	90	0	10	23	100	0	0

*See Fig. 1. Extracts were prepared with a phosphate-buffered medium (0.25M Sucrose, 0.2M NaCl, 0.01M $MgCl_2$, 0.002 M Na_2 • EGTA, NaH_2PO_4-Na_2HPO_4, pH 6.4).

portion retained CSF activity for 48 h under the same conditions (Table II). However, all samples, either from gelatinous or fluid portion of cytoplasm, when subjected to the second centrifugation immediately after the eggs were crushed, were found to retain CSF activity for at least 48 h. These results suggest that CSF inactivation is brought about by its interaction with heavier molecules which are mainly distributed in the gelatinous portion of the cytoplasm.

Curiously, however, we observed that CSF activity could be detected in extracts which had previously lost CSF activity (Meyerhof and Masui, 1977). This secondary emergence of CSF activity occurs within several days following its disappearance regardless of ionic conditions in the extracts. We have tentatively designated the factor responsible for this secondary appearance of CSF activity "secondary CSF" (2°CSF) to distinguish it from "primary CSF" (1°CSF) which was detected in fresh extracts.

The secondary CSF was found to have the same effect on chromosome behavior as primary CSF when introduced into cleaving blastomeres. However, 2°CSF showed much higher stability than 1°CSF. The activity of 2°CSF was found to be preserved for at least a few weeks of cold storage, and it could arrest cleavage even if injected into eggs at 15 min after insemination (Meyerhof and Masui,, 1977). These intriguing observations of the appearance of 2°CSF need clarification before any significance is attached to them. However, it might be possible that 1°CSF once inactivated recovers its activity to become 2°CSF after the cellular components involved in its inactivation are degraded.

III. CYTOSTATIC FACTOR IN OTHER ANIMALS

A. *Methodology*

The existence of CSF in the oocyte of other animals has not been demonstrated until recently. Evidence to the contrary was reported in the frog, *Rana temporaria* and the sturgeon, *Acipenser stellatus* (Chulitskaia and Feulgengauer, 1977). Perhaps positive evidence for the presence of CSF is difficult to obtain, first because of its ephemeral activity and second because eggs are easily activated in some species during storage or manipulation. For example, in our experiences, *Xenopus laevis* eggs ovulated *in vivo* activate spontaneously in diluted Ringer solution. Also mouse eggs when freed from the cumulus using hyaluronidase often develop parthenogenetically (Graham, 1974). Therefore, for testing of

CSF activity in cytoplasmic samples it is essential to prevent donor eggs from activation. However, conditions required for suppressing egg activation vary among species. For example, while activation of *Rana pipiens* eggs by cortical injuries is reversibly suppressed by immersing them in an acid phosphate buffer (Ziegler and Masui, 1973), the same buffer induces egg activation in *Xenopus laevis* (Wolf, 1974). In *Xenopus* activation is suppressed in 10% Ca^{2+}-free Ringer solution, while it fails to stabilize *Rana* eggs (Lohka, 1978). Further difficulty in obtaining a reliable test for CSF arises when the cortical change observed following activation becomes an unreliable criterion for activation of internal cytoplams (de Roeper and Barry, 1976). This is particularly true when internal cytoplasm of unfertilized eggs is transferred to recipients for CSF test, since it is highly probable that mechanical agitation of the cytoplasm causes its activation without affecting the cortex. Therefore, in our experiments involving *Xenopus laevis,* operations on donor eggs for CSF were carried out in Ca^{2+}-free solution, and also donor eggs were injected with EGTA prior to withdrawing cytoplasm in order to suppress their internal activation. This method was also effective in preventing degeneration of recipient blastomeres following CSF transfer (Meyerhof and Masui, 1979b).

Experiments to obtain evidence for existence of CSF in oocytes of non-amphibian species raises more serious technical problems, since cytoplasmic transfer itself becomes formidable. Perhaps cell fusion techniques will serve us for this purpose. Recently Balakier and Czolowska (1977) performed cell fusion induced by Sendai virus between mature oocytes and blastomeres of 2-cell embryos, showing that most of the fused blastomeres were arrested at metaphase. However, the general applicability of this technique remains uncertain in view of the fact that mature mouse oocytes are frequently activated by virus-induced cell fusion (Graham, 1970; Soupart *et al.*, 1978).

As seen above, so far no direct approach to the problem of CSF in non-amphibian species has been established. Therefore, we have taken an indirect approach as the preliminary step to solve the problem. In this approach we assume that if CSF exists in an egg of a type similar to that of the amphibian, i.e. having chromosomes arrested at metaphase II, then the egg can be stabilized or destabilized against activation stimuli under ionic conditions which stabilize or destabilize amphibian CSF respectively. Thus, we have studied the effects of divalent cations on ionophore-induced activation in mouse oocytes (Masui *et al.*, 1977).

B. *CSF in Xenopus laevis Oocytes*

As seen in Table III (Meyerhof and Masui, 1979b), cytoplasm from unactivated *Xenopus* eggs induced cleavage arrest with a high frequency when injected into blastomeres of *Xenopus* 2-cell embryos. The arrested blastomeres were found to contain a single mitotic apparatus arrested at metaphase. Reciprocal transfer of cytoplasm between *Rana* and *Xenopus* eggs, however, gave different results when donor and recipient were changed. Whereas *Xenopus* egg cytoplasm injected into *Rana* 2-cell embryos caused cleavage arrest immediately, *Rana* egg cytoplasm failed to arrest cleavage in *Xenopus* embryos. The latter result may be explained by a relatively low activity of *Rana* CSF or a high resistance of *Xenopus* blastomeres to Rana CSF.

Development of CSF activity in *Xenopus* oocytes during the course of maturation was examined by injecting *Xenopus* oocyte cytoplasm into blastomeres of *Rana pipiens* 2-cell embryos. It was found that *Xenopus* oocytes exhibited CSF activity after GVBD took place, but no CSF activity was detected in *Xenopus* oocytes when they were activated by electric shock, or by phosphate buffer (Meyerhof, 1978; Meyerhof and Masui, 1979b). Similarly cytoplasm from blastomeres of 2-cell embryos

TABLE III

Interspecific Transfer of Cytoplasm from Mature Eggs into Cleaving 2-Cell Embryos in Presence and Absence of EGTA

Donor	Recipient*	EGTA** in donor (mM)	No. of cases	Stage of Arrested Embryos: 1-2 cell (%)	4-8 cell (%)	no arrest (%)
Rana	Rana	0	71	37	9	54
		5	172	94	3	3
Xenopus	Xenopus	0	17	23	12	65
		5	50	94	4	2
Xenopus	Rana	5	70	100	0	0
Rana	Xenopus	5	27	0	4	96
Rana 2-cell	Rana	0	76	0	1	99
		5	67	4	2	94
Xenopus 2-cell	Rana	0	40	0	0	100
Xenopus 2-cell	Xenopus	5	12	0	0	100

Rana pipiens recipients were injected with 60 nl of cytoplasm and were cultured for 18h. *Xenopus laevis* recipients were injected with 30 nl of cytoplasm and were cultured for 6 h.

**Donor eggs were injected with EGTA to give an internal concentration of 5 mM, which gave rise to 0.15 mM in the recipient.

did not exhibit CSF activity (Table III). These observations indicate a close similarity between the properties of CSF of two different species of amphibians.

C. *Effects of Divalent Cations on Mouse Oocyte Activation*

In this study, in order to prevent egg activation, cumuli were removed from oviducts 13 h after female mice (Swiss Webster) had been given human chorionic gonadotropin (5iu), and oocytes were freed from the cumuli by hyaluronidase treatment at 14 h. The oocytes were briefly washed in Ca^{2+}–Mg^{2+}–free Whitten medium (Whitten, 1969) containing 1mM EDTA before being subjected to treatments with ionophore A23187. Under these conditions no more than 10% of oocytes were found to be activated during 8 h culture in Whitten medium in the absence of ionophore.

When eggs were exposed to Mg^{2+}–free or CA^{2+}– and Mg^{2+}–free Whitten medium containing ionophore A23187 at a concentration of $2\mu M$ for 2h, most of the eggs were activated, resumed meiosis during the exposure to the ionophore, and reached the pronucleus stage by 6 h after transfer into normal Whitten medium (Table IV). When the oocytes were cultured longer, they synthesized DNA and began cleaving (Fig. 9), although these eggs were unable to continue cleavage. On the other hand when eggs were cultured for 8 hours but not exposed to ionophore, only a very few, 9.7% on the average, resumed meiosis. Therefore, it is clear that exposure to the ionophore in the absence of Mg^{2+} ions causes eggs to activate. In contrast, when eggs were exposed to the ionophore in media containing Mg^{2+} ions, the frequencies with which eggs were activated were very low, and similar to those found in untreated eggs (Table IV).

This remarkable effect of Mg^{2+} ions on the ionophore-induced activation led us to the following possible interpretations; the effect might have been brought about either by the Mg action to inactivate the ionophore or by that to stabilize oocytes against the action of the ionophore. In order to test the validity of these interpretations, we carried out an experiment in which oocytes were continuously treated with the ionophore at varying doses for 8 h in the presence or absence of Mg^{2+} ions. As seen in Fig. 10, eggs treated in the presence of Mg^{2+} ions were activated with significantly lower frequencies than those treated in its absence at all the ionophore doses examined. It was also found that the activation frequency was not further increased by increasing the dose of the ionophore above $1\mu M$, even though the maximum level reached in

TABLE IV

*Effects of Divalent Cations on Ionophore-Induced Activation in Mouse Eggs**

					MEIOTIC PROGRESS %				
Ionophore μM	Ca mM	Mg mM	Time after Ionophore h	Oocytes	Met. II	Ana. II	Tel. II	Pronuclei	Total Activation %
2	1.7	1.2	2	39	74	0	18	8	26
			4	51	64	0	8	28	36
			8	43	63	0	0	37	37
0	1.7	1.2	8	70	94	0	0	6	6
2	2.9	0	2	47	19	2	57	22	81
			4	48	19	0	2	79	81
			8	96	23	0	0	77	77
0	2.9	0	8	51	88	0	0	12	12
2	0	2.9	2	52	77	2	19	2	23
			4	56	69	4	7	20	31
			8	53	83	0	2	15	17
0	0	2.9	8	42	100	0	0	0	0
2	0	0	2	47	24	2	74	0	76
			4	49	15	2	6	77	85
			8	48	25	0	2	73	75
0	0	0	8	56	91	7	2	0	9

*Eggs were isolated from females 13 h post HCG, and treated with a modified Whitten medium, which contained ionophore A 23187 (2 μM) and divalent cations at the concentrations indicated, for 2 h and then cultured for 2 to 6h in Whitten medium at 37°C in a 5% CO_2 atmosphere. The eggs were rinsed with Ca- and Mg free Whitten medium containing 1 mM EDTA when transferred into different medium. Results were scored at the end of culture.

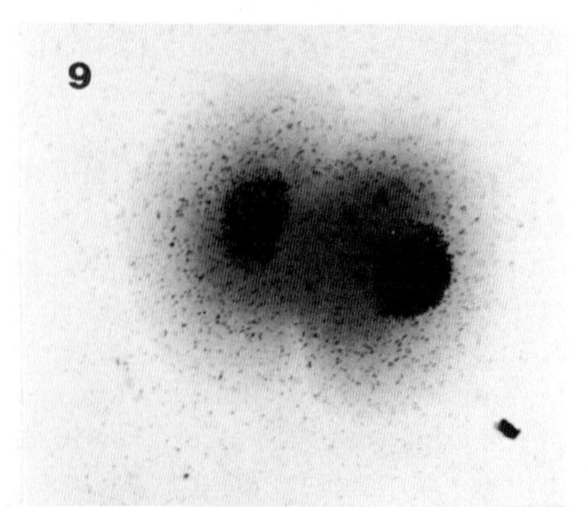

Fig. 9. Demonstration of DNA synthesis in a 2-cell mouse embryo developed parthenogenetically from eggs isolated from females 13 h post HCG and treated with Mg-free Whitten medium containing ionophore A23187 (2μM). After the ionophore treatment the embryo was cultured for 24 h in Whitten medium containing 1μC/ml H^3-thymidine at 37°C in a 5% CO_2 atmosphere, fixed in acetic acid-ethanol (1:3) mixture and dried on a slide. The specimen was thoroughly washed in cold TCA and processed for the autoradiography involving 1 week exposure to NTB2 (Kodak).

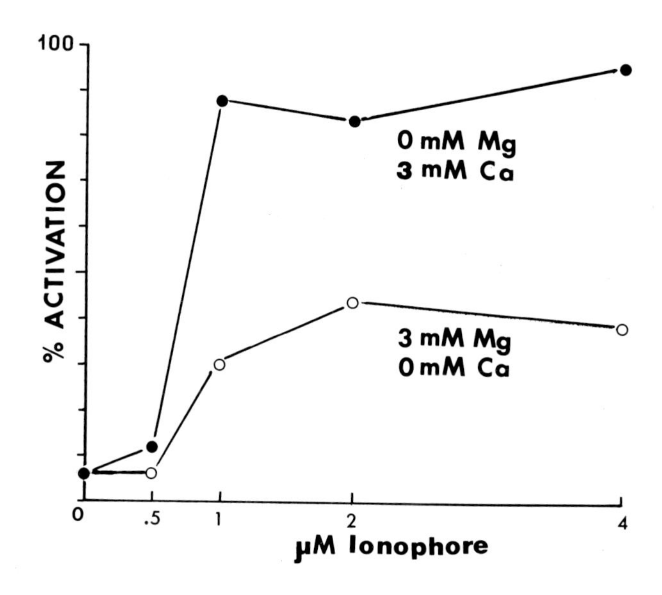

Fig. 10. Activation of mouse eggs by an exposure to ionophore A23187 at varying concentrations in the presence or absence of Mg ions. Eggs isolated from females 13 h post HCG were treated with Ca- or Mg-free Whitten medium containing the ionophore (0.4 μM) for 8 h continuously at 37°C in a 5% CO_2 atmosphere. The results were scored at the end of the treatment.

the Mg^{2+}-containing medium was still much lower than that reached in the Mg^{2+}-free medium. Therefore, it is apparent that any reduction in ionophore activity caused by presence of Mg^{2+} ions is not a limiting factor for activation, but rather that it is the Mg^{2+} action on the egg that interferes with activation. The action of Mg^{2+} ions to stabilize eggs against activation can be counteracted by Ca^{2+} ions. When eggs were exposed to the ionophore in media containing Ca^{2+} as well as Mg^{2+} at various concentrations, an antagonistic relation between these two ions became apparent (Fig. 11). Thus, the effect of Ca^{2+}, which promotes activation, can probably be attributed to its action counteracting the effect of Mg^{2+}, which stabilizes the egg. The mechanism proposed here may explain why eggs are readily activated by the ionophore in media lacking both Ca^{2+} and Mg^{2+}.

In conclusion, it may be stated that while Mg^{2+} ions enhance the stability of mouse oocytes against ionophore-induced activation, Ca^{2+} ions counteract the Mg^{2+} action and promote activation. Thus, an analogy may be seen between the Ca^{2+} and Mg^{2+} effects on amphibian

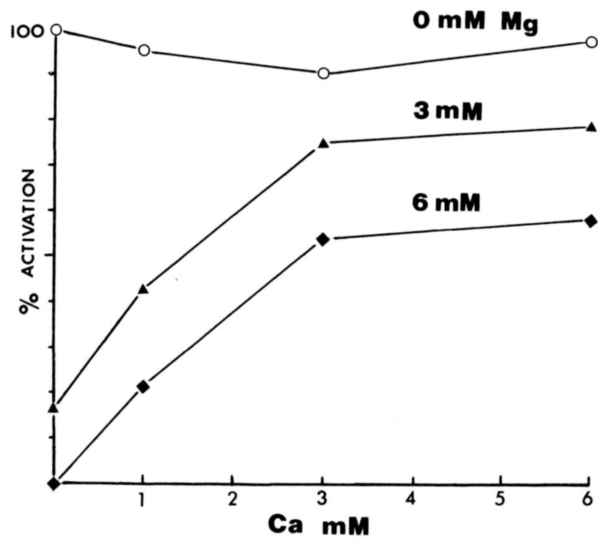

Fig. 11. Activation of mouse eggs by ionophore treatment at different levels of Mg and Ca ions. The eggs isolated from females 13 h post HCG were exposed to a modified Whitten medium, which contained ionophore A23187 (2μM) and Mg SO_4or $CaCl_2$ at various concentrations, for 2 h at 37°C and then cultured for 6 h in Whitten medium at 37°C in 5% CO_2 atmosphere. The results were scored at the end of the culture.

CSF and those on mouse eggs which have been made permeable to these ions by the action of ionophore A23187.

IV. DISCUSSION

A. *Nature of CSF*

In the foregoing sections of this paper we have presented evidence for the existence of the oocyte-cytoplasmic factor, CSF, which appears to be responsible for the meiotic arrest of the oocyte prior to fertilization. Therefore, its inactivation by insemination or by parthenogenetic activation may be considered to be a cause of the resumption of meiosis as well as of the initiation of mitosis. This CSF, which meets the criteria proposed in an earlier section in this paper has been called 1°CSF to distinguish it from other inhibitory substances.

It can be predicted that any substance which meets the criteria for CSF must be unstable in character, since its activity should be lost swiftly in response to activation signals. Therefore, it would not be surprising if the activity of our 1°CSF was unstable. It is perhaps for this reason that we have encountered difficulties in the stablization of this factor *in vitro*, and as a result its physicochemical nature is largely unknown, except that it is associated with very large molecules. This knowledge has come from the observation that CSF activity disappears from the supernatant of fresh extracts of unfertilized eggs following centrifugation beyond a certain range (4 h at 105,000 x g).

Though we know little of 1°CSF, some information concerning 2°CSF is available. This factor appears later in extracts from which 1°CSF activity has previously disappeared, and is stable. Its activity is also associated with very large molecules or molecular complexes since the activity can be eluted in the void volume from BioGel 5m and 15m (BioRad, Ltd.) columns. It was previously reported that CSF, perhaps 2°CSF because its activity was stable, could be precipitated in a 30% saturated $(NH_4)_2SO_4$ solution, and was susceptible to heat inactivation at 55°C and to RNAase treatment, but not to trypsinization (Masui, 1974). Our lack of knowledge of the relationship between 1° and 2°CSFs, however, does not allow us to use this information for further speculation on the nature of CSF *per se*.

B. *The Mode of CSF Action*

The cleavage-arresting effect of CSF on blastomeres can usually be observed in *Rana* when a blastomere of 2-cell embryos was injected with 60nl of egg cytoplasm or cytoplasmic extract from unfertilized eggs (30nl in Xenopus). The volume of cytoplasm injected represents about 5% of the total volume of the recipient blastomere, since the total volume of a single egg is about 2.5μl (1.2μl in *Xenopus*). Since, using our procedure of extraction, 2.5 ml of extract can usually be obtained from 1500 *Rana* eggs, the volume of extract injected contains only 3.4% of the extract contributed from a single egg and is probably equivalent to 7% of the corresponding portion of the cytoplasm in a single recipient blastomere. These figures clearly show that only a very small quantity of CSF, relative to the volume of the recipient cell, is sufficient for its effect.

Furthermore, since the bulk of cytoplasm in CSF-arrested blastomeres is able to induce metaphase chromosome condensation in nuclei exposed to it (Myherhof and Masui, 1979a), the action of CSF is not localized, but appears to spread throughout the cytoplasm of the recipient cell. In this instance, however, the possibility that amplification of CSF is the mechanism underlying the propagation of its effect may be ruled out, because no CSF activity could be detected in the cytoplasm of the recipient (Myerhof and Masui, 1979a). Perhaps CSF may act as a trigger to change certain conditions in the local cytoplasm into which CSF is first introduced. This initial change induced in one area of the cytoplasm may then be propagated throughout the rest of the cytoplasm. However, it is also possible that diffusion of CSF itself is responsible for changing the state of the entire cytoplasm of the recipient, in which case CSF may be too dilute to be detected by our assay method.

C. *Relationship of CSF to Other Cytoplasmic Factors*

Recently it was reported that cytoplasm of cleaving blastomeres develops maturation promoting factor (MPF) during mitosis (Wasserman and Smith, 1978). This cytoplasmic factor had previously been demonstrated by injecting mature oocyte cytoplasm into ovarian oocytes to initiate maturation of the recipient (for review, see Masui and Clarke, 1979). It was also shown that both mature oocytes and CSF-arrested blastomeres contained a cytoplasmic factor able to induce chromosome condensation to the metaphase state in interphase nuclei (Masui *et al.*, 1979); this factor has been named chromosome condensation factor (CCF). Thus, it is apparent that mature oocyte, cleaving blastomeres at metaphase, and CSF-arrested blastomeres share common cytoplasmic factors at least to some extent. The effects on chromosome behavior of these three different types of metaphase cells are similar.

However, it is important to remark the fact that the unfertilized egg or mature oocyte is the only cell among these three that possesses CSF. Thus, CSF occupies a unique position among the three. On the other hand, CCF is ubiquitous, for it appears in all three metaphasic cells, while MPF can be detected in all but CSF-arrested blastomeres.

The above observations make less likely the possibility that the three cytoplasmic factors can be represented by one molecular entity. Yet, common to all these cytoplasmic factors is the fact that they are sensitive to Ca^{2+} ions, but supported by Mg^{2+} ions, and that they appear during oocyte maturation and disappear shortly after egg activation (Masui, *et al.*, 1977).

Our experiments with *Xenopus laevis* and mouse eggs suggest the possibility that the rule we described above may be widespread. Furthermore, demonstrations of the efficacy of cytoplasmic factors across taxonomic boundaries (Brun, 1973; Reynhout and Smith, 1974; Meyerhof and Masui, 1979b), suggest the existence of species-non-specific cytoplasmic factors that regulate chromosome behavior. In this, we may find a general significance of the study of cytoplasmic factors in amphibian oocytes.

ACKNOWLEDGEMENTS

The authors thank Jean Bennett, University of California, Berkeley, for the critical-point drying and electronmicrographs of specimens, and Hugh J. Clarke, University of Toronto, for reading the manuscript. This research was supported by grants from the National Cancer Institute of Canada and the National Research Council of Canada. The scanning electron microscope was provided by the National Science Foundation grant GB-38359 to the Electron Microscope Laboratory of the University of California, Berkeley.

REFERENCES

Balakier, H. and Czolowska, R. (1977). *Exp. Cell Res.* **110**, 466-469.
Bataillon, E. and Tchou-Su. (1930). *Arch. Biol.* **40**, 441-553.
Brun, R. (1973). *Nature New Biology* **243**, 26-27
Chambers, E. L. (1974). *Biol. Bull.* **147**, 471, (abstract).
Chulitskaia, E. V. and Feulgengauer, P. E. (1977). *Ontogenez* **8**, 305-307.
de Roeper, A. and Barry, J. M. (1976). *Exp. Cell Res.* **100**, 411-415.
Epel, D. (1975). *Amer. Zool.* **15**, 507-522
Epel, D. (1978). *Curr. Top. Develop. Biol.* **12**, 186-246.
Graham, C. (1970). The Fusion of Cells with One-and-Two-Cell Mouse Embryos. *In* "Heterospecific Genome Interaction". (V. Defendi, ed.) Wistar Inst. Symp. Monogr. **9**, 20-35.
Graham, C. (1974). *Biol. Rev.* **49**, 399-422.
Gurdon, J. B. (1968). *J. Embryol. Exp. Morph.* **20**, 401-414.
Heilbrunn, L. V., Chaet, A. B., Dunn, A. and Wilson, W. L. (1954). *Biol. Bull.* **106**, 158-168.
Hollinger, T. G. and Schuetz, A. W. (1976). *J. Cell Biol.* **71**, 395-401.
Johnson, J. d., Epel, D. and Paul, M. R. (1976). *Nature* **262**, 661-664.
Lohka, M. J. (1978). A Study of the Ionic Control of Egg Activation in *Rana pipiens.* M. Sc. Thesis, University of Toronto.
Masui, Y. (1974). *J. Exp. Zool.* **187**, 141-147.
Masui, Y. and Clarke, H. J. (1979). *Int. Rev. Cytol.* **57**, 185-282.
Masui, Y. and Markert, C. L. (1971). *J. Exp. Zool.* **177**, 129-146.
Masui, Y., Meyerhof, P. G., Miller, M. A. and Wasserman, W. J. (1977). *Differentiation.* **9**, 49-57.
Masui, Y., Meyerhof, P. G. and Ziegler, D. H. (1979). *J. Ster. Biochem.* **11**, 715-722.
Mazia, D. (1937). *J. Cell Comp. Physiol.* **10**, 291-304.
Mazia, D., Shatten, G. and Steinhardet, R. A. (1975). *Proc. Nat. Acad. Sci. U.S.A.* **72**, 4469-4473.
Menkin, V. (1959). *J. Exp. Zool.*, **140**, 441-170.

Meyerhof, P. G. (1978). Studies on Cytoplasmic Factors from Amphibian Eggs which Cause Metaphase and Cleavage Arrest. Ph.D. Thesis, University of Toronto.

Meyerhof, P. G. and Masui, Y. (1977). *Develop. Biol.* **61**, 214-229.

Meyerhof, P. G. and Masui, Y. (1979a). *Exp. Cell Res.*, **123**, 345.

Meyerhof, P. G. and Masui, Y. (1979b). *Develop. Biol.*, **72**, 182-187.

Monroy, A. (1965). Biochemical Aspects of Fertilization. *In* "The Biochemistry of Animal Development" (R. Weber, ed.) Vol. I pp 73-135. Academic Press, New York.

Rebhun, L. I. (1977). *Int. Rev. Cytol.* **49**, 1-54

Reynhout, J. K, and Smith, L. D. (1974). *Develop. Biol.* **38**, 394-400.

Ridgeway, E. G., Gilkey, J. C. and Jaffe, L. E. (1977). *Proc. Nat. Acad. Sci. U.S.A.* **74**, 623-627.

Schuel, H. (1978). *Gamete Res.* **1**, 299-382.

Shen, S. S. and Steinhardt, R. A. (1978). *Nature* **272**, 253-254.

Soupart, P., Anderson, M. L. and Repp, J. E. (1978). *Theriogenology* **10**, 102. (Abstract).

Steinhardt, R. A., Epel, D., Carroll, E. J. and Yanagimachi, R. (1974). *Nature* **252**, 41-43.

Wasserman, W. J. and Smith, L. D. (1978). *J. Cell. Biol.* **78**, R15-22.

Whitten, W. K. (1969). The Effect of Oxygen on Cleavage of Mouse Eggs *In Vitro. Proc. 2nd Ann. Symp. Soc. Study of Repro.* p. 29.

Wolf, D. P. (1974). *Develop. Biol.* **40**, 102-115.

Ziegler, D. H. and Masui, Y. (1973). *Develop. Biol.* **35**, 283-292.

Ziegler, D. H. and Masui, Y. (1976). *J. Cell Biol.* **68**, 620-628.

Zucker, R. S., Steinhardt, R. A. and Winkler, M. M. (1978). *Develop. Biol.* **65**, 285-295.

III. The Cell Surface in Normal and Abnormal Development

Cell Configuration, Substratum and Growth Control

Judah Folkman

Department of Surgery
Children's Hospital Medical Center
Harvard Medical School
Boston, Massachusetts 02115

Robert W. Tucker

Department of Medical Oncology and
Cell Growth and Regulation
Sidney Farber Cancer Institute
Harvard Medical School
Boston, Massachusetts 02115

I. INTRODUCTION

It has long been assumed that the major operational difference between normal cells and neoplastic cells is that neoplastic cells grow more rapidly. There are now so many exceptions to this idea that it is no

ISBN 0-12-612984-3

longer valid. A more fruitful conceptual framework may be reached by focusing attention on the phenomenon of crowding. Neoplastic cells seem able to continue their growth at levels of crowding that would shut off the growth of normal cells. In cell culture, multiplication of normal cells is inhibited at high cell densities whereas the proliferation of neoplastic cells is not. This phenomenon has been termed "density dependent inhibition" of cell growth by Stoker and Rubin (1967). Ponten (1975) has used the term "crowding index". The mechanism of density dependent inhibition of cell growth remains as a central question in cell biology. Most investigators have studied diffusible or humoral factors that might mediate this growth control in normal cells. For example, Holley (1975) has proposed a humoral hypothesis which suggests that high cell density limits the availability of medium components such as growth factors present in the serum.

However, it has only recently been understood that cells (unlike bacteria) respond also to spatial information about their crowdedness. This spatial information seems to be transmitted to them during close packing of cells by changes in shape or configuration of the cell. Shape change is then transmitted to the cytoplasm and the nucleus to act as a regulator of growth.

The important role of cell shape in growth control became apparent during an investigation of "anchorage dependence". Most primary fibroblasts can proliferate if anchored to glass or plastic, but do not grow when suspended in the medium (Stoker *et al.*, 1968). While fibroblasts suspended in the medium are spheroidal, those attached to the substratum are well spread or flat. If small glass fibrils are dispersed in the medium, floating fibroblasts will attach to the fibrils and will proliferate if the fibrils are long enough (Maroudas *et al.*, 1973). However, if the fibrils are less than 20 microns long, proliferation ceases despite the fact that the cells are still anchored. On the long fibrils cells are stretched out, while on the short fibrils, cells are foreshortened. A third observation is that while many cells do not adhere to bacteriologic plastic, those that do, proliferate very slowly or not at all. Again, cells that attach to bacteriologic plastic are not well spread and are more rounded than cells attached to tissue culture plastic.

Based upon these observations, we proposed a hypothesis that appropriate cell shape is critical for DNA synthesis by normal cells; as cells become round, DNA synthesis is shut down (Folkman and Greenspan, 1975; Folkman, 1976, 1977).

More recently we have reported quantitative evidence in support of this hypothesis (Folkman and Moscona, 1978). This quantitative data has been generated from a new method that permits cell shape to be precisely varied so that cells in culture can be held at any one of a graded series of quantitative cell shapes. We have shown that the adhesiveness of plastic tissue culture dishes can be permanently reduced in a graded manner by applying increasing concentrations of poly(2-hydroxyethyl methacrylate) (poly(HEMA)).

II. CELL SHAPE OF SPARSE CELLS IS CONTROLLED BY SUBSTRATUM ADHESIVITY

When an alcoholic solution of poly(HEMA) is pipetted into plastic culture dishes at a constant volume of 95 $\mu l/cm^2$, a thin, hard, sterile film of optically clear polymer remains tightly bonded to the plastic surface after the alcohol is evaporated at 37°C. A stock solution of poly(HEMA)* (12 percent) is made by dissolving 6 grams of the powder in 50 ml of 95% ethanol, and turning the mixture slowly overnight at 37°C in a 100 cc bottle on a roller apparatus, to completely dissolve the polymer. This results in a clear, viscous solution which is then centrifuged for 30 minutes at 2500 rpm to remove the remaining, barely visible undissolved particles. This stock is then diluted with 95% alcohol to make dilutions from 10^{-1} (i.e., 1.00 ml poly(HEMA) + 9.0 ml ethanol) to 10^{-4}. If the 2.1 cm^2 wells of Falcon 3008 culture plates are used, each microwell is filled with 200 μl of the alcoholic solution. For 35 mm Falcon plates (No. 3001), 917 μl are used per plate. The plates are dried in a 37°C warmroom on a level bench, free of vibrations for 48 hours with the lids in place. A brief exposure to ultraviolet light before use does not harm the poly(HEMA) film, and can be used to sterilize the plates. After drying, these films varied between 35 and 0.0035 μm thickness, depending on the original dilution used. The higher the concentration, the thicker the film, and the less the adhesivity of the plate. Thus, plastic coated with 10^{-1} poly(HEMA) was the least adhesive and cells plated on it were spheroidal. By contrast, 10^{-4} poly (HEMA) was the most adhesive and cells on it were well spread, almost like cells on uncoated plastic.

In our original work (Folkman and Moscona, 1978), cell shape was quantitated by measuring two dimensions. The longest cell diameter was measured at 200X with a Nikon Profile projector. The other dimension, the height of the cell, was measured to $\pm$ 0.1 μm by focusing on the body

*Poly(HEMA) is obtained as "Hydron" from Hydron Laboratories, 783 Jersey Avenue, New Brunswick, New Jersey 08902. Poly(HEMA) from other sources is less soluble and may not be as pure.

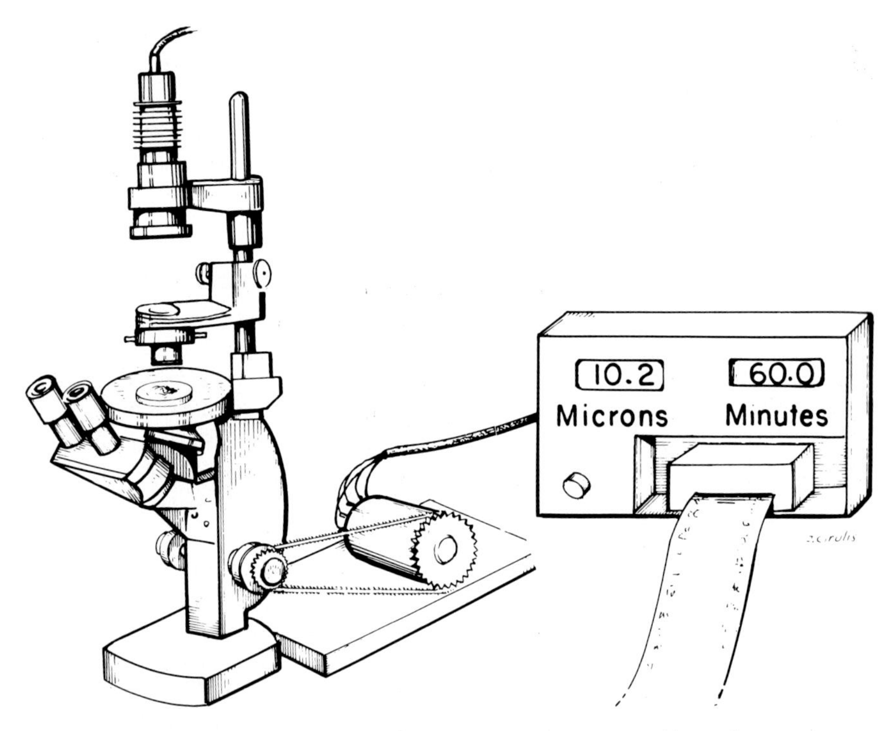

Fig. 1. The micrometer of an inverted Nikon Phase Microscope is connected by a pulley to an angular transducer which is fed to a digital read-out device, calibrated to read directly in microns. The system is accurate to ± 0.1 μm.

of each cell with an inverted Nikon Phase microscope to which the micrometer was connected by a pulley to an angular transducer, coupled to a digital read-out device. (Fig. 1). After cells are plated on a substratum, they flatten out in proportion to the adhesivity of the substratum. Once equilibrium is reached, the mean shape does not change (over a 7-day observation). Endothelial cells flatten the most rapidly after plating, i.e., equilibrium is reached after about one hour (Fig. 2); by contrast, chondrocytes take the longest time to reach shape equilibrium on plastic, i.e., 2½ days.

Recently, we have found it more efficient to measure cell area. This can be done rapidly by transferring the image of a field of cells from an inverted phase microscope to a small television screen. Cell outlines are then traced on transparent paper or plastic. The cell outlines are then converted to area (μm^2) by tracing them on a MOP-3 Zeiss Digital Image Analyzer.

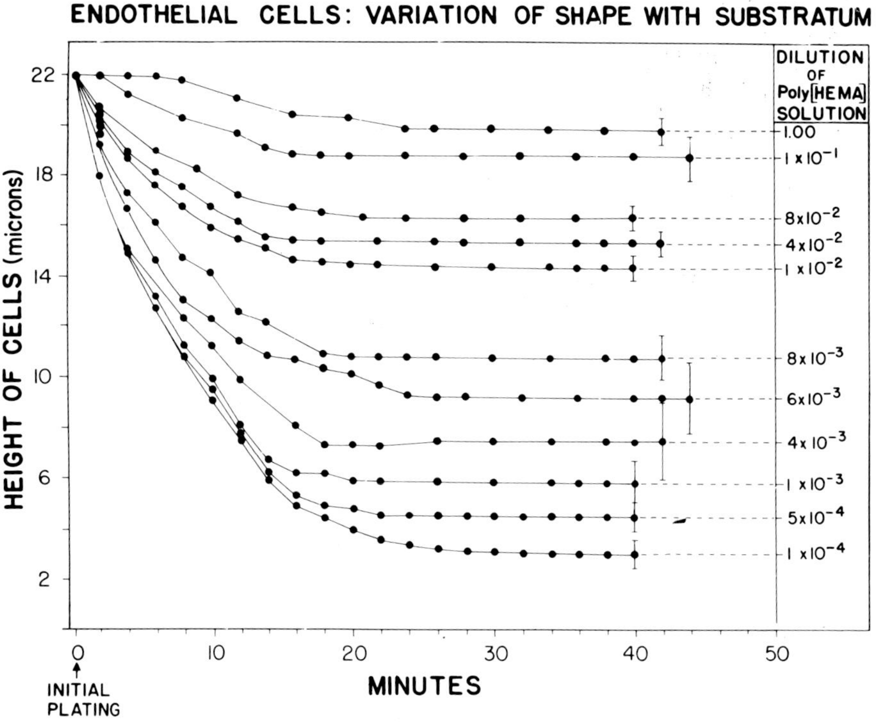

ENDOTHELIAL CELLS: VARIATION OF SHAPE WITH SUBSTRATUM

Fig. 2. The time required for flattening of cells after plating poly(HEMA) substrata of various adhesivities. Focal depth measurements were made on bovine aortic endothelial cells from time zero after plating. Cells that remained completely round, or just barely attached, had a mean focal depth measurement of 22 μm above the surface of the plate. By contrast, cells that were completely flattened (1 X 10^{-4} poly(HEMA)) reached approximately 3 μm. 1 X 10^{-4} poly(HEMA) was equivalent to tissue culture plastic. Each point represents the mean height of 5 cells measured to ± 0.1 μm by focusing on the body of the cell. The zero point is calibrated by focusing on a fine scratch made on the poly(HEMA) surface with a cotton swab. (Folkman and Moscona, 1978).

III. DNA SYNTHESIS VARIES WITH CELL SHAPE

As cells became elongated and more spread, increasing numbers of them incorporated ^{3}H-thymidine. Conversely as cells became more foreshortened and more rounded on substrata of decreasing adhesivity, DNA synthesis was diminished or stopped completely. This was true for fibroblasts, endothelial cells and chondrocytes. The cell lines included WI-38 (human cells), human chondrocytes, human skin fibroblasts, bovine aortic endothelial cells, mouse A-31 cells and mouse 10T½ cells. In

Fig. 3, this relationship is demonstrated for aortic endothelial cells. Cell shape is quantitated by measuring the mean cell height. The flattest cell focuses at 3 μm above the surface while the most rounded cell is focused at a mean height of 20 μm. In Fig. 4, mean cell area is used to quantitate the cell shape of 10T½ cells. In all of these experiments, the cells were subconfluent or sparse so that they did not contact each other. These cells, thus, were not crowded. Their shape was controlled by adhesivity of the substratum.

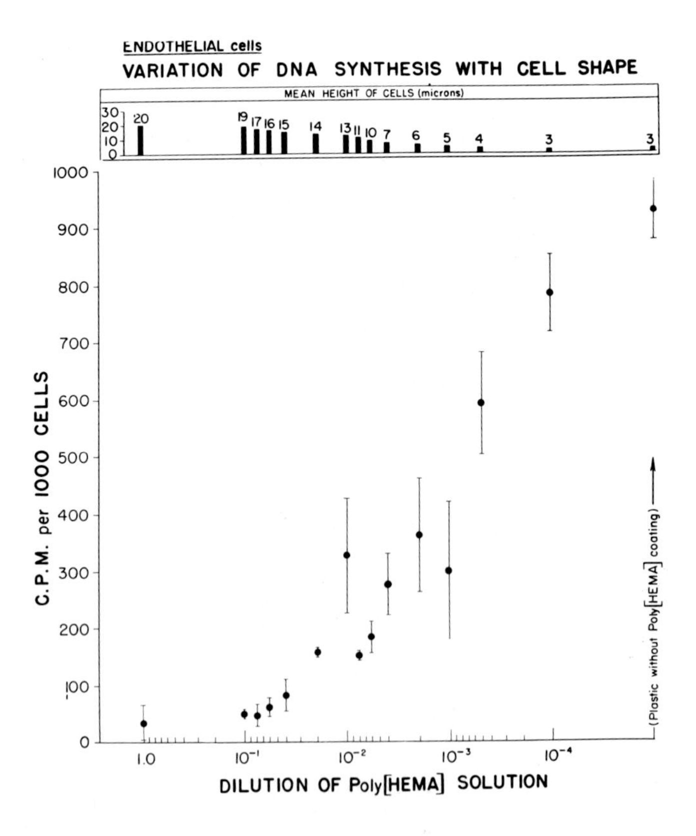

Fig. 3. Variation of DNA Synthesis with Cell Shape in Endothelial Cells. Aortic endothelial cells were plated on poly(HEMA) substrata and plastic as in the previous figure, at 10,000 cells/cm², in a Falcon No. 3008 culture plate. Four wells were used for each poly(HEMA) dilution. Cell height was measured in the first well. Cells in the second well were counted at the end of the experiment (48 hours). Cells in the third and fourth well were exposed to ^{3}H-thymidine (1 μCi/ml for 42 hours) and CPM/1,000 cells was determined on precipiated DNA. (Folkman and Moscona, 1978).

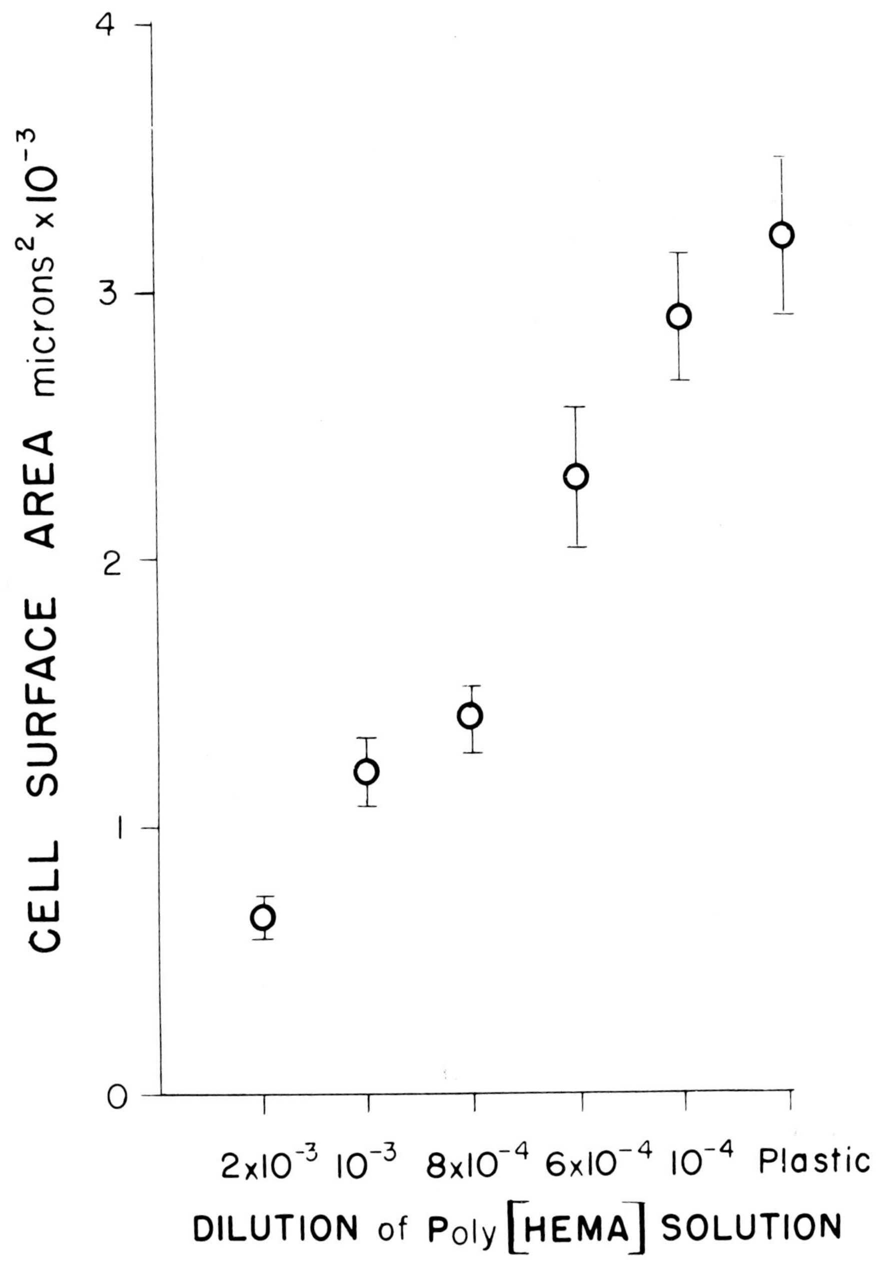

Fig. 4a. Area of Spreading of Cells onSubstrata of Varying Adhesivity. Mouse 10T½ cells were plated at 10^3 cells/cm^2. Each point represents the mean area for 100 cells, 24 hours after plating. The error bars represent standard error of the means. (Brouty-Boye *et al.*, 1979).

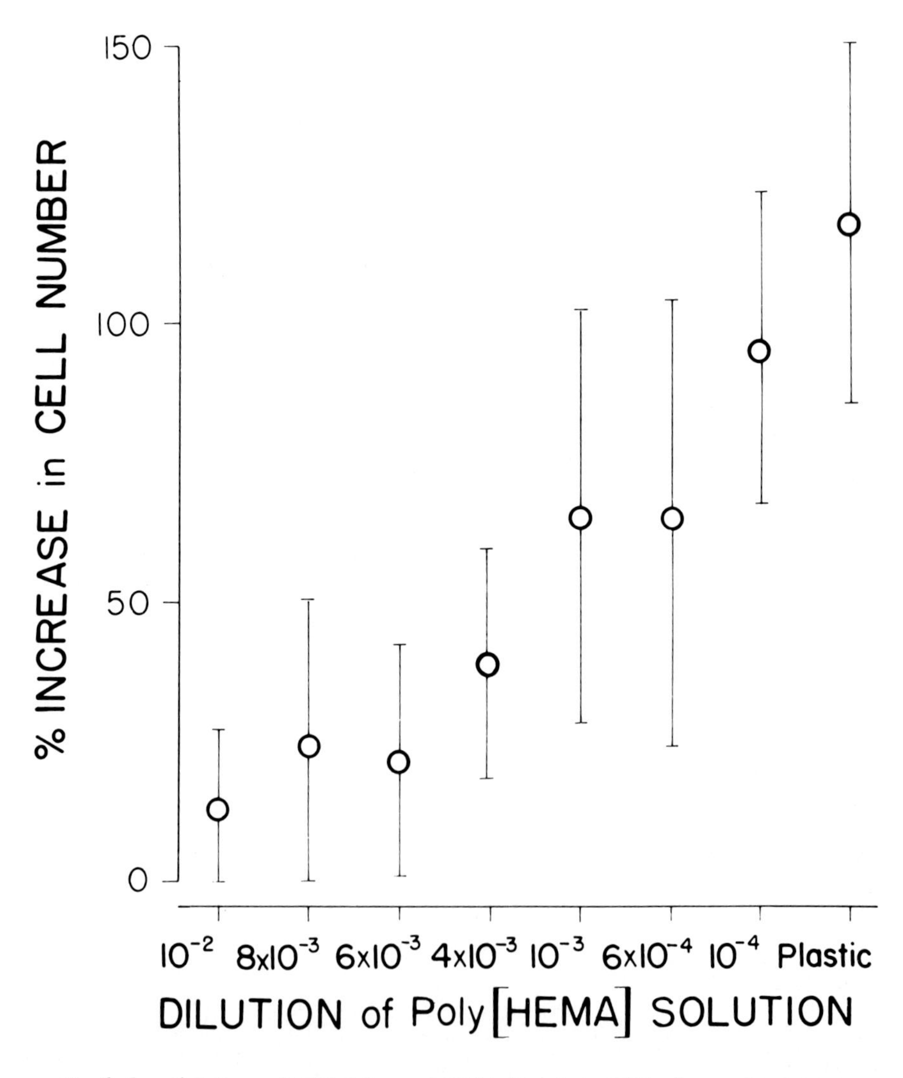

Fig. 4b. Growth Patterns of 10T½ Cells on poly(HEMA) Substrata. 10T½ cells were plated at the same concentrations as for Fig. 4a. Adherent cells were counted in a Coulter counter after trypsinization on the first and second days after plating. Cell growth was expressed as the percent increase in the cell number in 24 hours:

$$\left(\frac{\text{cell number day 2} - \text{cell number day 1}}{\text{cell number day 1}}\right) \times 100$$

These non-transformed mouse cells increase their growth rate in sparse culture as they are permitted to increase their spreading on substrata of greater adhesivity. These cells behave similarly to the endothelial cells described in Fig. 3. (Brouty-Boye *et al.*, 1979).

IV. CROWDING OF CELLS ALSO CHANGES CELL SHAPE

As cells grow to confluence on a highly adhesive substrate such as tissue culture plastic, each cell becomes more cuboidal or rounded due to close cell packing. Spreading area is markedly reduced as cell density increases (Zetterberg and Auer, 1970). Also, in our own experiments as cells grew to confluence, they came out of focus. This indicated that gradual lateral compression of a cell by its neighbors forced it to increase its height. The most confluent cells were the tallest, and also incorporated the least amount of ^{3}H-thymidine. When WI-38 cells were plated sparsely on poly(HEMA) and held at the same height as their confluent counterparts on plastic, the level of incorporation of ^{3}H-thymidine for the two populations was nearly identical (Fig. 5). In other

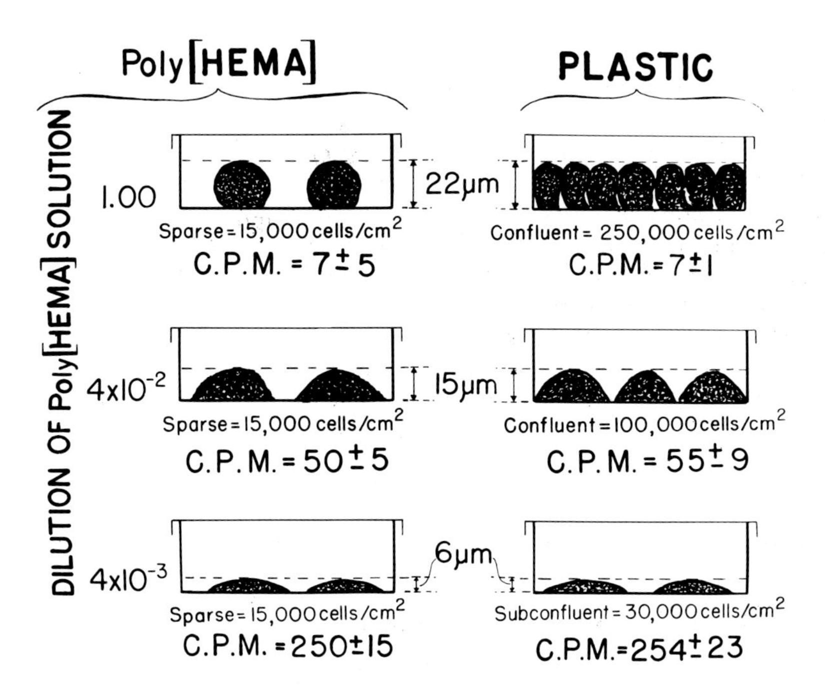

Fig. 5. Effect of Crowding on Cell Shape and DNA Synthesis. WI-38 cells were plated at the densities indicated on uncoated plastic (Falcon No. 3008) and also on a series of poly(HEMA) substrata of graded thickness. Cell height, number and ^{3}H-thymidine incorporation (CPM) were measured as in Fig. 3. When poly(HEMA) plates of different adhesivity containing sparsely plated cells were selected for cells whose height most closely matched the height of cells on plastic at various densities, the incorporation of ^{3}H-thymidine was more equivalent for these pairs of plates than for any other combination of plates. Similar results were obtained with endothelial cells.

words, sparse cells held at rounded shapes approached levels of DNA synthesis that closely matched that of confluent cells. On the poly(HEMA), cell shape was driven toward the spheroidal configuration by a reduction in substrate adhesiveness. The cells were not in contact with each other. By contrast, the cells on plastic were on a maximally adhesive substratum, but could not flatten because they were being crowded by their neighbors. The important point is that the resulting reduction in DNA synthesis was similar in both situations. The results were similar when this experiment was repeated with endothelial cells. The implication is that the mechanism of inhibition of cell growth at high cell density is partly mediated by cell shape, (i.e., restricted cell spreading). Thus, density dependent inhibition of growth may result from the close packing of cells by their neighbors, producing cell shapes that do not permit DNA synthesis.

This idea was further corroborated by a "wounding" experiment. When wounds are made in a confluent monolayer of cultured cells, the surviving cells at the wound edge and those in several rows behind the edge, show an increased incorporation of ^{3}H-thymidine (Dulbecco and Stoker, 1970). After the cells have closed the gap, DNA synthesis drops to the low level of the previous confluent layer. When we made a wound in the confluent monolayer of mouse A-31 cells on plastic, focal depth measurements initiated immediately after wounding showed that the first row of cells at the wound edge began to flatten within 30-60 minutes, as they gradually spread out onto the exposed substratum (Fig. 6). Within one hour the first row of cells had flattened by 54% of their original height, the next row by 42%, and the third row by 30%. Beyond the fifth row, cells were still at the same height as the remainder of the confluent layer. This explains why the most spread cells at the very edge

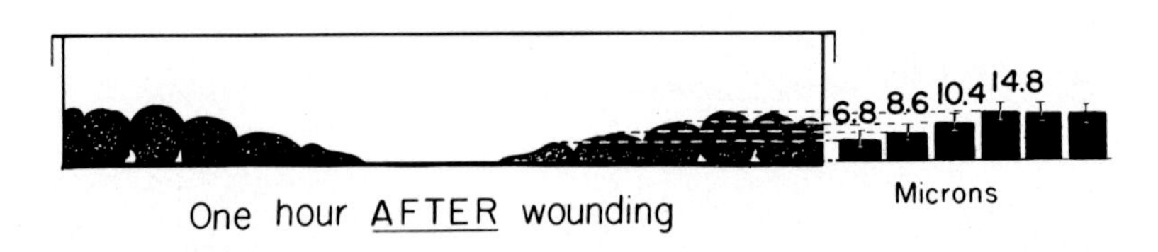

Fig. 6. Change in Cell Shape after Wounding of Confluent Monolayer. BALB/3T3 (A-31) cells (early passage) were allowed to grow to early confluence on tissue culture plastic (Falcon 3001), on which a fine scratch had previously been made for cell height calibration. The height of cells over the whole plate before wounding was 14.8 ± 0.5 μm. One hour after the monolayer had been wounded with a glass knife, cell heights were measured again. Cells at the wound edge had spread out and flattened down to 6.8 μm. Cells in the next interior row had also begun to flatten, but to a lesser extent and without loss of contact. (Folkman and Moscona, 1978).

of the wound would have the highest proliferation rate, the next interior row slightly less, the next interior row slightly less, until finally, about the fourth or fifth row the confluent cells would remain quiescent, unaffected by the wound. Here, again, the visco-elastic properties of cells permit shape changes that result from close packing during crowding, or unpacking of the cells when new space is provided for spreading. Furthermore, this phenomenon explains why colonies of cells on plastic may demonstrate inhibition of DNA synthesis in their centers.

V. CELL SHAPE MODULATES SERUM STIMULATION OF GROWTH

Since cell shape and DNA synthesis are so tightly coupled, what then is the influence of serum growth factors? It is well known that the addition of fresh serum, or of purified growth factors (Mierzejeweski and Rozengurt, 1977), or the increased velocity of medium flow (Stoker, 1973) can sometimes overcome the restriction of growth imposed by high cell density. Recent experiments (Tucker and Folkman, 1979a) suggest that for non-transformed cells, cell shape determines the sensitivity of the cells to serum or other growth factors. Thus, higher concentrations of growth factors are required for cells that move toward spheroidal configuration. By contrast, very low concentrations of serum are required for the same level of growth in flat cells. Balb/c 3T3 cells were held at a variety of different shapes by sparse plating on plastic dishes coated with different concentrations of poly(HEMA). The serum concentration of the media was varied from 0.5% through 30%, and total cell counts were carried out after 72 hours. Cells that were well spread on highly adhesive plastic substrata increased their cell number as serum concentration was increased (Fig. 7). As cells were foreshortened on a substrate of lower adhesivity, serum had less and less stimulatory effect. Finally at a completely round configuration, even 30% serum could not significantly stimulate growth.

This data explains previous reports in the literature that cells in suspension have a very high serum requirement for growth (Paul *et al.*, 1974). This also explains why cells grown in suspension but attached to beads, would have a reduced serum requirement for the same level of growth. Furthermore, these results may explain why cells at high density on plastic become refractory to growth stimulation by platelet growth factors (Vogel *et al.*, 1979), and also how other growth factors interact with cells in culture. For a given growth factor, there may be an optimal cell configuration for maximal stimulation by the factor (Gospodarowicz *et al.*, 1978).

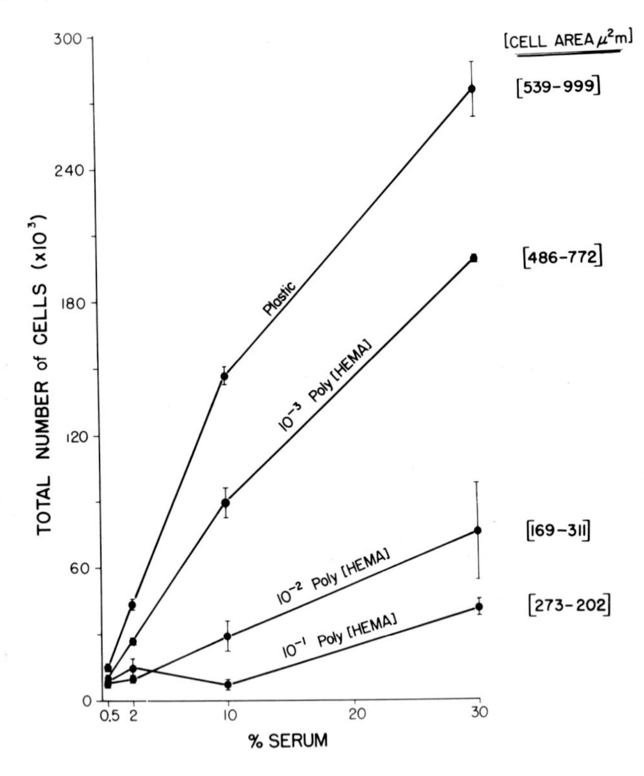

Fig. 7. Effect of Cell Shape on Serum Stimulation of Growth. 25,000 BALB/c 3T3 cells were plated on Falcon dishes (35 mm diameter), previously coated with various concentrations of poly(HEMA). Serum concentration was varied from 0.5% through 30%. For each point, the areas of 100 cells were measured. The range of areas for each substratum was plotted. After 72 hours, another cell count was carried out for each dish. Cells are sparsely plated. As spreading is restricted by the substrata of decreasing adhesivity, the growth response to serum is also diminished. (Tucker and Folkman, 1979).

VI. METABOLIC CHANGES ARE ASSOCIATED WITH CELL SHAPE CHANGES

A central question is, "How would information from changes in cell configuration be transmitted to the nucleus?". By what mechanism can spatial information be transduced to growth regulatory information? While nothing is known about this, Penman and his colleagues have made a first approach (Benecke *et al.*, 1978). They showed that when 3T6 cells that had previously been flat, were removed from anchorage and allowed to float in suspension culture there was a rapid suppression of mRNA appearance in the cytoplasm followed by a slow, but extensive decline in protein synthesis (Fig. 8). Furthermore, when the cells were allowed to touch down to a plastic substratum and begin spreading, protein synthesis recovered rapidly and there was a much slower recovery in message production (Fig. 9).

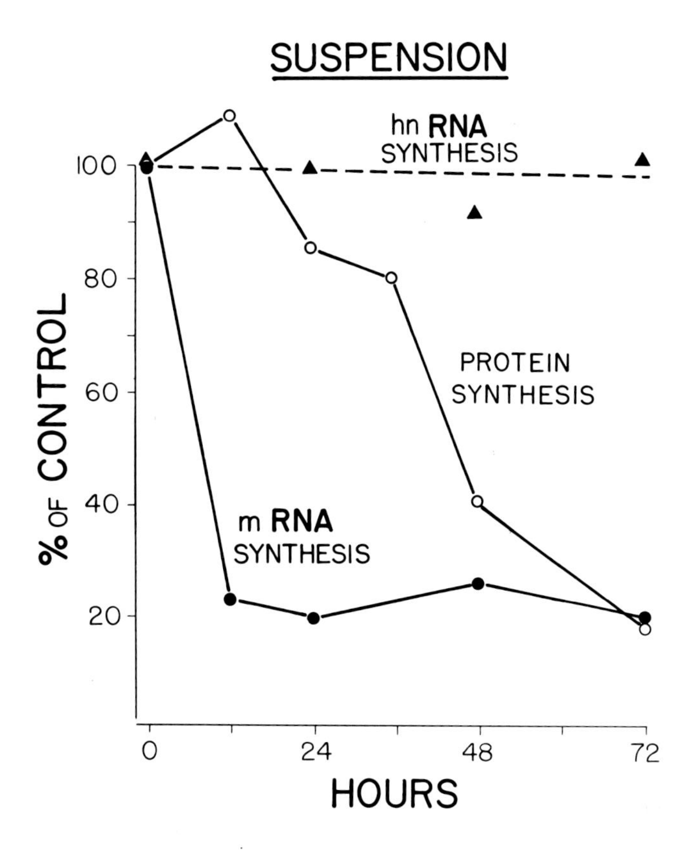

Fig. 8. 3T6 cells were suspended in methocel medium. At the times indicated, 50 ml of suspended cells were harvested and re-suspended in 9.0 ml of methocel-free medium. 6 ml of the suspension was treated with actinomycin D, and then labelled with ^{3}H-uridine for 60 minutes. The nuclear RNA as well as the cytoplasmic poly(A) RNA were analyzed on an SDS-sucrose gradient and centrifuged for 14 hours at 30,000 rpm. 3 ml of the cell suspension were labelled with ^{35}S-methionine for 30 minutes. Cells were then precipitated with 10% TCA and collected on Watman GF/A filters. The normalization of the different time points was based on counting an aliquot of each nuclear fraction in a phase contrast microscope. (Benecke *et al.*, 1978).

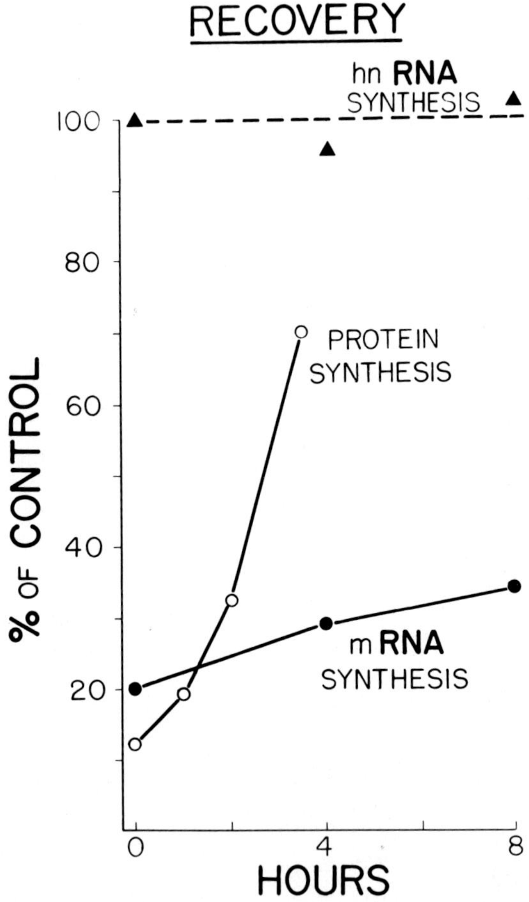

Fig. 9. The same procedure was used as in Fig. 8, except that at 72 hours, suspended cells were removed from methocel medium and transferred to tissue culture dishes. At the times indicated, the rates of hnRNA, mRNA, and protein synthesis were determined as described. (Benecke *et al.*, 1978).

VII. TRANSFORMED AND NEOPLASTIC CELLS PROLIFERATE IRRESPECTIVE OF CELL SHAPE.

It is well known that some transformed cells can proliferate without anchorage (Shin *et al.*, 1975). Furthermore, Tucker and Folkman (1979b) have found that when transformed cells are changed toward the spheroidal configuration by decreasing substratum adhesiveness, they continue to proliferate at high rates in contrast to their non-transformed counterparts which stop proliferating at round configurations. In other words, transformed and neoplastic cells have in some way escaped the subtle controls that are dictated by cell shape. A hallmark of malignancy is continual proliferation despite cell crowding. Therfore, the capacity of a neoplastic cell to overcome the effects of close cell packing and thus to continue DNA synthesis, may be an essential component of the malignant state. Furthermore, for certain cells, crowding may be necessary to maintain the neoplastic phenotype (Brouty-Boye *et al.*, 1979).

VIII. DISCUSSION

Having advanced the argument that cell shape plays an important role in growth control, we wish to caution that "cell shape" is a misleading term. We use it for want of a better word, and because it is the only parameter that can be measured so far. While measurements of cell shape appear tightly coupled to growth control, it should not be misconstrued that shape itself has anything to do with the mechanism of its effect. It is more likely that some as yet unknown event associated with change in cell shape brings about the appropriate biochemical change that regulates growth. For example, one highly improbable speculation is that a very small change in tension of certain parts of the cytoskeleton might lead to large changes in the rate of protein synthesis. Until there is a better conceptual framework of how the cell actually interprets shape information, we shall have to be satisfied with measuring cell shape.

The mechanism by which poly(HEMA) reduces the adhesivity of the polystyrene in tissue culture plastic is not entirely known. Poly(HEMA) is a hydrophilic hydrogel of neutral charge. It may act by reducing the negative electrostatic charge of the polystyrene, but other mechanisms are tenable. For example, it is possible that tiny spicules of plastic, below the resolution of scanning electron microscopy, protrude as multiple contact points to which cells stick. Increasing concentrations of poly(HEMA) would then act to decrease the number of available contacts for the cells. The important conclusion is that whatever the mechanism by which poly(HEMA) reduces adhesivity of polystyrene, cells respond by a change in shape.

These experiments suggest that several different phenomena of cell culture may be linked because they are all mediated by cell shape. Thus, "anchorage dependence", "density dependent inhibition of growth", and the response to "wounding", may be explained by the changes which they bring about in cell shape. For example, as a culture of cells becomes confluent, the pressure generated by each mitosis is attenuated throughout the monolayer. This pressure would be sufficient to force changes in cell shape. In this way, each nucleus in a large carpet of cells would receive information about the level of crowdedness.

Another puzzling phenomenon is now also explainable. It has been commonly observed in tissue culture, that if certain late-passage fibroblasts capable of forming 3-4 layers in culture, are fed ^{3}H-thymidine, autoradiographs of cross-sections show bottom cells incorporating ^{3}H-thymidine compared to the upper layer cells which are not. One would expect the opposite, i.e., that the cells near the media would be proliferating faster than cells on the bottom of the pile. However, cells on the bottom are flat and attached to plastic. Cells on the top layer are more rounded because they lie on the dorsal surfaces of underlying cells. Cells on the top layer are analogous to cells on bacteriologic plastic, or on poly(HEMA) of low adhesivity.

Finally, there is an interesting link between cell shape and tumor angiogenesis. Because tumor cells seem to be able to escape the controls of cell shape on growth, they reach cell densities and a level of crowdedness not attainable by normal cells. The cost of such high cell density appears in the greatly increased demand for oxygen and nutrients and for a means of diffusing away wastes. Once neoplastic cells reach this crowded state, those that will form growing tumors appear to have a mechanism for inducing the host to supply new capillaries above and beyond the normal capillary density, i.e., tumor angiogenesis. The importance of angiogenesis capacity for tumor cells to surmount the diffusion problems of crowding has previously been reported (Folkman and Cotran, 1976).

ACKNOWLEDGEMENTS

Supported by grant CA14019-04 from the National Cancer Institute and a grant to Harvard University from the Monsanto Company.

REFERENCES

Benecke, B-J, Ben-Ze'ev, A. and Penman, S. (1978). *Cell* **14**, 931-939.

Brouty-Boye, D., Tucker, R. and Folkman, J. (1979). submitted.

Dulbecco, R. and Stoker, M.G.P. (1970). *Proc. Nat. Acad. Sci. U.S.A.* **66**, 204-210.
Folkman, J. (1976). *In* "Advances in Pathobiology, Cancer Biology II, Etiology and Therapy" (C. Fenoglio and D.W. King, eds.), Vol. 4, pp. 12-28. Stratten Intercontinental Press, New York.
Folkman, J. (1976). *Sci. Amer.* **234**, 58-73.
Folkman, J. (1977). *In* "Recent Advances in Cancer Research: Cell Biology, Molecular Biology, and Tumor Virology" (R.C. Gallo, ed.), Vol. 1, pp. 119-130. CRC Press, Cleveland, Ohio.
Folkman, J. and Cotran, R. (1976). *In* "International Review of Experimental Pathology" (G.W. Richter and M.A. Epstein, eds.), Vol. 16, pp. 207-248. Academic Press, New York.
Folkman, J. and Greenspan, H.P. (1975). *Biochim. Biophys. Acta* **417**, 211-236.
Folkman, J. And Moscona, A.A. (1978). *Nature* **273**, 345-349.
Gospodarowicz, D., Greenberg, G. and Birdwell, C.R. (1978). *Cancer Res.* **38**, 4155-4158.
Holley, R.W. (1975). *Nature* **258**, 487-490.
Maroudas, N.G., O'Neill, C. and Stanton, M.F. (1973). *The Lancet* **i**, 807-809.
Mierzejewski, K. and Rozengurt, E. (1977). *Nature* **269**, 155-156.
Paul, D., Henahan, M. and Walter, S. (1974). *J. Nat. Cancer Inst.* **53**, 1499-1503.
Ponten, J. (1975). *In* "Human Tumor Cells in Vitro" (G. Fogh, ed.), Chapter 7, pp. 175-206. Plenum Press, New York.
Shin, S., Freedman, V., Risser, R. and Pollack, R. (1975). *Proc. Nat. Acad. Sci. U.S.A.* **72**, 4435-4439.
Stoker, M.G.P. (1973). *Nature* **246**, 200-203.
Stoker, M.G.P., O'Neill, C., Berryman, S. and Waxman, V. (1968). *Int. J. Cancer* **3**, 683-693.
Stoker, M.G.P. and Rubin, H. (1967). *Nature* **215**, 171-172.
Tucker, R.W. and Folkman, J. (1979a)., unpublished data.
Tucker, R.W. and Folkman, J. (1979b)., unpublished data.
Vogel, A., Ross, R. and Raines, E. (1979). *J. Cell Biol.,* in press.
Zetterberg, A. and Auer, G. (1970). *Exp. Cell Res.* **62**, 262-270.

Lectins and Their Saccharide Receptors as Determinants of Specificity in the *Rhizobium*-Legume Symbiosis[1]

Frank B. Dazzo

Department of Microbiology and Public Health
Michigan State University
East Lansing, Michigan 48824

I. INTRODUCTION

Nitrogen fixation is the process of converting N_2 gas into ammonia, a biologically available form. Root nodules of various legumes constitute major N_2-fixing sites (Fred *et al.*, 1932). Members of the genus of bacteria called *Rhizobium* grow in these nodules in symbiosis with the legume and are the source of the N_2-fixing enzyme, nitrogenase. Many legumes are of agricultural and economic importance as food crops (e.g., soybeans, beans, peas, peanuts) and as forage (e.g., alfalfa and clover). N_2 fixed within the root nodule is made readily available to the plant as an inexpensive, renewable, and nonpolluting resource. Approximately 3.3 billion dollars worth of nitrogen is fixed annually by legumes in the United States (Wittwer, 1977).

[1]This is Michigan Agricultural Experiment Station Journal Article No. 9064

ISBN 0-12-612984-3

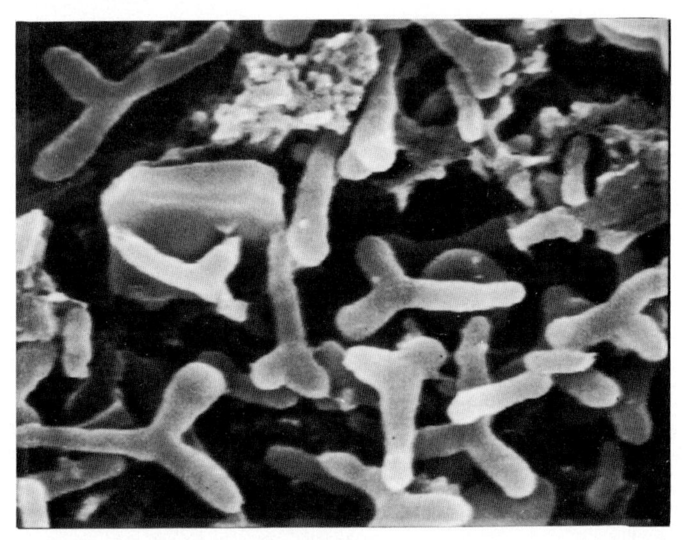

Fig. 1. The nitrogen-fixing symbiont of peas, *Rhizobium leguminosarum*, is released from the infection thread in the root nodule. Vegetative bacteria are transformed into the characteristic "Y" shaped bacteroids as shown in this scanning electron micrograph taken by Estelle Hrabak in the author's laboratory.

The infection process consists of several essential events which precede active N_2-fixation in the *Rhizobium*–legume symbiosis. These include mutual host–symbiont recognition of the *Rhizobium* species on the legume rhizoplane, specific rhizobial adherence to differentiated epidermal cells called root hairs, root hair curling, root hair infection, root nodulation, and transformation of vegetative bacteria into enlarged, pleomorphic bacteroids which fix N_2 (Fig. 1). Many steps of this infection process have become more clearly understood since the development of the slide culture technique of Fahraeus (1957), and its use in the production of an elegant time-lapse cinema at the light microscope level (Nutman *et al.*, 1973).

The symbiosis is characterized by a high degree of host specificity. For instance, *R. trifolii* infects and nodulates white clover roots but not alfalfa or soybean roots. Host specificity is expressed at a very early step of infection of the root hair by infective rhizobia prior to the penetration of the root hair cell wall (Napoli and Hubbell, 1975) and formation of the root hair infection thread (Li and Hubbell, 1969). This structure is a cellulosic tube which carries the rhizobia into the root and releases the bacteria within nodule cells.

The intent of this article is to draw attention to the interaction of legume lectins (carbohydrate-binding proteins) with their specific saccharide receptors on the bacterium and the plant, and the possible importance of this interaction in the expression of host specificity in the *Rhizobium*-legume symbiosis. All of the data which have been accumulated on the *Rhizobium*-clover symbiosis suggest that the lectin receptors on the bacterium and the plant host are chemically related saccharide structures which play an essential role in the development of a complementary macromolecular interface which initiates the selective adherence of the bacterium to the root hairs of the host. Other recent reviews on this rapidly developing subject are available (Bauer, 1977; Broughton, 1978; Dazzo, 1979a,b,c,d; Sequeira, 1978).

II. THE LECTIN-RECOGNITION HYPOTHESIS

Several investigators are studying the possible role of lectins in the specific recognition between legumes and rhizobia. The lectin-recognition hypothesis states that recognition at infection sites involves the binding of specific legume lectins to unique carbohydrates found exclusively on the surface of the appropriate rhizobial symbiont.

A. *The Soybean–Rhizobium japonicum Symbiosis*

Bohlool and Schmidt (1974) first suggested the role of lectins as determinants of host specificity by demonstrating that a fluorescein isothiocyanate (FITC)-labelled soybean lectin preparation only bound to strains of *R. japonicum* capable of nodulating soybean. Bauer and co-workers (Bauer, 1977; Bhuvaneswari *et al.*, 1977) have reconfirmed these observations with highly purified FITC- and tritium (^3H)-labelled soybean lectin, and have shown that the lectin-binding to *R. japonicum* is reversible by N-acetyl-galactosamine, a specific inhibitor of binding by this soybean lectin. Wolpert and Albersheim (1976) reported similar lectin-binding specificities with rhizobial lipopolysaccharide (LPS) preparations although they did not determine if the lectin-binding was reversible with the respective sugar inhibitors.

Bohlool and Schmidt (1976) and Tsien and Schmidt (1977) have found a distinctive polarity to exponentially growing cells of some *R. japonicum* strains. They found that an extracellular polysaccharide material which was fibrillar in nature accumulated near one end of the cell. One of the cell poles binds soybean lectin, and they proposed that cell polarity might explain the striking polar end-on attachment of some strains of *Rhizobium* to roots and other surfaces.

Bauer *et al.* (Bauer, 1977; Bhuvaneswari *et al., 1977)* found that the affinity of soybean lectin for most *R. japonicum* strains was biphasic, suggesting multiple receptors. Bal *et al.* (1978) and Calbert *et al.* (1978) showed that encapsulated cells of *R. japonicum* strains selectively bind soybean lectin. A rapid decline of soybean lectin in developing soybean seedling roots has been described (Pueppke *et al.,* 1978).

However, Pull *et al.* (1978) have suggested that the N-acetylgalactosamine-specific, 120,000-dalton soybean seed lectin (SBL) is probably not required for the initiation of the soybean-*Rhizobium japonicum* symbiosis. They identified five lines of soybean which lack the N-acetyl galactosamine specific 120,000-dalton soybean lectin in their seeds, yet their roots were still nodulated by *Rhizobium.* However, this study did not determine whether SBL was present in the roots and, more specifically, on infective sites at the soybean root hair surface. This determination is most critical since it is the root hair differentiated from the epidermis which acquires the cell surface components capable of recognizing *Rhizobium* (discussed below).

There are other questions to be asked of these data which have been taken as evidence countering the lectin-recognition hypothesis. No experiments were performed to determine if those lines of soybean lacking seed lectin have indeed retained or lost their selectivity for *R. japonicum.* Do the bacteria in the nodulation studies gain entry into old roots by a different route (e.g., through void spaces created by epithelial desquamation or lateral root emergence) which may bypass the specific infection of root hairs as is normally observed? Could soybean lectin be induced in roots in the presence of the rhizobial symbiont? Could a structural mutation in these soybean lines lead to a more labile protein in the plant which normally retains its ability to bind *Rhizobium,* but is inactivated during extraction and the alkaline conditions (pH 9) of FITC-coupling? Another critical question is whether the 120,000-dalton lectin is *the* one of several lectins in this angiosperm which is required for the infection of soybean roots by *Rhizobium.*

B. *The Tropical Legume–Rhizobium Symbioses*

The major exception to the general rule of restricted host range in *Rhizobium*-legume combinations is associated with the tropical legumes. These legumes are considered the primitive ancestors to the temperate legumes (Norris, 1959), and are nodulated by the promiscuous strains of *Rhizobium* described as the "cowpea miscellany."

The mechanism of rhizobial infection of several tropical legumes (e.g., peanut, *Arachis hypogaea,* and joint vetch, *Aeschynomene americana)* is distinctly different from the root hair infection of temperate legumes (Allen and Allen, 1940; Napoli *et al.*, 1975b; Chandler, 1978). Nodules on many tropical legumes form only where lateral roots emerge. Root hairs also develop in these areas but they lack infection threads. Ultrastructural studies show that intrusion of the rhizobia into roots of *Aeschynomene* occurs through voids in the epidermis created by desquamation or lateral root emergence. In *Arachis,* the rhizobia enter the root at the junction of the root hair and the epidermal and cortical cells (Chandler, 1978).

This condition suggests that recognition processes governing specificity in these more primitive legumes are less specific and have broad boundaries. Consistent with this hypothesis is the observation that the lectin, Concanavalin A (ConA), from the tropical legume *Canavalia ensiformis* (jackbean), can bind in a biochemically specific manner (α-methyl-mannoside inhibitable) with surface determinants produced by many rhizobia, regardless of their ability to nodulate jackbean (Dazzo and Hubbel, 1975b; Chen and Phillips, 1976; and Kamberger, 1979). Why Wolpert and Albersheim (1976) failed to detect this promiscuous ConA-binding to rhizobia remains open to question.

There are other reports of anomalous interactions of lectins with rhizobia which do not correspond with their nodulation properties (Brethauer and Paxton, 1977; Broughton, 1978; Chen and Phillips, 1976; and Law and Strijdom, 1977). A plausible explanation might be that these legumes may contain several lectins of different carbohydrate specificity (Kauss and Glaser, 1974), and therefore lectins having possible functions other than *Rhizobium* recognition may have been the ones examined. Many investigators have considered only "classical" lectins which agglutinate erythrocytes, and therefore, may have overlooked hitherto unrecognized legume proteins which may recognize unusual saccharides on *Rhizobium* which are not present on erythrocytes. Other possible explanations for the anomalous binding reactions include the use of excessively high seed lectin concentrations (up to 13 mg/ml of solution) without due attention to levels which are more physiological, improper bacterial growth conditions, or use of media containing yeast extract which affects the apparent composition of *Rhizobium* polysaccharides (Humphrey *et al.*, 1974). Sugar-inhibited controls to rule out nonspecific binding of lectin were not included in some studies. Many of the anomalous negative reactions could be due to the curious, transient appearance of the lectin receptors on the symbiont rhizobia during normal culture growth (Bauer, 1977; Dazzo *et al.*, 1979, discussed below).

C. *The Clover-Rhizobium trifolii Symbiosis*

We are testing the lectin-recognition hypothesis in *Rhizobium trifolii*-clover combinations because the sequence of events in the infection process in this symbiosis has been described (Napoli and Hubbell, 1975). Our approach has been to combine ultrastructural, biochemical, immunochemical, genetic and quantitative light and fluorescent microscopic techniques to examine roots directly for legume proteins which bind specifically at very low concentration to unique surface receptors of *Rhizobium* without being biased towards only considering "classical" lectins that agglutinate erythrocytes. (In fact, white clover phytohemagglutins were unknown at the time.) Our basic working hypothesis is that host specificity involves an interaction of complementary macromolecules on the surfaces of both symbionts. Since these macromolecules are antigenic, we considered it feasible to initiate studies which could identify components of the host-symbiont interface by taking advantage of the techniques of immunochemistry. The first clue which suggested that specific macromolecules may be involved in infection came from serological studies which showed antigenic differences between infective and related noninfective strains of *R. trifolii* using hyperimmune antisera obtained by a persistent immunization schedule (Dazzo and Hubbell, 1975a). More detailed studies demonstrated that the surfaces of infective *R. trifolii* and clover epidermal cells contain a unique antigen which is immunochemically cross-reactive (Dazzo and Hubbell, 1975c). The immunological cross-reactivity suggests their structural relatedness. This antigen is referred to as the cross-reactive antigen (CRA). The CRA contains receptors which bind to a lectin called trifoliin, which has recently been purified from clover seeds and seedling roots and partially characterized (Dazzo *et al.*, 1978).

Quantitation of trifoliin activity was based on its ability to agglutinate a suspension of *R. trifolii.* An early effort was made to identify critical variables which influenced the quantitation of activity. The culmination of that effort resulted in over a 100-fold increase in sensitivity of the quantitative agglutination assay as compared with procedures described in the original report (Dazzo and Hubbell, 1975c). An important variable, as with the *R. japonicum*-soybean lectin interaction (Bauer, 1977), was found to be the age of the culture of cells (Dazzo *et al.*, 1979).

Trifoliin can be extracted from ground clover seeds in phosphate buffered saline (PBS) containing sodium ascorbate and insoluble polyvinylpyrrolidone (PVP), and then purified by a combination of gel filtration, ultrafiltration, and ion exchange chromatography (Dazzo *et al.*, 1978). A 310-fold purification is accomplished by these procedures, as measured by an increase in specific agglutinating activity (agglutinating units/mg protein). Purity of the trifoliin was indicated by the elution of a symmetrical peak of 280 nm absorbance from a Sephadex G200 column, the presence of a single sharp band in 6% and 8% polyacrylamide gels stained with Coomassie blue R-250 for protein and stained with periodate-Schiff's reagent for carbohydrate, and single immunoprecipitin bands in Ouchterlony immunodiffusion gels with trifoliin and antiserum prepared against crude seed extract or against purified trifoliin. Interestingly, an 11-fold increase in the total activity over the crude seed extract was obtained during purification and removal of low molecular weight compounds which presumably interfere with the agglutination reaction. Trifoliin, at concentrations as low as 0.1–0.2 μg protein/ml, specifically agglutinates the symbiont of clover, *R. trifolii.* This purified clover protein was composed of subunits with molecular weight in SDS-gels of approximately 50,000 daltons, an isoelectric point of 7.3, and contained approximately 6 μmole reducing sugar/mg protein. Trifoliin aggregates near its isoelectric point to form spherical particles of 10 nm diameter (interestingly, aggregates of soybean lectin are of the same size; Bauer, 1977). Trifoliin can be eluted from intact clover seedling roots in the presence of PBS containing 2-deoxyglucose, sodium ascorbate, and PVP. The protein could then be concentrated by ultrafiltration and purified to homogeneity by preparative slab gel electrophoresis. Trifoliin from clover seedling roots shared the following properties with trifoliin from seeds: same electrophoretic mobility in native and SDS-polyacrylamide gels, immunochemical reaction of identity using antiserum to trifoliin from seeds and specificity of agglutination towards *R. trifolii* (inhibited by 2-deoxyglucose.).

Dazzo and Hubbell (1975b) proposed a model to explain an early recognition event occurring on the clover root hair surface prior to infection. According to this hypothesis, trifoliin recognizes similar saccharide residues on *R. trifolii* and clover and cross-bridges them in a complementary fashion to form the correct molecular interfacial structure that initiates the preferential and specific adsorption of the bacteria to the root hair surface (Figs. 2 and 3).

The following is the experimental evidence to support this model, although it must be emphasized that the final proof has not been established. The CRA is only present in *Rhizobium* strains capable of infecting white clover (Dazzo and Hubbell, 1975c), and only these strains are agglutinated by trifoliin (Dazzo *et al.,* 1978). Clover root hairs preferentially adsorb infective *R. trifolii* (Dazzo *et al.,* 1976) and its CRA (Dazzo and Brill, 1977). The receptor sites on clover roots which immediately bind both *R. trifolii* and its CRA match the distribution of

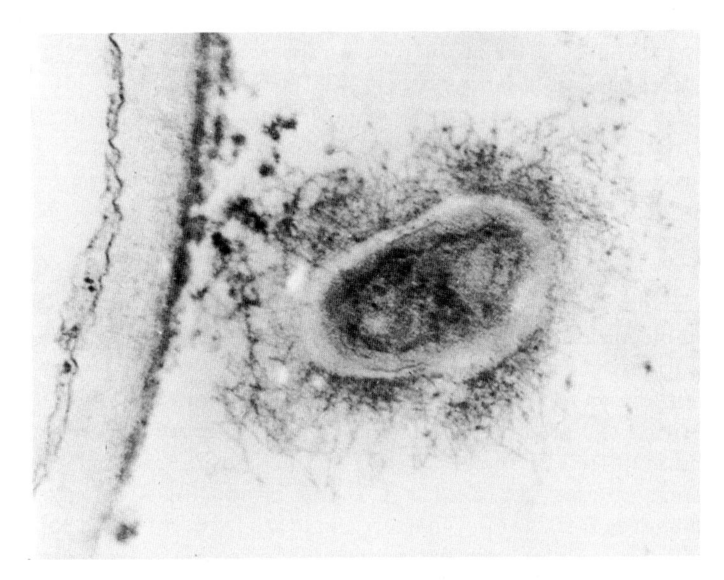

Fig. 2. Transmission electron micrograph of *Rhizobium trifolii* in the docking stage of Phase I adherence to a clover root hair. The fibrillar capsule of the bacterium is in contact with electron-dense globular aggregates on the outer periphery of the root hair cell wall (Dazzo and Hubbell, 1975c, and courtesy of Carolyn Napoli and the American Society for Microbiology).

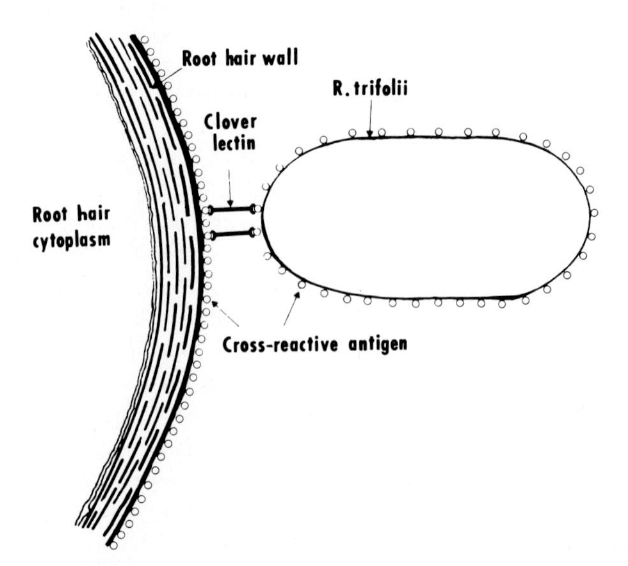

Fig. 3. Schematic diagram of the model to explain Phase I adherence. According to this model, antigenically cross-reactive saccharide receptors on *Rhizobium trifolii* and clover root hairs are specifically cross-linked with the clover lectin, trifoliin (Dazzo and D. Hubbell, 1975c) and courtesy of the American Society for Microbiology).

trifoliin on clover root surfaces (Dazzo *et al.*, 1978). These receptors are located at discrete sites which have differentiated on the epidermal root surface. They accumulate at root hair tips and diminish towards the base of the root hair (Figs. 4 and 5, Dazzo and Brill, 1977, 1979). By contrast, the CRA is uniformly distributed on both root hairs and undifferentiated epidermal cell walls in the root hair region of the clover seedlings (Dazzo and Brill, 1979). Trifoliin is multivalent in binding and agglutinating *R. trifolii* (Dazzo *et al.*, 1978). The sugar 2-deoxyglucose specifically inhibits agglutination of *R. trifolii* by trifoliin and anti-clover root antibody (Dazzo and Brill, 1977; Dazzo and Hubbell, 1975c), the binding of *R. trifolii* or its capsular polysaccharide to clover root hairs (Dazzo and Brill, 1977; Dazzo *et al.*, 1976), and specifically elutes trifoliin from intact clover roots (Dazzo and Brill, 1977; Dazzo *et al.*, 1978). As a negative control,

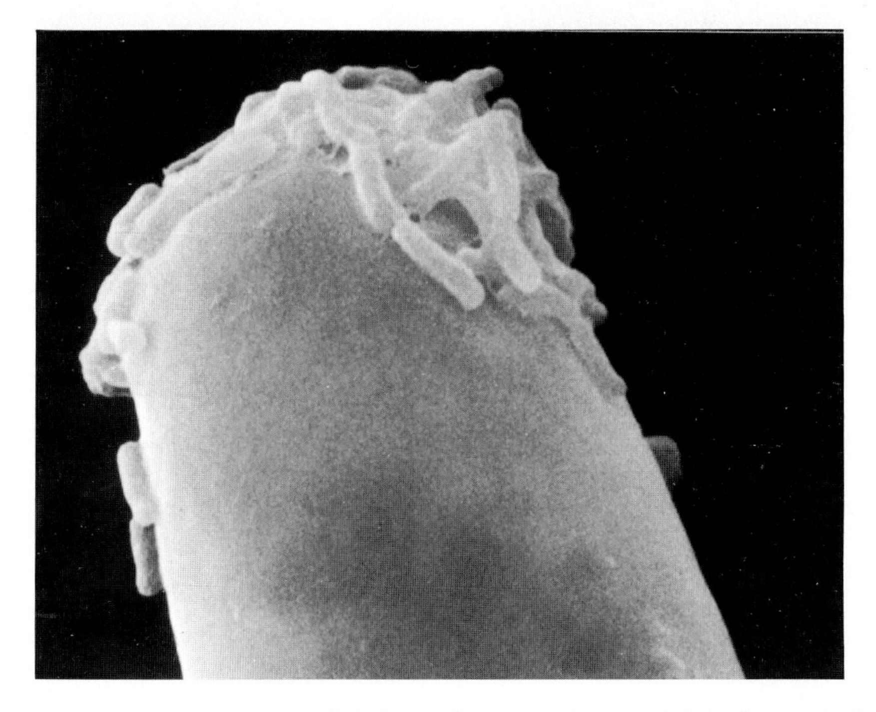

Fig. 4. Phase I selective adherence of *Rhizobium trifolii* 0403 to a clover root hair tip after 15 min of incubation as examined by scanning electron microscopy (Dazzo and Brill, 1979, and courtesy of the American Society for Microbiology).

Fig. 5. The exclusive relation between a legume and a nitrogen-fixing bacterium is demonstrated by the binding of *Rhizobium trifolii* to root hairs of its plant host clover. A fluorescent dye was first conjugated to the bacterial capsular polysaccharide. This labeled polysaccharide was then incubated with clover roots. Fluorescence of the roots showed that binding of the bacterial polysaccharide was restricted to the root hairs, the differentiated epidermal cells that serve as target cells for infection by *Rhizobium trifolii*. Similar results have been obtained with other rhizobia and their corresponding specific legume hosts (Dazzo and Brill, 1977, and courtesy of the American Society for Microbiology).

2-deoxyglucose does not inhibit adsorption of *R. meliloti* or its capsular polysaccharide to alfalfa root hairs (Dazzo and Brill, 1977; Dazzo *et al.*, 1976). This *Rhizobium*-host combination constitutes a different cross-inoculation group (Burton and Wilson, 1939), and displays specific alfalfa lectin-binding which is reversible by haptens other than 2-deoxyglucose (Kamberger, 1979; Leps and Brill, personal communication).

In such hapten inhibition studies it is believed that the sugar acts by combining with the site on the lectin or the antibody which is normally occupied by the polysaccharide. This implies a close but not necessarily identical structure of the inhibitor and antigenic determinants. 2-deoxyglucose may, therefore, only represent an analog of the native haptenic determinant on the surface polysaccharide of *R. trifolii* which binds specifically to trifoliin. However, it is important to note that some lectins undergo conformational changes when associated with saccharide binding (Reeke *et al.*, 1975), and so this possibility must also be considered in the interpretation of hapten inhibition studies.

Rhizobium–cells, if precoated with their corresponding plant lectin, will adhere in higher numbers than non-coated cells to their host roots (Dazzo *et al.*, 1976; Solheim, 1975). Quantitative adsorption studies with seedling roots of several legumes indicate that the CRA related to *R. trifolii* is unique to the specific host clover, and it is not detected immunochemically on the surface of roots of alfalfa or joint vetch as examples of other cross-inoculation groups (Dazzo and Brill, 1979).

Several experiments suggest that trifoliin and the cross-reactive anti-clover root antibody bind to the same or similar overlapping determinants on *R. trifolii* (Dazzo and Brill, 1979). First, the antibody and the lectin bind specifically to the same isolated polysaccharide from *R. trifolii.* Second, this interaction is specifically inhibited by 2–deoxyglucose. Third, the genetic markers on *R. trifolii* which bind trifoliin and the antibody co-transform into *Azotobacter vinelandii* with 100% frequency (Bishop *et al.*, 1977). And fourth, monovalent Fab fragments of IgG from anti-clover root antiserum strongly block the binding of trifoliin to *R. trifolii.* Considered collectively, these studies suggest that *R. trifolii* and clover roots have similar saccharide receptors for trifoliin. However, the definitive test of their identity as antigenically related structures will require knowledge of the minimal saccharide sequence which binds trifoliin.

The model would predict that all infective strains of *R. trifolii,* regardless of their antigenic disparity, should possess the same saccharide sequence cross-reactive with clover. It was very important to test this prediction, since it is well known that the serology of surface polysaccharides on *Rhizobium* is very strain-specific because of compositional differences in the immunodominant determinants (Carlson *et al.*, 1978; Dudman, 1977). Passive hemagglutination and its inhibition, nevertheless, showed that the saccharide sequence responsible for the antigenic cross-reactivity with clover roots was conserved in otherwise immunochemically distinct surface polysaccharides of *R. trifolii* strains (Dazzo and Brill, 1979). These results emphasize the existence of compositional differences in surface polysaccharides which are unrelated to their involvement in lectin-recognition processes, and also indicate that, as predicted (Dazzo and Hubbell, 1975a,c), the saccharide sequence responsible for lectin-binding and antigenic cross-reactivity need not be the immunodominant structure of the polysaccharide.

III. THE LECTIN RECEPTOR ON *RHIZOBIUM*

A major controversy exists concerning the nature of the carbohydrate receptor for the lectin on the *Rhizobium* cell. The lectin receptors were identified as capsular polysaccharide (Dazzo and Hubbell, 1975c), lipopolysaccharide (LPS) (Wolpert and Albersheim, 1976), and glycan (Planqué and Kijne, 1977). Kamberger (1979) has recently shown both host-specific LPS-lectin interactions (*R. melioloti*-alfalfa) and exopolysaccharide-lectin interactions (*R. japonicum*-soybean). In agreement with the results of Dazzo and Hubbell (1975c), these have all been characterized as surface, acidic heteropolysaccharides (Carlson *et al.*, 1978; Dazzo and Hubbell, 1975c; Planqué and Kijne, 1977). Electron microscopic examination of *R. trifolii* revealed a very prominent fibrillar capsule which surrounds the cell and stains positive for acidic polysaccharide with ruthenium red (Fig. 6, Dazzo and Brill, 1979). Fluorescence microscopic techniques indicated greater reactivity of anti-clover root CRA and purified trifoliin from seeds and roots with

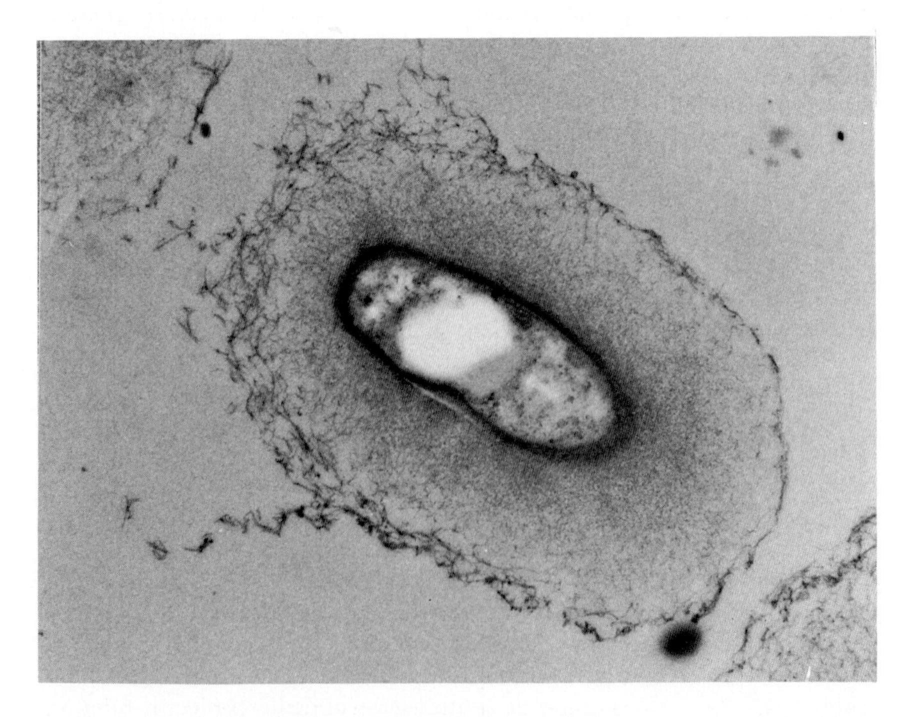

Fig. 6. Staining of *Rhizobium trifolii* 0403 with ruthenium red reveals the dense fibrillar capsule surrounding cells which can bind the clover lectin, trifoliin (Dazzo and Brill, 1979, and courtesy of the American Society for Microbiology).

encapsulated as compared with unencapsulated *R. trifolii* cells which had been separated by differential centrifugation (Dazzo and Brill, 1977; Dazzo and Hubbell, 1975c; Dazzo *et al.*, 1978). The same observation was subsequently made with *R. japonicum* and soybean lectin (Bal *et al.*, 1978; Calbert *et al.*, 1978).

In a recent study (Dazzo and Brill, 1979), the capsule of *R. trifolii* 0403 was extracted from cells harvested at early stationary phase, and then fractionated by ion exchange and gel filtration chromatography. Three polysaccharides were obtained. The major polysaccharide (85% of the total carbohydrate equivalents) was an acidic heteropolysaccharide which lacked components of the R-core (KDO, heptose) and endotoxic lipid A in LPS isolated from the same organism. The second polysaccharide (12% of the total carbohydrate equivalents) had properties consistent with LPS. It was immunogenic, contained endotoxic lipid A, KDO, heptose, and had an immunodominant carbohydrate antigen linked to the native structure through a weakly acid-labile linkage characteristic of the KDO linkage in the R-core of LPS. The third polysaccharide was minor (3% of the total carbohydrate equivalents), and was smaller in size than the second polysaccharide. It contained saccharide markers which were found in neither of the other two polysaccharides, and it was non-antigenic.*

Both trifoliin and the anti-clover root antibody reacted through 2-deoxyglucose-reversible binding with the first two polysaccharides described above; binding to the minor polysaccharide has not been observed. Both trifoliin-positive polysaccharides carried antigenic determinants in common, but on electrophoretically distinct molecular species. Electrophoresis in SDS-PAGE suggested that the first polysaccharide described above was larger in size than the second, yet acid hydrolysates of these two polysaccharides yielded the same paper chromatograms when developed with alkaline silver nitrate. These data are consistent with the hypothesis that the major polysaccharide in the capsule of *R. trifolii* 0403 at this culture age is an endotoxin-free acidic heteropolysaccharide which immunochemically conforms to the same O-antigen found in the LPS isolated from cells at this age. Because of its larger size and absence of heptose, this major polysaccharide in the capsule is not considered a precursor or partial degradation product of LPS.

---

*The invariable presence of this small, neutral polysaccharide in the capsule of *R. trifolii* suggests that it normally associates with the acidic polysaccharide, perhaps to perpetuate the capsular structure on the bacterial cell surface (Dazzo and Brill, 1979). If this hypothesis were correct, then it might explain the puzzling observation that some rhizobial strains only accumulate "microcapsules" rather than the well-developed capsules typical of *R. trifolii* 0403 and other strains (Dazzo and Brill, 1979).

Polysaccharides which bear this relation to LPS have been described for several Gram negative bacteria in the family Enterobacteriaceae (Anacker *et al.*, 1966; Rudbach, *et al.*, 1967; Jann *et al.*, 1970; Jann and Westphal, 1975; Luderitz, 1977). They are believed to arise under certain conditions by the transfer of the acidic capsular heteropolysaccharide to the core-lipid A and replace the O-antigen side chain in the bacterial lipopolysaccharide (Jann *et al.*, 1970; Jann and Westphal, 1975; Luderitz, 1977).

A clearer understanding of these data may be possible by a brief review of the process of lipopolysaccharide biosynthesis which has been described for *Salmonella* and *Escherichia coli* (Luderitz, 1970; Osborne *et al.*, 1972). The biosynthesis of the O-antigen chain of LPS and capsular polysaccharides occurs by a complex series of reactions with participation of a C_{55} polyisoprenoid-monophosphate carrier and membrane bound glycosyl transferases. Sugars are transferred by specific transferases from their activated sugar nucleotides into oligosaccharide units of O–antigen covalently attached to antigen carrier polyisoprenoid lipid (ACL). These oligosaccharide units are polymerized in a block fashion by a polymerase with the side chain (attached to ACL) being lengthened by one repeating unit at each growing step. The final O-specific chain is transferred by a translocase onto the R–core which is anchored to the lipid A on the outer envelope of the Gram negative bacterium. Other enzymes (e.g., acetyl transferases) transfer additional substituents (e.g., O-acetyl branch groups) onto the O-specific chain after the LPS is completed. The highly polymerized polysaccharides and the repeating somatic O-antigen of LPS are probably synthesized by the same mechanism (Jann *et al.*, 1970; Jann and Westphal, 1975; Luderitz, 1977). The lipid A- and heptose free polysaccharide is believed to accumulate when control of the O–antigen polymerase by the translocase is not functioning (Jann *et al.*, 1970; Jann and Westphal, 1975; Luderitz, 1977).

Recently, detailed analysis of nodulating and relating non-nodulating mutant strains of *Rhizobium* have revealed structural changes in their LPS which may relate to their nodulating properties (Maier and Brill, 1978; Saunders *et al.*, 1978). An analysis of nodulating and non-nodulating mutant strains of *R. japonicum* showed that the antigenic difference on their cell surfaces resided in the somatic antigens, which were acidic heteropolysaccharides which could be extracted and purified as either linked to the core-lipid A of LPS or as the free, unlinked polysaccharide (Maier and Brill, 1978). Discrete, silver nitrate-reactive components were missing in acid hydrolysates of LPS and the acidic heteropolysaccharide from the *R. japonicum* non-nodulating mutant strains. Such phenotypes could arise from mutations in glycosyl transferases, enzymes which catalyze the interconversion of various sugars and their phosphorylated derivatives (e.g., epimerases), or subsequent modifying enzymes (e.g., methyl transferases). In another study (Saunders, *et al.*, 1978), mutant strains of *R. leguminosarum* were selected by filtration techniques to yield cells which do not produce large gummy colonies on agar. One of these mutant strains, EXO-1, does not nodulate peas, and 27% of its total LPS mass is anthrone-reactive carbohydrate, as compared to 63% of the total LPS mass from the wild-type *R. leguminosarum* (Saunders *et al.*,1978). The glycosyl and antigenic compositions of the O-antigen of the mutant and wild-type strains do not seem to be different, and both strains are lysed by the same bacteriophages. The hypothesis advanced by the authors was that the mutant strain EXO-1 has reduced its production of extracellular polysaccharide and not of LPS (Saunders *et al.*, 1978). However, an alternative hypothesis is that the EXO-1 phenotype could be due to a defective O-antigen polymerase which would fail to polymerize in a block fashion the repeating O-antigen on the ACL. When this O-antigen oligosaccharide is transferred to the R-core lipid A via the translocase reaction, an LPS with reduced O-antigenic polymerization would result. Bacteriophage with receptors for the O-antigens could still recognize those defective LPS structures, their immunodominant antigens would be immunochemically identical when isolated, and their glycosyl composition would be the same or similar, exactly matching the phenotype of these non-nodulating *R. leguminosarum* mutant strains described by Saunders *et al.* (1978).

IV. RHIZOBIAL ADHESION TO LEGUME ROOT HAIRS

Contact between the symbionts on the root surface constitutes one of the early, critical "recognition" interactions prior to successful infection and development of the symbiotic state. Quanitative microscopic assays (Dazzo, 1979a; Dazzo *et al.*, 1976) and numerous electron microscopic studies (Dart, 1971; Dazzo, 1979a,c and d; Dazzo and Brill, 1979; Dazzo and Hubbell, 1975c; Napoli, 1976; Napoli and Hubbell, 1975) have revealed multiple mechanisms of rhizobial adherence to clover root hairs. A nonspecific mechanism allows all rhizobia to attach in low numbers (2–4 cells per 200 μm root hair length per 12 hr). In addition, a specific mechanism in the *Rhizobium trifolii*-clover symbiosis allows selective adherence in significantly (P = 0.005) larger numbers (22–27 cells per 200 μm root hair length per 12 hr). Electron microscopy (Dazzo and Hubbell,

1975c) disclosed that the initial bacterial attachment step consisted of contact between the fibrillar capsule of *R. trifolii* and electron-dense globular aggregates lying on the outer periphery of the fibrillar clover root hair cell wall (Fig. 2). This "docking" stage is called *Phase I adherence* and occurs within a few hours after inoculation of the host with rhizobia.

What cell surface molecules are involved in Phase I adherence? The first clue that the clover lectin, trifoliin, may be involved came from the observation that 2-deoxyglucose specifically inhibited the attachment of *R. trifolii* to clover root hairs (Dazzo *et al.*, 1976), and it reduced this high level of bacterial adhesion to that characteristic of background. Subsequent studies showed that 2-deoxyglucose specifically facilitated the elution of trifoliin from the intact clover root (Dazzo and Brill, 1977; Dazzo *et al.*, 1978). A summary of the 2-deoxyglucose effect is presented in Fig. 7. Kijne (personal communication) has also found hapten-facilitated elution of root lectin from peas.

The results of three experiments indicate that trifoliin and antibody to the clover root cross-reactive antigen bound to the same *R. trifolii* saccharide determinants which bound these bacteria to clover root hairs

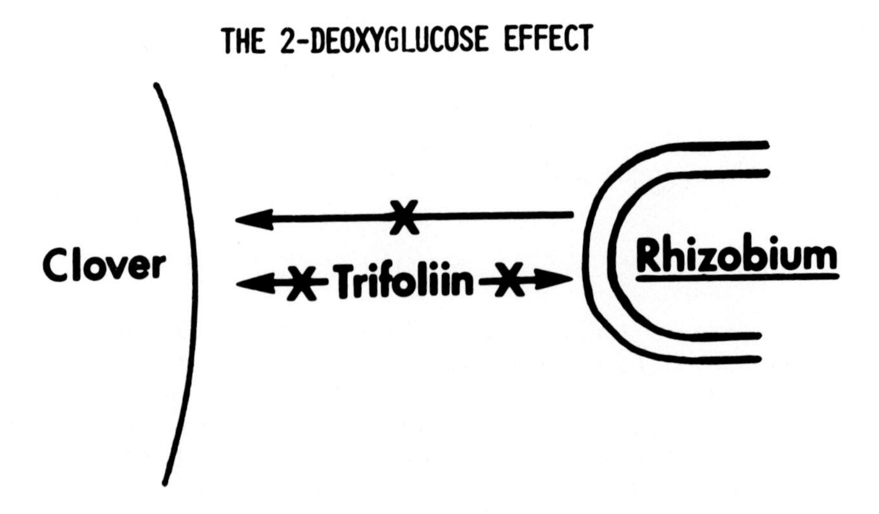

Fig. 7. Summary of the effects of 2-deoxyglucose on Phase I adherence in the *Rhizobium*-clover symbiosis. This hapten sugar inhibits the binding of *R. trifolii* and its capsular polysaccharide to root hairs, and facilitates the elution of trifoliin from these symbiont surfaces.

(Dazzo and Brill, 1979). First, Fab fragments of anti-clover root antibody adsorbed on *R. trifolii* blocked the bacterial adhesion to clover root hairs. Second, only the *A. vinelandii* hybrid transformants which carried the trifoliin receptor bound to clover root hairs. Third, competition assays using fluorescence microscopy indicated that the *R. trifolii* polysaccharides which carried the trifoliin receptor had the highest affinity for clover root hairs. Critical negative controls in these clover-root binding experiments included Fab fragments of preimmune antibody, 2-deoxyglucose, hybrid transformants of *A. vinelandii* which carry surface antigens from *R. japonicum* which are important for nodulation of soybean, and polysaccharides from *R. meliloti*.

At this time, thoughts on the function of the CRA on the clover root are only speculative. Presumably it is a component of the plant cell wall. Indeed, the pea lectin binds to the pectin fraction of the pea root cell walls (Kijne, personal communication). Perhaps a normal, physiological function of these lectins is to cross-link polysaccharide structures during tip growth of the root hair cell wall (see Kauss and Glaser, 1974). The lectin receptor would then serve as the "address" for trifoliin during its mobilization to the root cell wall. Selective pressures imposed by invasive pathogens may have favored polysaccharide and lectin-binding diversities of the host. The advantage of N_2 fixation may have favored the selection of mechanisms which accumulate high numbers of rhizobia on root hair tips of young seedlings which would be necessary for successful infection and symbiotic development. It has been proposed that common antigenic constituents between pathogen and host may possibly foil defense mechanisms associated with plant hypersensitivity (DeVay and Adler, 1976; Chen and Phillips, 1976). Designing the same complementary sites on the polypeptide chains of this lectin, which would recognize both the plant cell wall structures and the appropriate rhizobial polysaccharides, would be considered a genetically conservative venture. Functional multivalency could be accomplished by aggregation of identical subunits as products of one gene.** Perhaps the CRA could serve as a scavenger to salvage trifoliin which has eluted from clover roots, and would provide a specific binding receptor on clover roots for the enhanced adherence of trifoliin-coated *Rhizobium trifolii* which is observed (Dazzo *et al.*, 1976; Raa *et al.*, 1977; Solheim, 1975). In addition, the presence of a lectin receptor in the host cell wall may possibly provide an opportunity of the plant to regulate the *in situ* activity of trifoliin.

**SDS-PAGE suggests that a single subunit combines to form the 10 nm, multivalent agglutinating aggregate of trifoliin (Dazzo *et al.*, 1978).

If the pH of the clover rhizosphere is above 4, then successful completion of Phase I adherence would require that the bacteria overcome the repulsive energy barrier created by polyanionic charges on the root and bacterial surfaces. A recent analysis of clover root exudate has given a clue to how this may possibly be accomplished. Automated amino acid analysis showed that lysine was the major free amino acid present in the aseptically collected root exudate (Smucker and Dazzo, unpublished observation). It would be interesting to know if the abundance of lysine in the clover rhizosphere may provide sufficient electropositive potential to dampen the repulsive electro-negative energies during docking. At pH of 4 or less, the CRA on *R. trifolii* is destroyed (Dazzo and Hubbell, 1975c).

Soybean root cells in culture selectively adhere to homologous *R. japonicum* cells, and bacterial adherence is inhibited by pectinase treatment of the root cells (Reporter *et al.*, 1975). Encapsulated cells of *R. japonicum* bind to soybean lectin and soybean root hairs (Bal *et al.*, 1978; Calbert *et al.*, 1978). The lectin from bean *(Phaseolus vulgaris)* binds specifically with surface polysaccharides of the bean symbiont, *R. phaseoli* (Wolpert and Albersheim, 1976), and there is indirect evidence that this lectin may be on the bean root epidermis at sites where binding by rhizobia and symbiotic infection occur (Hamblin and Kent, 1973). Alfalfa root hairs carry the receptor sites which specifically bind polysaccharides from *R. meliloti*, the alfalfa symbiont (Dazzo and Brill, 1977). Similarly located receptor sites for polysaccharides from *R. japonicum* are found on soybean root hairs (Hughes and Elkan, personal communication). Fewer cells of *R. japonicum* and *R. leguminosarum* adhere to roots of non-nodulating soybean and pea varieties, respectively, than to roots of their normal nodulating soybean and pea varieties (Degenhardt *et al.*, 1976; Elkan, 1962). Non-nodulating mutant strains of *R. leguminosarum*, which are defective in extracellular polysaccharide production (Saunders *et al.*, 1978), adhere in smaller numbers to pea root hairs as compared with the wild-type nodulating strains (C.A. Napoli, personal communication). Clearly, selectivity in rhizobial adherence to legume roots may operate in several species.

Other mechanisms (collectively called Phase II adherence) firmly anchoring the bacteria to the root hair surface may operate during later preinfective stages after the early period of docking (Dazzo, 1979a). During Phase II adherence, fibrillar materials associated with the adherent bacteria are visualized under the scanning electron microscope (Dart, 1971; Dazzo, 1979c,d; Fig. 8, compare with Fig. 4). These appendages are associated with the adherent bacteria after prolonged incubation with the clover root. Similar appendages appear on *R. trifolii*

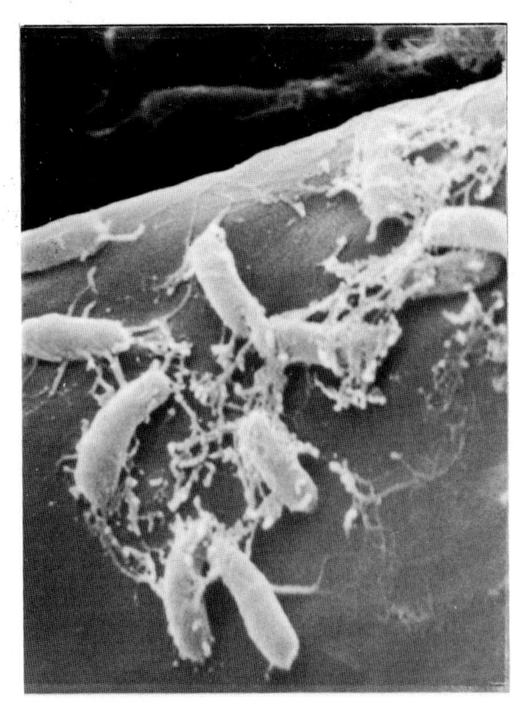

Fig. 8. Phase II adherence of *Rhizobium trifolii* 0403 to a clover root hair after 2 days incubation as examined by scanning electron microscopy. Note the fibrillar appendages associated with the adherent bacteria.

grown in aseptically collected clover root exudate (Napoli, 1976).

The composition of these aggregated microfibrils characteristic of Phase II adherence is unknown. One possibility is that they are the cellulose microfibrils produced by many rhizobial species (Deinema and Zevenhuizen, 1971; Napoli *et al.*, 1976a). These neutral exopolysaccharides tend to flocculate cells, often when they are in stationary phase of growth in broth culture. Another possibility is that they may consist of bundles of low molecular weight, extracellular β-1, 2 glucans (Zevenhuizen, 1976). Since these two types of neutral exopolysaccharides are produced by many species of rhizobia, they alone are unlikely to account for specific recognition of legumes (Napoli *et al.*, 1975a; Zevenhuizen, 1978). Nevertheless, they may be important in maintaining the firm contact between the bacterium and the host root hair necessary for triggering the tight root hair curling (Shepherd's crook formation) and successful infection (Napoli *et al.*, 1975a).

Other data also suggest multiple mechanisms of bacterial adherence to roots. Nissen (1971) differentiated specific from nonspecific mechanisms of bacterial adsorption to barley roots by their sensitivity to acid pH and requirements for calcium and plant-derived compounds for energy. Werner *et al.* (1975) found that multiple mechanisms of adsorption to legume roots can be separated by their different sensitivities to dissociation by large salt concentrations. Chen and Phillips (1976) found that the kinetics of rhizobial adsorption to pea roots progressed at a rapid rate initially, and later more slowly. They also identified the root hair tips as important sites of rhizobial attachment but were unable to correlate rhizobial adsorption to roots with host-symbiont selectivity. However, data from viable counts and scintillation counts of radiolabelled bacteria on roots may not be reliable since both techniques count many unadsorbed bacteria which are not washed away, as well as bacteria which adhere to undifferentiated epidermal cells. Other complications involve the entrapment of flocs of cells, and the variability of surface areas among individual plant roots. Furthermore, Kotarski and Savage (1976) have shown that the use of radiolabelled cells for adherence studies is complicated by the narrow range of linearity in radioactive counts per minute vs. cell number, and the transfer of radiolabelled metabolites from the bacteria to the host tissue. Use of fluorescently labelled cells for quantitative adherence studies also yields questionable data since the bacterial cell surfaces have been undesirably modified by a coating of antibody-dye complexes prior to incubation with the root. On the other hand, the direct microscopic assay is reproducible (Dazzo, 1979a; Dazzo *et al.*, 1976), and has a high "signal-to-noise" ratio since the only bacterial cells which are scored are those which have their native surface in physical contact with root hair cell walls of standardized length and surface area and with a uniform state of physiological development.

Although adherence of infective rhizobia to target root hairs is a prerequisite for infection, several observations indicate that other undefined events must occur to initiate infection. Heterologous rhizobia can adhere in small numbers to root hairs but do not infect them (Dazzo, 1979a; Dazzo *et al.*, 1976), and very few of the root hairs to which infective rhizobia adhere eventually become infected. Genetic hybrids (Dazzo and Brill, 1979) of *Azotobacter vinelandii* (which carry the trifolliin saccharide receptor on their surface as a result of intergeneric transformation with DNA from *R. trifolii*, Bishop, *et al.*, 1977) have acquired the ability to adhere specifically to clover root hairs but do not infect them.

V. REGULATION OF RECOGNITION IN THE *RHIZOBIUM*-LEGUME SYMBIOSIS

How is the recognition process regulated? Do both symbionts participate in regulating the components of the recognition process? Answers to these important questions are beginning to emerge. It has been known for many years that fixed nitrogen (e.g., NH_4^+, NO_3^-) is one of the many environmental factors which limits the development and the success of the *Rhizobium*-legume symbiosis in nature (for a review, see Fred, *et al.*, 1932). Recently, we developed an immunocytofluorimetric assay which allows us to examine the regulation by fixed nitrogen ions of trifoliin present on the seedling root (Dazzo and Brill, 1978). The measurement was based on the quantitation of fluorescent light intensity (as photovolts/mm^2) emitted from intact root surfaces following their incubation with anti-trifoliin antiserum and examination by indirect immunofluorescence (Dazzo *et al.*, 1978). This assay was very specific for trifoliin on clover (Dazzo *et al.*, 1978).

We found that the immunologically detectable levels of trifoliin and the specific binding of *R. trifolii* to root hairs decreased in a parallel fashion as the concentrations of either NO_3^- or NH_4^+ were increased in the rooting medium (Fig. 9). These experiments support the hypothesis that trifoliin is involved in binding *R. trifolii* to clover root hairs. The levels of these fixed ions which inhibited attachment were well below the levels which stunted seedling growth. Interestingly, the levels of trifoliin and rhizobial attachment were increased at a relatively low concentration of NO_3^-. Thus, low levels of NO_3^- enhanced the recognition process and high levels shut it off (Dazzo and Brill, 1978). Recent ligand-binding studies using radiolabelled $^{13}NO_3^-$ have detected no direct interactions between NO_3^- and trifoliin, and this anion does not interfere with the direct interaction of trifoliin and its specific antibody (Dazzo, unpublished observations). Garcia *et al.*, (submitted for publication) have observed also that adherence of *Azospirillum brasilense* to root hairs of the grass, pearl millet, is suppressed when the roots are grown in critical concentrations of NO_3^-. Upon examination by scanning electron microscopy, there was observed a distinct degranulation in the surface topography of the root hair cell grown in NO_3^--containing medium. These observations have opened new avenues of investigation which may ultimately provide a solution to the important problem of how fixed nitrogen ions regulate the developmental events of the *Rhizobium*-legume symbiosis.

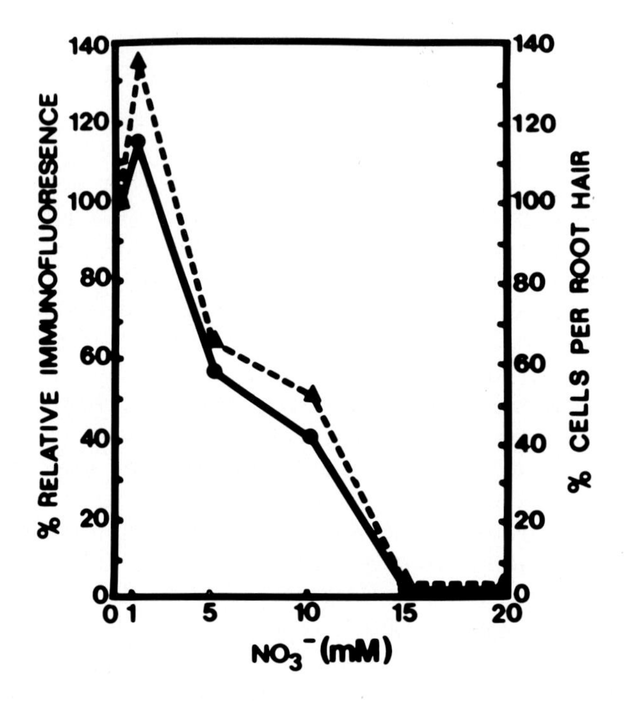

Fig. 9. The effect of NO_3^- on adsorption of *Rhizobium trifolii* 0403 to root hairs (solid line) and on immunologically detectable trifoliin (dotted line) in the root hair region of clover seedlings. Bacterial adsorption was measured by direct counting and trifoliin was measured by cytofluorimetry using indirect immunofluorescence. Values from roots grown in N-free nutrient solution are taken as 100%, and represent 980 photovolts/mm^2 and 21 cells/root hair 200 μm in length. Points along the curve are means from 10-15 root hairs or seedling roots, standard deviations vary within 10% of the means. Values are corrected for non-specific rhizobial adsorption, root autofluorescence, and non-specific adsorption of conjugated goat antirabbit γ globulin (Dazzo and Brill, 1978, and courtesy of the American Society of Plant Physiologists).

The selective ability of *R. trifolii* to adhere to clover root hairs should be influenced by conditions that affect the accumulation of trifoliin on the host root surface and the saccharide receptor on the bacterium. Evidence supporting this hypothesis comes from data described above on the fixed-nitrogen regulation of levels of trifoliin on clover roots. Other studies also support this hypothesis by showing that, under certain growth conditions, the transient appearance of trifoliin receptors on *R. trifolii* may influence the ability of these bacteria to attach to clover root hairs (Dazzo *et al.*, 1979). Cells grown on agar plates of defined media were most susceptible to agglutination by trifoliin when they were harvested at five days of growth. The antigenic determinants cross-reactive with clover roots were exposed for only short periods on the bacteria as broth cultures left lag phase and again as they entered stationary phase (Fig. 10). Clover roots adsorbed the bacteria in greatest quantities when they were harvested from plate cultures incubated for five days and from broth cultures in early stationary phase. These studies draw attention to an important explanation as to why many anomalous lectin-binding results have been obtained. The results also indicate that the regulation of these receptors may be very complex.

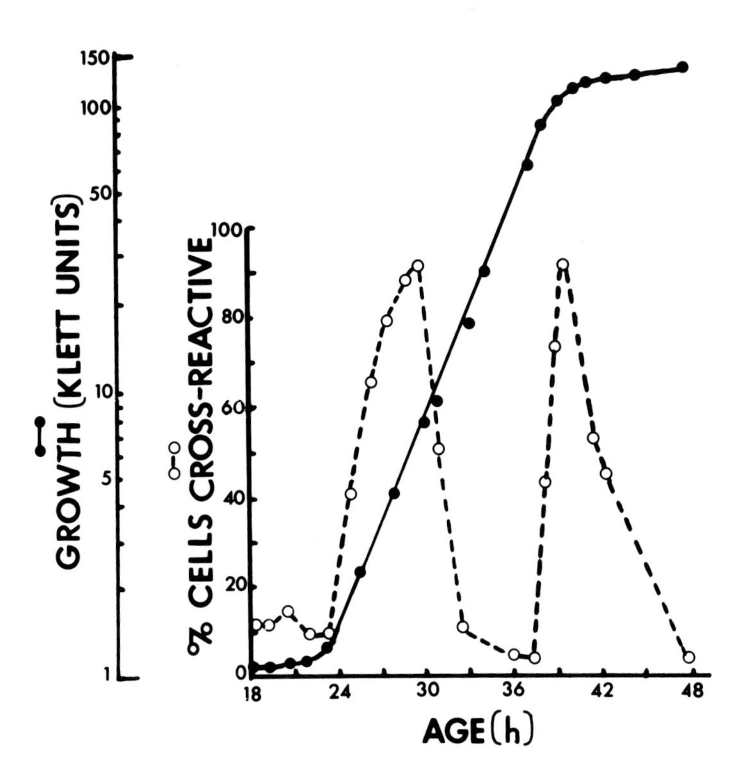

Fig. 10. The effect of culture age on the percentage of *Rhizobium trifolii* 0403 cells reactive with anti-clover root antiserum. Cells were grown in a chemically defined broth at 30°C with shaking. Culture growth (closed circles, log scale) was measured with a Klett-Summerson colorimeter at 660 nm and the cross-reactive antigen (open circles, linear scale) by indirect immunofluorescence (Dazzo *et al.*, 1979, and courtesy of Springer-Verlag).

Some of the intergeneric hybrids of *A. vinelandii-R. trifolii* deposited the CRA in a patchy distribution on their cells surfaces (Bishop *et al.*, 1977). These "patchy" hybrids offer potentially useful intermediates for studies of the regulation of the synthesis and compartmentation of this polysaccharide.

Bauer and co-workers (Bauer, 1977; Bhuvaneswari and Bauer, 1978; Bhuvaneswari *et al.*, 1977) have done seminal studies on the regulation of the lectin receptor on *R. japonicum.* They observed a transient appearance and disappearance of the receptor on *R. japonicum* which specifically binds soybean lectin (SBL). While most strains had their highest percentage of SBL-positive cells and the greatest number of SBL-binding sites per cell in the early and mid-log phases of growth, one strain accumulated the lectin receptor as cells left exponential growth and entered stationary phase. The proportion of galactose residues in the capsular polysaccharide is high at a culture age when the cells bind the galactose-reversible soybean lectin (Mort and Bauer, 1978). A decline in lectin-binding activity accompanying cultural aging is concurrent with a decline in galactose content and a rise in 4-0-methyl galactose residues. The latter methylated sugar has low affinity for the soybean lectin.

Interestingly, some *R. japonicum* stains, which lack soybean lectin receptors when grown in bacterial culture media, would acquire this property when grown in root exudate (Bhuvaneswari and Bauer, 1978).

It is possible that the effect of culture age on expression of lectin receptors may influence the competitiveness and ecological behavior of different rhizobial strains in soil and on the roots of their legume hosts. The delay in appearance of the capsule surrounding cells of *R. trifolii* TA1 (Dudman, 1968; Humphrey and Vincent, 1969) may improve their competitiveness for nodulation of clover roots in field soils (Dudman, 1968). Inoculum prepared from early stationary-phase cells of *R. trifolii* NA 30 gives rise to more clover root hair infections than equivalent inocula prepared from exponentially growing or late stationary-phase cells (Napoli, 1976). The one strain of *R. japonicum* (3I1b123) found to accumulate the SBL-binding capsule during early stationary phase (Bhuvaneswari *et al.*, 1977) has been recognized as the most frequently found serogroup in soybean nodules from many soils of the central United States (Damirgi *et al.*, 1967; Ham *et al.*, 1971).

Albersheim and Wolpert (1977) have reported that lectins are enzymes which specifically degrade the LPS of their symbiont rhizobia. It is uncertain from their studies whether seed glycosidases could have contaminated the lectin preparations during their purification by affinity chromatography using saccharide-coupled columns. Albersheim has subsequently retracted this published report (retraction unpublished).

The possibility that lectins are glycosidases is potentially important, since it may provide clues as to how selective, rhizobial adherence may lead to softening of the root hair cell wall and subsequent infection (Ljunggren and Fahraeus, 1961) and also as to how the vegetative bacteria loose most of their LPS and transform into pleomorphic bacteroids (VanBrussel *et al.*, 1977). More recently, Hankins and Shannon (1978) reported an exo-α-galactosidase activity associated with mung bean lectin isolated from imbibed seeds.

With regard to genetic regulation, exciting evidence is beginning to emerge that genetic elements important to surface polysaccharides and symbiotic recognition are encoded on plasmids of *Rhizobium* (Johnson *et al.*, 1978; Zurkowski and Lorkiewicz, 1978; Nuti *et al.*, 1979; Van der Schaal and Kijne, 1979). The report that a fast-growing strain of *R. trifolii* has all the necessary genes to enable the organism to nodulate soybean effectively (O'Gara and Shanmugham, 1978) has been retracted.

In my opinion, the most fundamental question regarding the regulation of recognition is the following: how does the recognition process of mechanical maneuvering of *Rhizobium* at the root hair surface interface with the informational exchange between symbionts which then triggers the very specific events of shepherd's crook formation and penetration of the root hair cell wall to result in successful infection? Correct answers to this question should yield the molecular basis for host specificity in the *Rhizobium*-clover symbiosis.

VI. IMPLICATIONS OF THE RESEARCH

Studies of host recognition processes in this symbiosis are timely because of the awareness that nitrogen is the most commonly limiting nutrient to crop productivity, and legumes can offset that major limitation by entering efficient N_2-fixing symbioses.

An understanding of the mechanism and control of host recognition would not only help to elucidate the developmental events in the legume symbiosis, but also indicate ways in which *Rhizobium* and the host plant may be manipulated genetically to increase the range of agricultural crops which can enter efficient N_2-fixing symbioses.

There are other benefits to be gained from better understanding of cell-recognition processes between microorganisms and their plant hosts. For example, these data will relate directly to studies of the physiological recognition process of plant hosts by phytopathogens, with the eventual goal of reducing the burden of plant pathogenesis on agricultural crops by the induction of disease resistance (Graham *et al.*,

1977). An important attribute of many microbial pathogens of plants and animals is their ability to adhere to and colonize the host tissue (Gibbons, 1977). This appears to be one of the dominating factors determining successful colonization. Indeed, the selectivity of bacterial attachment appears to be the major underlying basis which accounts for the tissue, organ, and host tropisms of a variety of microorganisms (Gibbons, 1977). Interactions which display selective adherence also serve as useful models for studies of the biochemistry of cellular recognition, cell surface receptors, fertilization, and compatibility-incompatibility of self and nonself tissues which are topics central to developmental biology.

ACKNOWLEDGEMENTS

I wish to thank Dr. David Hubbell of the University of Florida and Dr. Winston Brill of the University of Wisconsin for the leadership they provided during the earlier development of this research. Critical comments on this manuscript by Estelle Hrabak and Dr. Harold Sadoff of Michigan State University are also appreciated. Portions of this research were supported by the Michigan Agricultural Experiment Station, the College of Natural Science and the College of Osteopathic Medicine, Michigan State University, and by NSF Grant PCM 78-22922 and USDA CGO Grant 78-00099.

REFERENCES

Albersheim, P., and Wolpert, J. (1977). *In* "Cell Wall Biochemistry Related to Specificity in Host-Plant Pathogen Interactions" (B. Solheim and J. Raa, eds.), pp. 373-376. Universitetsforlanget, Norway.

Allen, O.N. and Allen, E.K. (1940). *Bot. Gaz.* **102**, 121-142.

Anacker, R.L., Bickel, W.D., Haskins, W.T., Milner, K.C., Ribi, E. and Rudbach, J.A. (1966). *J. Bacteriol.* **91**, 1427-1433.

Bal, A.K., Shantharam, S. and Ratnam, S. (1978). *J. Bacteriol.* **133**, 1393-1400.

Bauer, W.D. (1977). *Basic Life Sci.* **9**, 283-297.

Bhuvaneswari, T.V. and Bauer, W.D. (1978). *Plant Physiol.* **62**, 71-74.

Bhuvaneswari, T.V., Pueppke, S.G. and Bauer, W.D. (1977). *Plant Physiol.* **60**, 486-491.

Bishop, P.E., Dazzo, F.B., Appelbaum, E.R., Maier, R.J. and Brill, W.J. (1977). *Science* **198**, 938-940.

Bohlool, B.B. and Schmidt, E.L. (1974). *Science* **185**, 269-271.

Bohlool, B.B. and Schmidt, E.L. (1976). *J. Bacteriol.* **125**, 1188-1194.

Brethauer, T.S. and Paxton, J.D. (1977) *In* "Cell Wall Biochemistry Related to Specificity in Host-Plant Pathogen Interactions" (B. Solheim and J. Raa, eds.), pp. 381-389. Universitetsforlanget, Norway.

Broughton, W.J. (1978). *J. Appl. Bact.* **45**, 165-194.

Burton, J.C. and Wilson, P.W. (1939). *Soil Science* **47**, 293-303.
Calbert, H.E., Lalonde, M., Bhuvaneswari, T.V. and Bauer, W.D. (1978). *Can. J. Microbiol.* **24**, 785-793.
Carlson, R.W., Sanders, R.E., Napoli, C. and Albersheim, P. (1978). *Plant Physiol.* **62**, 912-917.
Chandler, M.R. (1978). *J. Exp. Bot.* **29**, 749-755.
Chen, A.T. and Phillips, D.A. (1976). *Physiol. Plant.* **38**, 83-88.
Damirgi, S.M., Frederick, L.R. and Anderson, I.C. (1967). *Agron. J.* **59**, 10-12.
Dart, P.J. (1971). *J. Exp. Bot.* **22**, 163-168.
Dazzo, F.B. (1979a). *In* "Adsorption of Microorganisms to Surfaces" (G. Bitton and K.C. Marshall, eds.), J. Wiley and Sons, New York, in press.
Dazzo, F.B. (1979b). *Amer. Soc. Microbiol. News* **45**, 238-240.
Dazzo, F.B. (1979c). *In* "Advances in Legume Science" (A.J. Bunting and R. Summerfield, eds.), University of Reading Press, England., in press.
Dazzo, F.B. (1979d). *In* "Proceedings of the Steenboch-Kettering International Symposium on Nitrogen Fixation" (W.H. Orme-Johnson and W.E. Newton, eds.), University Park Press, Maryland., in press.
Dazzo, F.B. and Brill, W.J. (1977). *Appl. Environ. Microbiol.* **33**, 132-136.
Dazzo, F.B. and Brill, W.J. (1978). *Plant Physiol.* **62**, 18-21.
Dazzo, F.B. and Brill, W.J. (1979). *J. Bacteriol.* **137**, 1362-1373.
Dazzo, F.B. and Hubbell, D.H. (1975a). *Appl. Microbiol.* **30**, 172-177.
Dazzo, F.B. and Hubbell, D.H. (1975b). *Plant Soil* **43**, 713-717.
Dazzo, F.B. and Hubbell, D.H. (1975c). *Appl. Microbiol.* **30**, 1017-1033.
Dazzo, F.B., Napoli, C.A. and Hubbell, D.H. (1976). *Appl. Environ. Microbiol.* **32**, 168-171.
Dazzo, F.B., Urbano, M.R. and Brill, W.J. (1979). *Curr. Microbiol.* **2**, 15-20.
Dazzo, F.B., Yanke, W.E. and Brill, W.J. (1978). *Biochim. Biophys. Acta* **539**, 276-286.
Dengenhardt, T., LaRue, T. and Paul, E. (1976). *Can. J. Bot.* **54**, 1633-1636.
Deinema, M.H., and Zevenhuizen, L.P.T. (1971). *Arch. Mikrobiol.* **78**, 42-57.
DeVay, J.E. and Adler, H.E. (1976). *Annu. Rev. Microbiol.* **30**, 147-168.
Dudman, W.F. (1968). *J. Bacteriol.* **95**, 1200-1201.
Dudman, W.F. (1977). *In* "A Treatise on Dinitrogen Fixation" (R. Hardy and W. Silver, eds.), Vol. IV, pp. 487-508. J. Wiley and Sons, New York.
Elkan, G. (1962). *Can. J. Microbiol.* **8**, 79-87.
Fahraeus, G. (1957). *J. Gen. Microbiol.* **16**, 374-381.
Fred, E.B., Baldwin, I.L. and McCoy, E. (1932). "Root Nodule Bacteria and Leguminous Plants." University of Wisconsin Press, Wisconsin.
Gibbons, R.J. (1977). *In* "Microbiology-1977" (D. Schlessinger, ed.), pp. 395-406, American Society for Microbiology, Washington, D.C.
Graham, T.L., Sequeira, L. and Huang, T.R. (1977). *Appl. Environ. Microbiol.* **34**, 424-432.
Ham, G.E., Frederick, L.R., and Anderson, I.C. (1971). *Agron. J.* **63**, 69-72.
Hamblin, J. and Kent, S.P. (1973). *Nature New Biol.* **245**, 28-30.
Hankins, C.N. and Shannon, L.M. (1978). *J. Biol. Chem.* **253**, 7791-7797.
Humphrey, B. and Vincent, J.M. (1969). *J. Gen. Microbiol.* **59**, 411-425.
Humphrey, B., Edgley, M. and Vincent, J.M. (1974). *J. Gen. Microbiol.* **81**, 267-270.
Jann, B., Jann, K., Schmidt, G., Orskov, I. and Orskov, F. (1970). *Eur. J. Biochem.* **15**, 29-39.
Jann, K. and Westphal, O. (1975). *In* "The Antigens" (M. Sela, ed.), Vol. III, pp. 1-125. Academic Press Inc., New York.
Johnston, A.W., Beynon, J.L., Buchanan-Woolaston, A.V., Setchell, S.M., Hirsch, P.R. and Beringer, J.E. (1978). *Nature* 276, 634-636.
Kamberger, W. (1979). *Arch. Microbiol.* **121**, 83-90.
Kauss, H. and Glaser, C. (1974). FEBS Lett. **45**, 304-307.

Kotarski, S.F. and Savage, D.C. (1978). Proc. Amer. Soc. Microbiol. Meeting, Las Vegas, Nevada, *1978*, p. 15 (B11.)
Law, I.J., and Strijdom, B.W. (1977). *Soil Biol. Biochem.* **9**, 79-84.
Li, D. and Hubbell, D.H. (1969). *Can. J. Microbiol.* **15**, 1133-1136.
Ljunggren, H. and Fahraeus, G. (1961). *J. Gen. Microbiol.* **26**, 521-528.
Luderitz, O. (1970). *Agnew. Chem. Internat. Edit.* **9**, 649-663.
Luderitz, O. (1977). *In* "Microbiology-1977" (D. Schlessinger, ed.), pp. 239-246. American Society for Microbiology, Washington, D.C.
Maier, R.J. and Brill, W.J. (1978). *J. Bacteriol.* **133**, 1295-1299.
Mort, A. and Bauer, W.D. (1978). Ann. Meeting Amer. Soc. Plant Physiol. 1978, p. 325.
Napoli, C.A. (1976). "Physiological and Ultrastructural Aspects of the Infection of Clover *(Trifolium fragiferum)* by *Rhizobium trifolii* NA 30." Ph.D. Thesis, University of Florida. 103 pp.
Napoli, C.A. and Hubbell, D.H. (1975). *Appl. Microbiol.* **30**, 1003-1009.
Napoli, C.A., Dazzo, F.B. and Hubbell, D.H. (1975a). *Appl. Microbiol.* **30**, 123-131.
Napoli, C.A., Dazzo, F.B. and Hubbell, D.H. (1975b). Proc. 5th Austr. Leg. Confr. Brisbane, Australia.
Nissen, P. (1971). *In* "Information Molecules in Biological Systems," (L.G. Ledoux, ed.), pp. 201-212. Elsevier North Holland Press, Amsterdam.
Norris, D.O. (1959). *J. Aust. Inst. Agric. Sci.* **25**:202.
Nuti, M.P., Cannon, F.C., Prakash, R.K., Lepidi, A.A. and Schilperoort, R.A. (1979). Proc. Amer. *Rhizobium* Confr. VII.
Nutman, P.S., Doncaster, C.C. and Dart, P.J. (1973). "Infection of Clover by Root-Nodule Bacteria." British Film Institute, London.
O'Gara, F. and Schanmugam, K.T. (1978). *Proc. Nat. Acad. Sci. U.S.A.* **75**, 2343-2347.
Osborne, M.J., Cynkin, M.A., Gilbert, J.M., Muller, L. and Singh, M. (1972). *Meth. Enzymol.* **28**, 583-601.
Planqué, K., and Kijne, J.W. (1977). *FEBS Lett.* **73**, 64-66.
Pueppke, S.G., Keegstra, K., Ferguson, A.L. and Bauer, W.D. (1978). *Plant Physiol.* **61**, 779-784.
Pull, S.P., Pueppke, S.G., Hymowitz, T. and Ord, J.H. (1978). *Science* **200**, 1277-1279.
Raa, J., Robertson, B., Solheim, B. and Tronsmo, A. (1977). *In* "Cell Wall Biochemistry Related to Specificity in Host-Plant Pathogen Interactions" (B. Solheim and J. Raa, eds.), pp. 11-30. Universitets forlaget, Oslo.
Reeke, G.N., Becker, J.W., Cunningham, B.A., Wang, J.L., Yahara, I. and Edelman, G.M. (1975). *Adv. Exp. Med. Biol.* **55**, 13-33.
Reporter, M., Raveed, D. and Norris, G. (1975). *Plant Sci. Lett.* **5**, 73-76.
Rudbach, J.A., Anacker, R.L., Haskins, W.T., Milner, K.C. and Ribi, E. (1967). *J. Immunol.* **98**, 1-7.
Saunders, R.E., Carlson, R.W. and Albersheim, P. (1978). *Nature* **271**, 240-242.
Sequeira, L. (1978). *Ann. Rev. Phytopathol.* **16**, 453-481.
Solheim, B. (1975) *In* "Specificity in Plant Disease," NATO Advanced Study Institute, 1975.
Tsien, H.C. and Schmidt, E.L. (1977). *Can. J. Microbiol.* **23**, 1274-1284.
Van Brussel, A.A., Planqué, K. and Quispel, A. (1977). *J. Gen. Microbiol.* **101**, 51-56.
Van der Schaal, I. and Kijne, J. (1979). Proc. Amer. *Rhizobium* Confr. VII.
Werner, D., Wilcockson, J. and Zimmerman, E. (1975). *Arch. Microbiol.* **105**, 27-32.
Wittwer, S.H. (1977). *Basic Life Sci.* **9**, 515-519.
Wolpert, J.S. and Albersheim, P. (1976). *Biochem. Biophys. Res. Commun.* **70**, 729-737.
Zevenhuizen, L.P.T. (1978). Proc. 12th Int. Congr. Microbiol., Munich, p. 75.
Zurkowski, W. and Lorkiewicz, Z. (1978). *Genet. Res. Camb.* **32**, 311-314.

Matrix Glycoproteins in Early Mouse Development and in Differentiation of Teratocarcinoma Cells

Jorma Wartiovaara

*Department of Medical Biology, University of Helsinki,
Siltasvuorenpenger 20 A,
00170 Helsinki 17, Finland*

Ilmo Leivo

*Department of Pathology, University of Helsinki,
Haartmaninkatu 3
00290 Helsinki 29, Finland*

Antti Vaheri

*Department of Virology, University of Helsinki,
Haartmaninkatu 3
00290 Helsinki 29, Finland*

I. INTRODUCTION

Extracellular matrix components have been suggested as important factors in cell differentiation and morphogenesis (Hay, 1973; Bernfield *et al.*, 1973; Slavkin and Greulich, 1975; Lash and Vasan, 1977). There is recent evidence that extracellular matrix molecules are involved also in the very early stages of embryogenesis (Zetter and Martin, 1978;

ISBN 0-12-612984-3

Adamson and Ayers, 1979; Wartiovaara *et al.*, 1979). Biochemical studies of matrix components in the early embryo are tedious because of the minute amounts of material available. During recent years specific antibodies have been raised against several defined matrix proteins (e.g. Gunson and Kefalides, 1976; Timpl *et al.*, 1977). This has provided a means to study the appearance and localization of the components in small structures. Another advancement in the field has been the establishment of teratocarcinoma cell lines that differentiate in culture and mimic events of early embryogenesis (Martin, 1975; Jacob, 1977). In the following, we first describe some of the salient features of extracellular matrix components of early mouse development, and of teratocarcinoma differentiation , and then summarize recent findings on matrix proteins in early mouse embryogenesis and teratocarcinoma differentiation.

A. *Components of the Extracellular Matrix and Their Localization*

The extracellular matrix contains collagenous and noncollagenous glycoproteins and glycosaminoglycans. In most connective tissues collagen is the major molecule. *Collagen* (for refs. see: Ramachandran and Reddi, 1976; Fessler and Fessler, 1978; Prockop *et al.*, 1979) exists as distinct genetic types (Table I), and has a very characteristic structure. Collagen molecules are composed of three polypeptide chains, known as α chains, folded into rod-like triple-helical molecules and fibrils. Type I collagen is the major connective tissue protein of a number of tissues including skin, bone, dentin and tendon. A small amount of type I α_1 trimer has been detected in embryonic chick tendons and calvaria. Type II collagen is the major collagen in cartilage, but has been detected also in the vitreous body and neural retinal tissues. Type III is present in large amounts in many adult tissues such as arteries, muscle, lung and liver. The ratio of type III/type I collagen decreases in skin after birth.

The biosynthesis of the interstitial collagens is characterized by multiple post-translational modifications (cf. Prockop *et al.*, 1979). Collagens are synthesized as procollagens. The intracellular processing includes removal of presequences, hydroxylation of proline and lysine residues, glycosylation of hydroxylysine residues and the extension peptides, and association and disulfide bonding of procollagen chains into triple helices. After secretion to the extracellular space, the extension peptides are cleaved off the procollagen molecules. The resulting collagen molecules associate into fibrils which obtain their characteristic tensile strength through covalent cross-linking.

TABLE I

Major Defined Matrix Components

Molecule	Chain composition	Distribution
Procollagens — Collagens		
Type I	$[\alpha_1(I)]_2\alpha_2$	Interstitial
I — trimer	$[\alpha_1(I)]_3$	Interstitial
II	$[\alpha_1(II)]_3$	Interstitial
III	$[\alpha_1(III)]_3$	Interstitial
IV	$[\alpha_1(IV)]_3$	Basement membranes
AB	$\alpha A\ (\alpha B)_2$	Basement membranes
Fibronectin	2 x 220 kd	Loose connective tissue matrix Basement membranes Body fluids
Laminin	n x 220 kd n x 440 kd	Basement membranes
Tropoelastin — Elastin	70 kd	Elastic fibers
Microfibrillin	270 kd	Elastic fibers elsewhere in matrix?
Glycosaminoglycans		Matrix and body fluids
Sulfated proteoglycans		
Hyaluronic acid		

The interstitial collagens have been structurally well characterized, and the main parts of their amino acid sequences have been determined. Affinity purified antibodies specific for the helical regions or propeptides of the different types of interstitial collagens have been raised and used in various studies (for refs. see Timpl *et al.*, 1977), such as localization of the antigens by immunofluorescence, quantitation by radioimmunoassay and immunoprecipitation.

In contrast, little information is available about the basement membrane collagens (Kefalides, 1978; Robert *et al.*, 1979). Type IV collagen has been identified as the major component of basement membranes found in a variety of tissues. Type IV collagen is nonfibrillar and nonstriated in contrast to the other collagens, and appears to be laid down as procollagen in an amorphous form. Another type of basement membrane collagen, composed of αA and αB polypeptides, has been detected in placenta, skin and a few other tissues. It now seems that basement membrane collagen is a group of several related proteins similar to the interstitial collagens. Antibodies have been raised against type IV and AB collagens, but it remains possible that they react with several distinct subtypes of basement membrane collagen.

A characteristic feature of the noncollagenous glycoprotein *fibronectin* (Vaheri and Mosher, 1978; Yamada and Olden, 1978), is that it is both a major insoluble protein in connective tissue and a major soluble protein in various body fluids such as plasma, amniotic fluid and cerebrospinal fluid. Several types of adherent cells, e.g. connective tissues cells and endothelial and certain epithelial cells, synthesize large amounts of fibronectin in cell culture conditions. Unlike normal cells, malignantly transformed cells do not usually lay down fibronectin in a pericellular matrix. Little is known about the mechanism of deposition of fibronectin in cell cultures or *in vivo*. Other characteristic features of fibronectin include interactions with collagen and with fibrin and heparin in the cold. Fibronectin seems to be relatively sensitive to cleavage by plasmin and certain other serine proteases. There is ample evidence that fibronectin functions as a cell adhesion protein (cf. Yamada and Olden, 1978), and it is possibly important also in cell migration *in vitro* (Ali and Hynes, 1978) and *in vivo* (Kurkinen *et al.*, 1979). Fibronectin has been detected in basement membranes, in the mesenchyme, in loose connective tissues, in chondroblasts, predentin and in the early stages of wound healing, but not in mature matrices such as cartilage, dentin or tendon (see Dessau *et al.*, 1978; Thesleff *et al.*, 1979; Stenman and Vaheri, 1978). These and other findings have suggested an organizing role for fibronectin in connective tissue formation. Soluble fibronectin may be quantitated, localized and precipitated with the aid of antibodies. Because of species cross-reactivity it should be noted, however, that for tissue culture studies the antibodies should be absorbed with e.g. agarose-conjugated bovine serum in order to avoid reactions with the bovine fibronectin present in calf serum.

Laminin (Timpl *et al.*, 1979) is a large noncollagenous glycoprotein that is detected in adult tissues in basement membranes only. Laminin was originally isolated from a mouse tumor basement membrane, and is both in structural and immunological terms distinct from fibronectin. Antibodies may be used to localize laminin but the relative insolubility of this glycoprotein complicates studies on immunological quantitation and precipitation. However, almost all antigenic reactivity resides in a large cystine-rich fragment obtained by pepsin digestion. Several types of adherent cells have been shown to synthesize laminin in culture. Certain types of cells, such as amniotic epithelial cells, also deposit laminin in pericellular matrix form. Reduction of disulfide bonds of laminin produces polypeptides, some of which co-migrate in gel electrophoresis with the subunits of fibronectin (220 kd). Hence, electrophoretic mobility and accessibility to surface labeling are insufficient criteria for identification of fibronectin. Another distinguishing property between laminin and fibronectin is that only the latter binds to gelatin-agarose.

Chung *et al.* (1979) recently identified an external glycoprotein, GP-2, in cultures of the mouse embryonal carcinoma-derived cell line M1536-B 3. There is now immunological evidence (Timpl *et al.*, 1979) that GP-2 is either laminin or a closely related protein.

The elastic fibers *in vivo* contain in addition to *elastin*(Sandberg, 1976) the *microfibrillar protein* (Muir *et al.*, 1976) that has an apparent molecular weight of 270,000, and has been suggested to be involved in the organization of the elastic fiber. Smooth muscle cells have been shown to synthesize in culture conditions both of these proteins, the elastin in the form of tropoelastin (Narayanan *et al.*, 1976; Jones *et al.*, 1979). Immunological studies on elastin and microfibrillin are in progress in several laboratories but so far little is known about the appearance and distribution of these proteins in embryonic or adult tissues.

Glycosaminoglycans, GAG (Rodén and Schwartz, 1975; Lindahl and Höök, 1978), with the exception of hyaluronic acid, occur *in vivo* mainly as proteoglycans, i.e. in covalent association with protein. Studies on the distribution of GAG in the adult and embryonic tissues have been based on chemical analyses, and on other tedious and relatively insensitive methods such as staining reactions before and after treatment of tissues with GAG-specific enzymes (for ref. see Slavkin and Greulich, 1975). Antibodies have been raised against the core proteins of cartilage proteoglycans, but the charged nature of the molecule complicates immunological localization.

B. *Early Development of the Mouse Embryo*

An outline of the events of mouse embryonic development relevant in this context is presented below. At the 8-16 cell stage of cleavage the previously loose spherical cells of the embryo assume a nonspherical compacted form. During this process an increase in intercellular apposition is seen. Changes occur in cell surface microvilli and cytoskeletal elements (Ducibella *et al.*, 1977). Also formation of juctional membrane complexes between the compacted cells takes place (Ducibella *et al.*, 1975; Magnuson *et al.*, 1977). At the 16-cell stage two cells become localized to the interior of the embryo, forming the future inner cell mass (ICM) (Barlow *et al.*, 1972; Graham and Lehtonen, 1979). In the blastocyst the ICM develops into ectoderm and endoderm, and basement membrane material is deposited between these germ layers. Later, basement membranes are present between all the three germ layers. Extracellular material has been described on the inner aspect of the trophectoderm already in the preimplantation blastocyst (Schlafke and

Enders, 1963; Nadijcka and Hillman, 1974). The Reichert's membrane is later laid down at this same site between the parietal endoderm and the trophoblast. Other extraembryonic basement membranes are formed in the amnion, the chorion and the visceral yolk sac (cf. Snell and Stevens, 1966). The mesoderm, the third germ layer, arises from the ectoderm in the primitive streak area of the 6 ½ day embryo. Early derivatives of the mesoderm include head and heart mesenchyme, the somite, and the allantois.

C. *Teratocarcinoma Differentiation as a Model*

Teratocarcinomas are malignant tumors of embryonic or germ cell origin. They are characterized by the presence of a distinct stem cell type termed embryonal carcinoma and its numerous differentiated non-malignant derivatives representing each of the primary germ layers (cf. Stevens, 1967; Pierce, 1967; Damjanov and Solter, 1974). Mouse teratocarcinoma has been used as a model system for the study of early mammalian development (for refs. see Sherman and Solter, 1975; Martin, 1975; Graham, 1977; Jacob, 1978). The differentiation of embryonal carcinoma cells *in vivo* and *in vitro* parallels in many aspects the early events of mammalian embryogenesis. Under certain conditions embryonal carcinoma cells, corresponding to the pluripotential cells of the early embryo, form aggregates termed embryoid bodies which are analogous to the inner cell mass of the 5 day mouse blastocysts (Fig. 1).

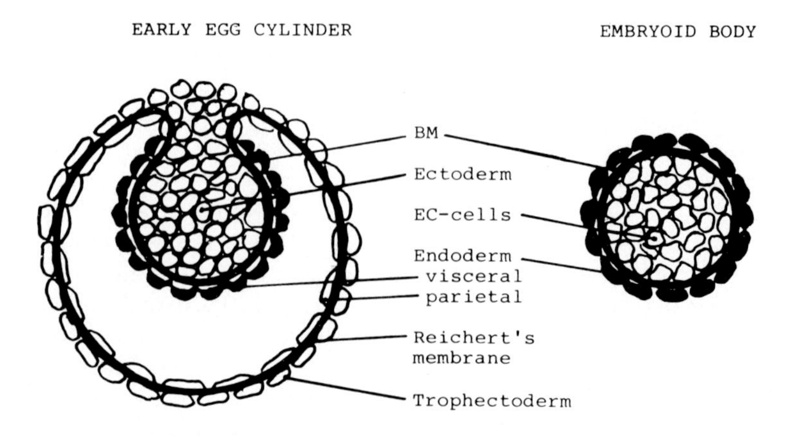

Fig. 1. Analogy between structures in mouse egg cylinder and in teratocarcinoma embryoid body. BM, basement membrane material; EC-cells, embryonal carcinoma cells.

The embryoid bodies differentiate in suspension into structures with an inner core of embryonal carcinoma cells surrounded by an outer rim of endodermal cells which is separated from the core by a layer of extracellular material (Martin and Evans, 1975a). If allowed to attach to a substratum, the embryoid bodies can differentiate further to form a variety of cell types. Clonal lines of mouse embryonal carcinoma cells have been isolated which, in contrast to the above described pluripotent cell lines, cannot differentiate or form embryoid bodies and are therefore termed nullipotent (Martin and Evans, 1975b). Recently, also some human teratocarcinoma cell lines have been isolated (Hogan *et al.*, 1977; Holden *et al.*, 1977; Zeuthen *et al.*, 1980). A detailed description of the teratocarcinoma system can be found elsewhere in this volume.

II. MATRIX PROTEINS IN EARLY EMBRYOGENESIS

A. *Preimplantation Development*

Immunofluorescence studies in early mouse embryos on the appearance and distribution of fibronectin (Wartiovaara *et al.*, 1979), laminin and different types of collagens (Leivo *et al.*, 1980) have recently been carried out in our laboratory. Monospecific antibodies against soluble fibronectin (Stenman *et al.*, 1977), type I and II collagens, type III procollagen (Timpl *et al.*, 1977), and type IV collagen and laminin (Timpl *et al.*, 1979) were used. A more intense fluorescence for the studied matrix proteins was obtained by pre-treating the samples with hyaluronidase.

In the earliest preimplantation stages of mouse development from ova to 4-8 cell stage, no specific staining was detected with antibodies against collagens I-IV, fibronectin or laminin (Leivo *et al.*, 1980). It should be mentioned, however, that the antibodies gave some background staining in these embryos, but blocked antibodies gave a similar staining, indicating non-specific adsorption.

The first specific staining for laminin is seen in the 3 day compacted morula of 16 to 32 cells. Cytoplasmic laminin fluorescence is apparent in most if not all of the cells at this stage (Fig. 2a), and laminin is also detected lining intercellular contours of the morulae (Fig. 2b). None of the other proteins studied could be detected at this stage, neither intra- nor extracellularly.

Notable changes occur as the blastocyst is formed. In the 3 day early blastocyst laminin is found in the two cell types forming the embryo at this stage: the cells of the ICM and the trophectoderm. Fluorescence for type IV collagen is first seen at this stage, and it is localized to the ICM

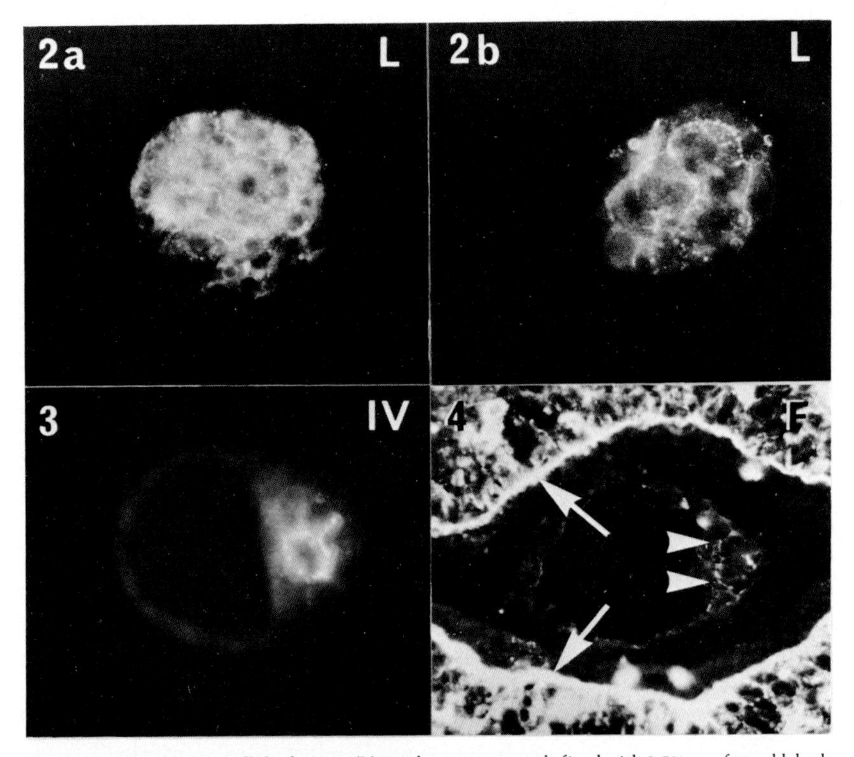

Fig. 2a. Staining for intracellular laminin (L) in 3 day mouse morula fixed with 3.5% paraformaldehyde and 0.05% nonionic detergent Nonidet P-40. Distinct cytoplasmic laminin fluorescence is seen in the cells. x 600.

Fig. 2b. Staining for extracellular laminin (L) in 3 day compacted morula fixed with 3.5% paraformaldehyde. Granular laminin fluorescence lines several intercellular contours. x 600. (From Leivo *et al.*, 1980, in press).

Fig. 3. Staining for type IV collagen (IV) in a 3 day blastocyst fixed for study of intracellular structures as in Fig. 2a. Distinct fluorescence is seen in the inner cell mass. x 600.

Fig. 4. Staining for fibronectin (F) in an implanting 4.5 day late blastocyst in a uterine crypt. Fixation in ethanol-glacial acetic acid (v/v, 99/1). Faint fluorescence is seen between the cells of the inner cell mass (arrowheads). Strong fluorescence is seen in the uterine epithelial basement membrane (arrows). x 600.

(Fig. 3), now in the process of differentiation into ectoderm and primitive endoderm (Enders *et al.*, 1978). Similar results with immunoperoxidase localization have been obtained by Adamson and Ayers (1979). Type IV collagen is sometimes seen also on the inner aspect of the trophectoderm facing the blastocoel cavity. In ultrastructural studies extracellular material has been described at this site well before the appearance of the

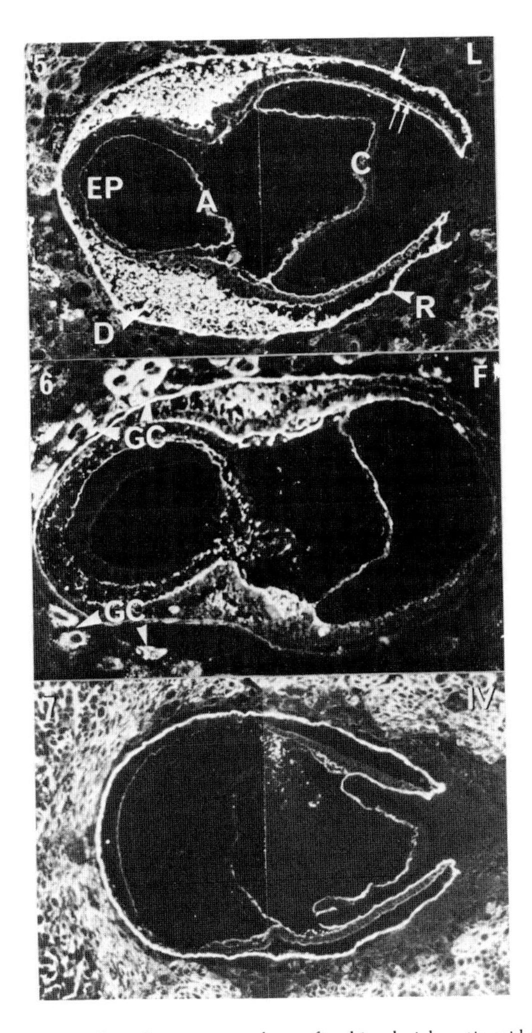

Figs. 5-7. Sagittal sections of 7.5 day mouse embryos fixed in glacial-acetic acid (v/v, 99/1). x 400.

Fig. 5. Staining for laminin. Fluorescence is seen between the ectodermal, mesodermal and endodermal cell layers both in the embryo proper (EP), and in the amnion (A) and chorion (C). Bright fluorescence is localized to the Reichert's membrane (R). Some fluorescence is also evident in the parietal (arrow) as well as in the visceral (double arrows) endoderm cells. Necrotic cell debris in the yolk cavity stains brightly (D).

Fig. 6. Staining for fibronectin. Fluorescence similar to that of laminin (above) is observed. Additionally, the trophoblastic giant cells (GC) surrounding the embryo stain brilliantly.

Fig. 7. Staining for type IV collagen. Staining similar to that seen above for laminin and fibronectin is seen in the embryo. The trophoblastic giant cells are negative.

true Reichert's membrane on the 6th day of development (Schlafke and Enders, 1963; Nadijcka and Hillman, 1974). Fibronectin is another matrix protein to appear at this stage. Similar to type IV collagen, fibronectin is first seen in the ICM of the blastocyst and on the inner surface of the trophectoderm (Fig. 4; see also Wartiovaara *et al.*, 1979). Zetter and Martin (1978) also described fibronectin in the 3-4 day ICM isolated by immunosurgery.

B. *Postimplantation Development*

After implantation the inner cell mass expands into the enlarging blastocoel cavity, and the first embryonic basement membrane is formed between the ectoderm and endoderm. At this egg cylinder stage, type IV collagen, laminin and fibronectin are deposited in the embryonic basement membrane as well as in the extra-embryonic Reichert's membrane that forms between the trophectoderm and the parietal endoderm. These three proteins can constantly be found in the parietal and visceral endoderm cells as soon as they are distinguished. Only fibronectin has a distinct apical distribution in the cells of the visceral endoderm (Wartiovaara *et al.*, 1979). Ultrastructural studies indicate that the apical cytoplasm of these cells contains vesicles with endocytotic function (King and Enders, 1970). It thus seems possible that the apical fibronectin might represent endocytosis of soluble fibronectin rather than synthesis. It is noteworthy that the ectoderm is negative for both fibronectin and type IV collagen. Laminin seems to be detectable to some extent also in ectodermal tissues in the stages studied.

Later, fluorescence for laminin, fibronectin and type IV collagen is seen in all the embryonic and extraembryonic basement membranes, including those of the chorion, the amnion and the visceral yolk sac (Figs. 5-7). These proteins are found also in mesodermal tissues and in the head and somite mesenchyme, and in the somites and allantois of the 8 day embryo (Leivo *et al.*, 1980).

During the first seven days of development no type I or type III collagen is seen in the embryo. At the 8 day 4-somite stage, however, type I and III collagens appear simultaneously in mesenchymal tissues and in basement membranes bounding mesenchyme (Figs. 8 and 9). These interstitial collagens now become detectable in the somites, and the head and heart mesenchyme, and in their basement membranes. Type II collagen, found early in chick notochord (von der Mark *et al.*, 1976), does not appear by this stage in the nascent mouse notochord. Interestingly, fibronectin, but not laminin or type IV collagen, is abundant in the

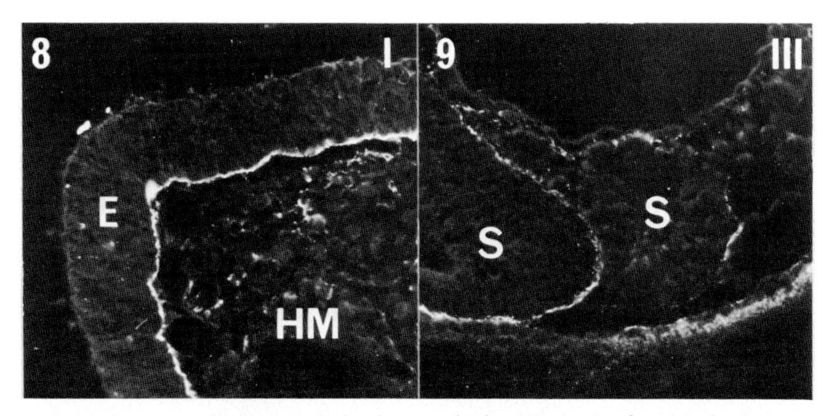

Fig. 8. Staining for type I collagen in the head region of 8-day, 3-4 somite embryo. Patchy fluorescence is observed in the head mesenchyme (HM) and the basement membrane between the mesenchyme and the epithelium (E). x 600. (From Leivo *et al.*, 1980).

Fig. 9. Staining for type III procollagen in the somite region of 8-day, 3-4 somite embryo. Fluorescence is seen surrounding the somites (S). x 600.

trophoblastic giant cells surrounding the embryo before the formation of the placenta (Fig. 6). These cells have invasive and phagocytic properties and secrete large amounts of the protease plasminogen activator (Sherman *et al.*, 1976; Strickland *et al.*, 1976).

III. MATRIX PROTEINS IN TERATOCARCINOMA DIFFERENTIATION

Several recent reports describe the matrix components in differentiating teratocarcinoma cells (Wartiovaara *et al.*, 1978a; 1978b;Zetter and Martin, 1978; Adamson and Ayers, 1979; Wolfe *et al.*, 1979).

Sectioned embryoid bodies of the mouse teratocarcinoma line OC15S1 were studied in our laboratory. Type IV collagen was detected by indirect immunofluorescence in the peripherally situated endoderm-like cells (Fig. 10). A similar pattern of fibronectin staining is seen in embryoid bodies of a number of cell lines including F9 and OC15S1 (Figs. 11 and 12). In older cystic embryoid bodies a bright fibronectin band is seen, corresponding to the basement membrane material found beneath the endoderm cells (Martin and Evans, 1975a), now forming the wall of the cyst (Fig. 12).

Differentiation of embryonal carcinoma cells in the endodermal

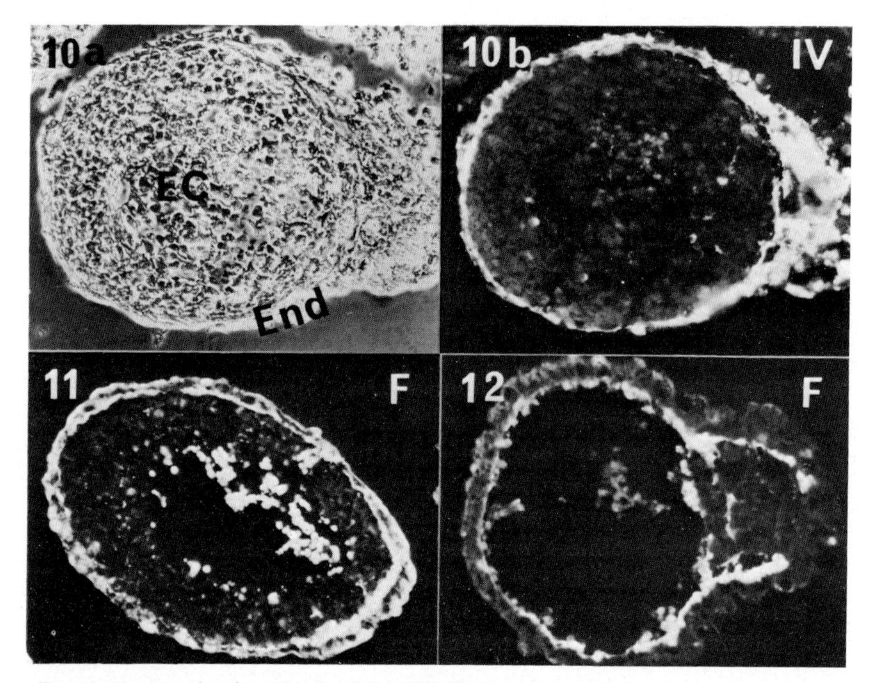

Fig. 10a. Section of embryoid body of line OC15S1 mouse teratocarcinoma grown for 6 days in suspension culture. EC, embryonal carcinoma cells; End, endoderm cells. x 650.

Fig. 10b. Staining of the same section as in Fig. 10a for type IV collagen. Fluorescence is seen in the peripherally located endoderm layer. x 650.

Fig. 11. Ten-day cystic embryoid body of mouse embryonal carcinoma line F9 stained for fibronectin. Fluorescence is seen in the peripheral endoderm layer and between embryonal carcinoma and endoderm layers. Necrotic cells in the central cyst are unspecifically stained. x 650.

Fig. 12. Fifteen-day cystic embryoid body of line OC15S1 where the EC core has largely degenerated away. Staining for fibronectin. Fibronectin fluorescence beneath the endoderm layer forms now a lining of the cyst. x 650.

direction takes place also in monolayer cultures under serum deprivation and at low cell density. Such monolayer cultures are well suited for immunofluorescence studies. When stained for laminin, both endoderm-like cells and embryonal carcinoma cells stain positively (Fig. 13). With anti-fibronectin (Fig. 14) and anti-type IV collagen (Fig. 15) antibodies, flattened endoderm-like cells were stained whereas embryonal carcinoma cells, characterized also by their high alkaline phosphatase activity (Fig. 14a), remained negative or stained very faintly. The flattened endoderm-like cells deposit both fibronectin (Fig. 14b), laminin

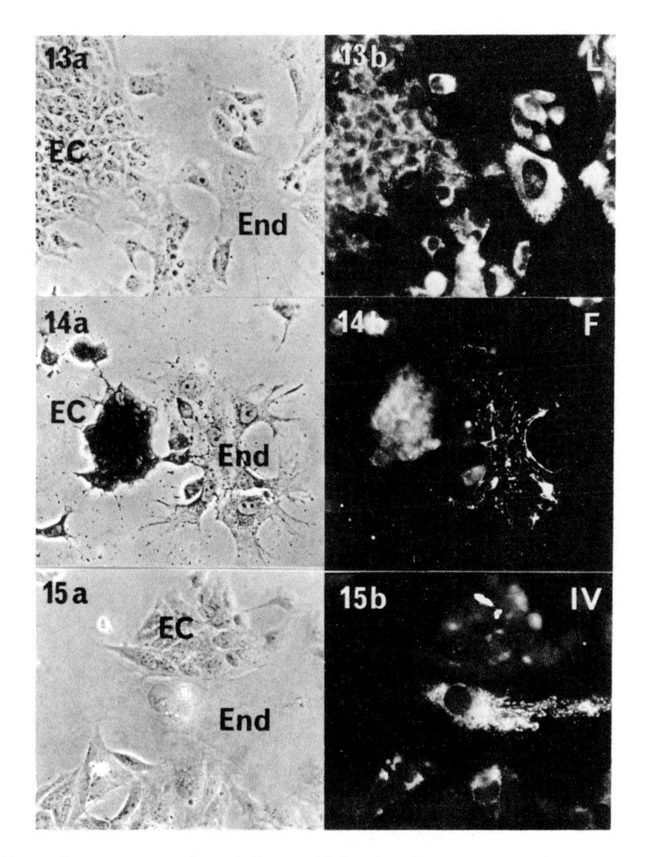

Fig. 13a. Three-day sparse monolayer culture of OC15S1 cells grown in 3% fetal calf serum, allowing endodermal differentiation. Fixation for intracellular fluorescence as in Fig. 2a. EC, embryonal carcinoma cells; End, flattened endoderm-like cells. x 250.

Fig. 13b. Staining for laminin of the same cells as in Fig. 13a shows strong cytoplasmic fluorescence both in embryonal carcinoma cells and in endoderm-like cells. x 250.

Fig. 14a. F9 cells grown as in Fig. 13. Fixation for extracellular fluorescence as in Fig. 2b. Diazo-staining for alkaline phosphatase gives strong precipitate in embryonal carcinoma (EC) but not in endoderm-like (End) cells. x 250.

Fig. 14b. Immunofluorescence for fibronectin in the same F9 cells as in Fig. 14a demonstrates strong fibrillar staining of endoderm-like cells. Some diffuse staining of EC-cells is seen. x 250.

Fig. 15a. OC15S1 grown and fixed as in Fig. 13. EC, embryonal carcinoma cells; End, endoderm-like cells. x 350.

Fig. 15b. Staining for type IV collagen of similar cells as in Fig. 15a shows strong fluorescence in the flattened endoderm-like cell. Faint fluorescence is seen in partly flattened cells and also in embryonal carcinoma cells. x 350.

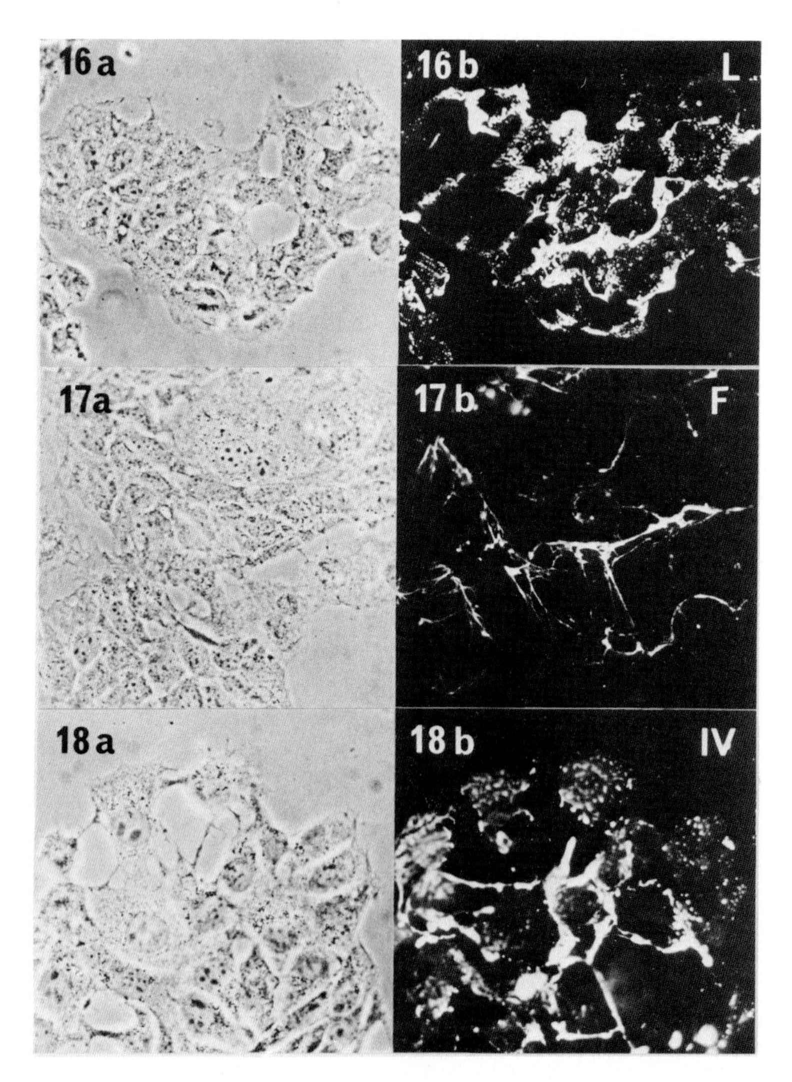

Fig. 16-18. Parietal yolk sac carcinoma (PYS-2) cells grown in 10% fetal calf serum and fixed for surface fluorescence as in Fib. 2b. x 300.

Fig. 16. Staining for laminin demonstrates intense fluorescence of extracellular matrix produced by the cells. In addition to a fibrillar pattern of fluorescence a punctate pattern is seen between the cells and the substratum.

Fig. 17. Staining for fibronectin results in distinct fluorescence of fibrillar type only.

Fig. 18. Staining for type IV collagen shows fluorescence both in strong fibrillar and in distinct punctate form.

and type IV collagen in a fibrillar pericellular matrix. Neither endoderm cells nor embryonal carcinoma cells stain positively for type III collagen.

Similar studies with monolayer cultures were performed with parietal yolk sac carcinoma PYS-2 cells introduced by Lehman *et al.* (1974). PYS-2 cells secrete large amounts of extracellular material analogous to Reichert's membrane that forms in the embryo from parietal endoderm cells. Staining for laminin, fibronectin and type IV collagen (Figs. 16-18) gave strong extracellular-type staining for all these matrix components; laminin and type IV collagen were the most abundant. In addition to the fibrillar pattern, laminin and type IV collagen formed also punctate substrate-attached deposits in the cell periphery (Figs. 16b and 18b). PYS-2 cells stained negative for type III and type I collagen. This result is in accordance with the lack of interstitial collagens in the Reichert's membrane *in vivo*.

The results with embryoid bodies and monolayer cultures of embryonal carcinoma cells and PYS-2 cells are analogous to the results with the early embryos. Endoderm cells seem to be associated with extracellular matrix material containing at least laminin, fibronectin and type IV collagen. In addition, the presence of laminin in embryonal carcinoma cells is in accordance with its presence in the embryonic cells after compaction and before endodermal differentiation of the inner cell mass.

IV. DISCUSSION AND CONCLUSIONS

The results presented in this paper on the appearance of matrix glycoproteins and their localization in the early mouse embryo and in cells of teratocarcinoma embryoid bodies are summarized in Tables II and III.

Of the matrix components studied, laminin was the first to be detected by immunofluorescence in the early embryo. The finding that laminin was detected already in the compacted morula might imply a role for this protein in the increased intercellular apposition seen at this stage (Ducibella *et al.*, 1977), but little attention has been given so far to the presence of extracellular material in morulae, although several ultrastructural aspects of morulae have recently been studied (Nadijcka and Hillman, 1974; Enders *et al.*, 1978). Analogous to the appearance of laminin in compaction seems to be its appearance in the aggregating cells of the mouse metanephrogenic mesenchyme during differentiation of kidney tubules (Ekblom *et al.*, 1980). After formation of kidney tubules, laminin is restricted to basement membranes, as is the case in adult

TABLE II

Appearance of Matrix Glycoproteins in Early Mouse Embryos

Stage	LAM	FN	IV	III	I	II
Ovum and zygote	–	–	–	–	–	–
2– to 8–cell cleavage	–	–	–	–	–	–
compacted morula	+	–	–	–	–	–
3-day blastocyst	+	+	+	–	–	–
7-day embryo	+	+	+	–	–	–
8-day embryo	+	+	+	+	+	–

LAM = laminin; FN = fibronectin; IV, III, I and II refer to the respective genetic types of (pro)collagens

TABLE III

Localization of Matrix Glycoproteins in Mouse Embryos and Teratocarcinoma Cells

Tissue or cell type	Non-collagenous		Collagenous		
	Laminin	Fibronectin	Type IV	Type III	Type I
Mouse embryo:					
Precompaction blastomeres	–	–	–	–	–
Compacted morula cells	+	–	–	–	–
Blastocyst inner cell mass	+	+	+	–	–
Ectoderm	+	–	–	–	–
Endoderm (visceral, parietal)	+	+	+	–	–
Mesoderm	+	+	+	–	–
Differentiating mesenchyme	+	+	+	+	+
Basement membranes					
— early postimplantation	+	+	+	–	–
— over 8 d mesenchymal	+	+	+	+	+
— Reichert's membrane	+	+	+	–	–
Trophoblast	+	+	+	–	–
Teratocarcinoma cells:					
Embryonal carcinoma	+	±	±	–	–
Endoderm	+	+	+	–	–
Parietal yolk sac carcinoma	+	+	+	–	–

tissues (Timpl *et al.*, 1979). The above findings, as well as the localization of laminin under PYS-2 cells (Fig. 16b) suggest a role for laminin in cell adhesion.

The appearance of fibronectin in the inner cell mass of the blastocyst, as reported previously (Zetter and Martin, 1978; Wartiovaara *et al.*, 1979), coincides with the formation of the first embryonic basement membrane. Localization of fibronectin to basement membranes and loose connective tissue is also characteristic for this protein later in development (Wartiovaara *et al.*, 1979; Thesleff *et al.*, 1979; Kurkinen *et al.*, 1979), as well as in adult tissues (Stenman and Vaheri, 1978). A conspicuous finding was the bright cytoplasmic fibronectin fluorescence in the trophoblastic giant cells surrounding the embryo. These cells are known to invade the uterine wall, become phagocytic and produce large amounts of proteolytic enzymes (Strickland *et al.*, 1976). It is of interest that monocytes, after conversion to phagocytic migratory macrophages, start to produce both plasminogen activator (Unkeless *et al.*, 1974) and to secrete fibronectin (Alitalo *et al.*, 1980).

Fibronectin, unlike laminin, was not found in the embryonic ectoderm. Also cultured cells of ectodermal origin have this characteristic, as has been reported for keratocytes, cervix epithelial cells, and various carcinoma cells (cf. Vaheri and Mosher 1978).

Of the collagenous matrix components, basement membrane collagen type IV is similarly located in the preimplantation embryo and in the embryoid bodies as fibronectin. In the implanted embryos the distribution of type IV collagen was similar to that of fibronectin (Table III), but it was not found in the trophoblastic giant cells. Our results on type IV collagen are in agreement with the immunoperoxidase findings of Adamson and Ayers (1979). In both studies this collagen was found in the endoderm, in the basement membranes of the embryo proper, and in those of the Reichert's membrane, the amnion and the visceral yolk sac. As pointed out in the Introduction, the basement membrane collagens probably form a group of several related proteins and it remains possible that they may have individual developmental characteristics.

Interstitial collagens, types I and III, seem to be associated with mesodermal differentiation in the early embryo. When the mesoderm, negative for the interstitial collagens, differentiates into the mesenchymal tissues of e.g. the head and the heart, both type I and type III collagens appear. At this stage of development these collagens seem to be present also in basement membranes lining the differentiating mesenchyme, a localization pattern not to be found in adult structures.

In conclusion, the different matrix components studied seem to have a characteristic developmental sequence in their appearance in the mouse embryo. They also show specificity as to the structures they become associated with. The abundance of extracellular matrix and its orderly structure in the developing embryo is notable. Each primary germ layer and its derivatives become compartmentalized by a limiting layer of extracellular material. In addition to giving the embryo an extracellular backbone, the extracellular material (ECM) may well serve as an adhesive substratum and play a role in cell movement. The constant changes in shape of the growing embryo require also remodeling of the extracellular scaffold. The induction of proteolytic activity capable of breaking down the ECM components has been detected in the early embryo (Sherman *et al.*, 1976; Strickland *et al.*, 1976; Bode and Dziadek, 1979). The teratocarcinoma cells seem to mimic the early embryo in respect to the formation of the ECM, as also demonstrated in recent biochemical studies (Chung *et al.*, 1979; Adamson *et al.*, 1979). The teratocarcinoma model may be useful also in analysis of the remodeling process of extracellular matrix material in early embryogenesis.

ACKNOWLEDGEMENTS

We thank Ms. Elina Waris for expert technical assistance during various stages of this work. The original studies were supported by grants from the Sigrid Jusélius Foundation, Helsinki, The Finnish Medical Research Council, the Paulo Foundation, Helsinki, and The National Institutes of Health, DHEW (grant no. CA-24605).

REFERENCES

Adamson, E.D. and Ayers, S.E. (1979). *Cell* **16**, 953-965.
Adamson, E.D., Gaunt, S.J. and Graham, C.F. (1979). *Cell* **17**, 469-476.
Ali, I.U. and Hynes, R.O. (1978). *Cell* **14**, 439-446.
Alitalo, K., Hovi, T. and Vaheri, A. (1980)., *J. Exp. Med.*, in press.
Barlow, P.W., Owen, D.A.S. and Graham, C.F. (1972). *J. Embryol. Exp. Morphol.* **27**, 431-445.
Bernfield, M.R., Cohn, R.H. and Banerjee, S.D. (1973). *Amer. Zool.* **13**, 1067-1083.
Bode, V.C. and Dziadek, M.A. (1979). *Develop. Biol.*, in press.
Chung, A.E., Jaffe, R., Freeman, I.K., Vergues, J.-P., Braginski, J.E. and Carlin, B. (1979). *Cell* **16**, 277-287.
Damjanov, I. and Solter, D. (1974). *Curr. Top. Pathol.* **59**, 69-130.
Dessau, W., Sasse, J., Timpl, R., Jilek, F. and von der Mark, K. (1978). *J. Cell Biol.* **79**, 342-355.
Ducibella, T., Albertini, D.F., Anderson, E. and Biggers, J.D. (1975). *Develop. Biol.* **45**, 231-250
Ducibella, T., Ukena, T., Karnovsky, M. and Anderson, E. (1977). *J. Cell Biol.* **74**, 153-167.
Ekblom, P., Alitalo, K., Vaheri, A., Timpl, R. and Saxén, L. (1980)., *Proc. Nat. Acad. Sci. U.S.A.* **77** in press.
Enders, A.C., Given, R.L. and Schlafke, S. (1978). *Anat. Rec.* **190**, 65-78.
Fessler, J.H. and Fessler, L.I. (1978). *Ann. Rev. Biochem.* **47**, 385-417.
Graham, C.F. (1977). *In* "Concepts in Mammalian Embryogenesis" (M.J. Sherman, ed.), pp. 315-394. M.I.T. Press, Cambridge, Mass.
Graham, C.F. and Lehtonen, E. (1979). *J. Embryol. Exp. Morphol.* **49**, 277-294.
Gunson, D.E. and Kefalides, N.A. (1976). *Immunology* **31**, 563-569.
Hay, E.D. (1973). *Amer. Zool.* **13**, 1085-1107.
Hogan, B., Fellous, M., Avner, P.R. and Jacob, F. (1977). *Nature* **270**, 515-518.

Holden, S., Bernard, O., Artzt, K., Whitmore, W. F. and Bennett, D. (1977). *Nature* **270**, 518-520.
Jacob, F. (1977). *Immunol. Rev.* **33**, 3-32.
Jacob, F. (1978). *Proc. Royal Soc. Ser. B.* **201**, 241-270.
Jones, P.A., Scott-Burden, T. and Gevers, W. (1979). *Proc. Nat. Acad. Sci. U.S.A.* **76**, 353-357.
Kefalides, N.A. (1978). "Biology and Chemistry of Basement Membranes". Academic Press, London.
King, B.F. and Enders, A.C. (1970). *Amer. J. Anat.* **129**, 261-288.
Kurkinen, M., Alitalo, K., Vaheri, A., Stenman, S. and Saxén, L. (1979). *Develop. Biol.* **69**, 589-601.
Lash, J.E. and Vasan, N.S. (1977). *In* "Cell and Tissue Interactions" (J.W. Lash and M.M. Burger, eds.), pp. 101-114. Raven Press, New York.
Lehman, J.M., Speers, W.C., Swartzendruber, D.E. and Pierce, G.B. (1974). *J. Cell Physiol.* **84**, 13-28.
Leivo, I., Vaheri, A., Timpl, R. and Wartiovaara, J. (1980). *Develop.Biol.*, in press.
Lindahl, U. and Höök, M. (1978). *Ann. Rev. Biochem.* **47**, 385-417.
Magnuson, T., Demsey, A. and Stackpole, C.W. (1977). *Develop. Biol.* **61**, 252-261.
Martin, G. (1975). *Cell* **5**, 229-243.
Martin, G.R. and Evans, M.J. (1975a). *Proc. Nat. Acad. Sci. U.S.A.* **72**, 1441-1445.
Martin, G.R. and Evans, M.J. (1975b). *In* "Teratomas and Differentiation" (M. J. Sherman and D. Solter, eds.), pp. 169-187. Academic Press, New York.
Muir, L.W., Bornstein, P. and Ross, R. (1976). *Eur. J. Biochem.* **64**, 105-114.
Nadijcka, M. and Hillman, N. (1974). *J. Embryol. Exp. Morphol.* **32**, 675-695.
Narayanan, A.S., Sandberg, L.B., Ross, R. and Layman, D.L. (1976). *J. Cell. Biol.* **68**, 411-419.
Pierce, G.B. (1967). *Curr. Top. Develop. Biol.* **2**, 223-246.
Prockop, D.J., Kivirikko, K.J., Tuderman, L. and Guzman,N.A. (1979). *New Engl. J. Med.* **30**, 13-24 and 77-85.
Ramachandran, G.N. and Reddi, A.H. (1976). "Biochemistry of Collagen". Plenum Press, New York.
Robert, A.M., Boniface, R. and Robert, L. (1979). "Biochemistry and Pathology of Basement Membranes". S. Karger, Basel.
Rodén, L. and Schwartz, N.B. (1975). *In* "Biochemistry of Carbohydrates" (W.J. Whelan, ed.), pp. 95-153. Butterworths, University Park Press.
Sandberg, L.B. (1976). *In* "International Review of Connective Tissue Research" (D.A. Hall and D.S. Jackson, eds.), vol. 7, pp. 159-210. Academic Press, New York.
Schlafke, S. and Enders, A.C. (1963). *J. Anat. (London)* **97**, 353-360.
Sherman, M.J. and Solter, D. (1975). "Teratomas and Differentiation". Academic Press, New York.
Sherman, M.J., Strickland, S. and Reich, E. (1976). *Cancer Res.* **36**, 4208-4216.
Slavkin, H.C. and Greulich, R.C. (1975). "Extracellular Matrix Influences on Gene Expression". Academic Press, New York.
Snell, G.D. and Stevens, L.C. (1966). *In* "The Biology of the Laboratory Mouse" (E.L. Green, ed.), pp. 205-245. McGraw-Hill Inc., New York.
Stenman, S. and Vaheri, A. (1978). *J. Exp. Med.* **147**, 1054-1064.
Stenman, S., Wartiovaara, J. and Vaheri, A. (1977). *J. Cell Biol.* **74**, 453-467.
Stevens, L.C. (1967). *Adv. Morphogen.* **6**, 1-31.
Strickland, S., Reich, E., and Sherman, M.J. (1976). *Cell* **9**, 231-240.
Thesleff, I., Stenman, S., Vaheri, A. and Timpl, R. (1979). *Develop. Biol.* **70**, 116-126.
Timpl, R., Wick, G. and Gay, S. (1977). *J. Immunol. Meth.* **18**, 165-182.

Timpl, R., Rohde, H., Gehron-Robey, P., Rennard, S.L., Foidart, J.M. and Martin, G.R. (1979). *J. Biol. Chem.*, **254**, 9933-9937.

Unkeless, J.C., Gordon, S. and Reich, E. (1974). *J. Exp. Med.* **139**, 834-850.

Vaheri, A. and Mosher, D.F. (1978). *Biochim. Biophys. Acta* **516**, 1-25.

Von der Mark, H., von der Mark, K. and Gay, S. (1976). *Develop. Biol.* **48**, 237-249.

Wartiovaara, J., Leivo, I., Virtanen, I., Vaheri, A. and Graham, C.F. (1978a). *Nature* **272**, 355-356.

Wartiovaara, J., Leivo, I., Virtanen, I., Vaheri, A. and Graham, C.F. (1978b). *Ann. N.Y. Acad. Sci.* **312**, 132-141.

Wartiovaara, J., Leivo, I. and Vaheri, A. (1979). *Develop. Biol.* **69**, 247-257.

Wolfe, J., Mautner, V., Hogan, B. and Tilly, R. (1979). *Exp. Cell Res.* **118**, 63-71.

Yamada, K.M. and Olden, K. (1978). *Nature* **275**, 179-185.

Zetter, B.R. and Martin, G.R. (1978). *Proc. Nat. Acad. Sci. U.S.A.* **75**, 2324-2328.

Zeuthen, J., Nørgaard, J.O.R., Aoner, P., Fellows, M., Wartiovaara, J., Vaheri, A., Rosén, A. and Giovanella, B.C. (1980). *Int. J. Cancer,* in press.

Use of Teratocarcinoma Cells as a Model System for Studying the Cell Surface During Early Mammalian Development

Gail R. Martin, Laura B. Grabel and Steven D. Rosen

Department of Anatomy
University of California, San Francisco
San Francisco, California 94143

ISBN 0-12-612984-3

I. INTRODUCTION

In recent years teratocarcinoma cells have become accepted as a model system for the study of early mammalian development. This acceptance is based on the close similarity between teratocarcinoma cells and normal embryonic cells, as assessed in two ways. First, under suitable conditions normal embryonic cells can readily become malignant teratocarcinoma stem cells and *vice versa*. Second, teratocarcinoma stem cells have numerous properties in common with normal embryonic cells: in particular, multipotency, but also morphological, biochemical and immunological characteristics. Much of the work that has been done in recent years has been aimed at defining the similarities between normal embryonic cells and malignant teratocarcinoma stem cells. At present we do not know which cells in the developing embryo the tumor cells resemble. It is therefore difficult to assess what particular aspects of embryogenesis it would be appropriate to study using teratocarcinoma cells. However, teratocarcinoma cells do offer the advantage of being available in large quantities and of being able to withstand cloning, culture and manipulation in ways that embryonic cells are not.

The main objective here is to present a general review of those studies in which teratocarcinoma cells have been used as a tool for examining the mammalian embryonic cell surface and its role in early development. There are available several recent, comprehensive reviews which deal with other aspects of teratocarcinoma cell biology (Jacob, 1977; Graham, 1977; Hogan, 1977; Martin, 1975, 1978; Solter and Damjanov, 1979).

A. *Teratocarcinomas and Their Relationship to the Normal Embryo*

Mouse teratocarcinomas are malignant tumors that are characterized by a variety of differentiated cell types and a distinctive cell type known as embryonal carcinoma (Fig. 1). The latter are the multipotent stem cells of the tumor, as evidenced by the fact that a single embryonal carcinoma cell can give rise to all the differentiated cell types that are observed in the tumors (Kleinsmith and Pierce, 1964). It is the presence of these multipotent undifferentiated embryonal carcinoma cells that is responsible for the malignancy (i.e. progressive growth and transplantability) of these tumors (reviewed by Pierce, 1967; Stevens, 1967; Damjanov and Solter, 1974). If no embryonal carcinoma cells are present, because they have either differentiated or died, the tumors are benign and are known as teratomas. This latter term, however, is often used to designate both the malignant and benign types of tumor.

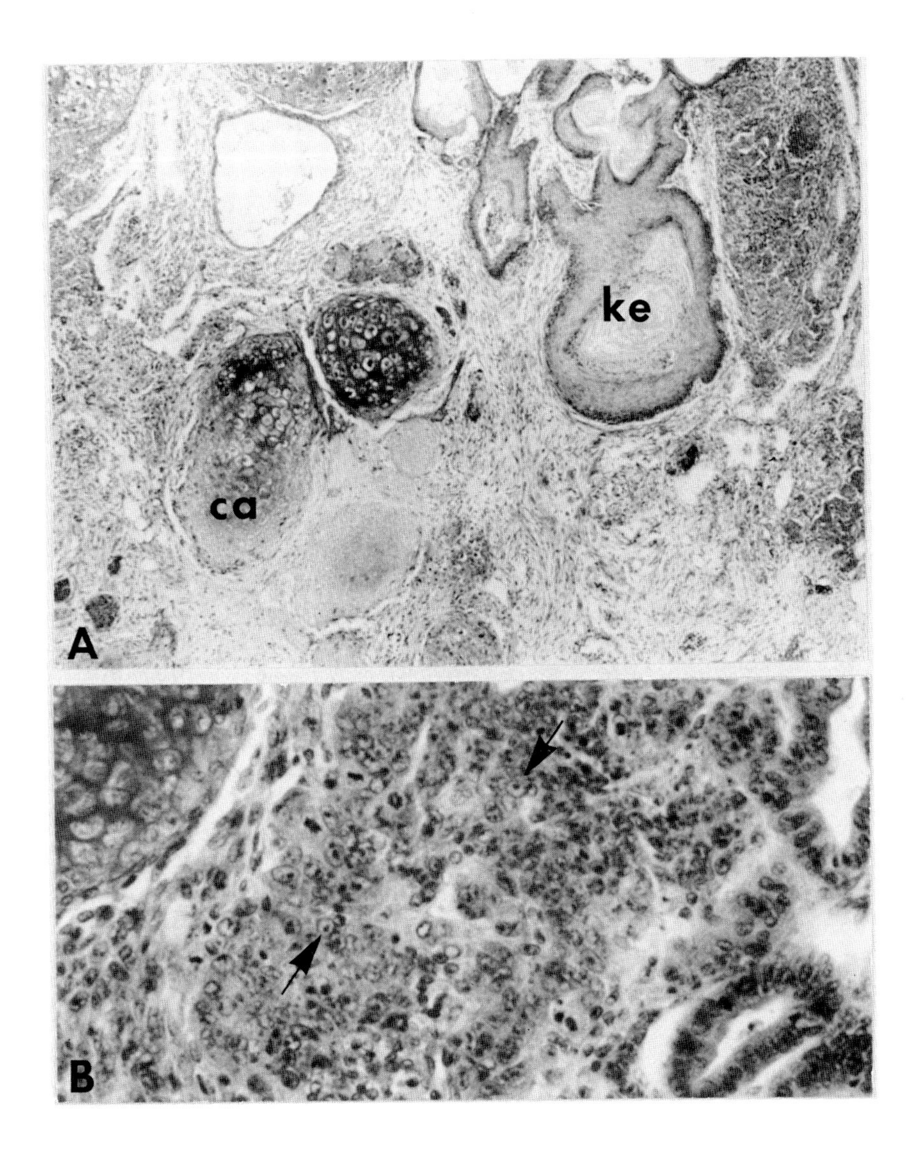

Fig. 1. Histological sections of a transplantable teratocarcinoma. (A) The tumor contains several differentiated cell types in a typically disorganized array. Cartilage (ca) and keratinizing epithelium (ke) are readily distinguished (approx. 35X). (B) Detail of embryonal cells found in such tumors (approx. 150X). The arrows point to particularly clear examples of this cell type: there is relatively little cytoplasm and a large nucleus often containing a single distinct nucleolus. These cells are always found in closely packed groups or "nests.' Contrast the embryonal carcinoma cells with cartilage in the upper left and neuroepithelial tubules on the right. Reproduced by permission from Martin (1978).

The close relationship between embryonal carcinoma cells and normal embryonic cells is underscored by the relative ease with which one can become the other. Teratocarcinomas and teratomas can be induced at very high frequency by transplanting a normal embryo to an extra-uterine site. The type of tumor which arises is a function of the age of the embryo as well as a variety of genetic and environmental factors. Thus, when mouse embryos of 8 or more days of development are transplanted beneath the kidney capsule they give rise only to teratomas (Damjanov *et al.*, 1971). It is thought that this is because, with the exception of germ cells, multipotent stem cells are no longer present in the embryo after the appearance of the organ primordia (at approximately 7.5 days of development). In contrast, stem cell-containing teratocarcinomas can be readily induced experimentally by extra-uterine transplantation of younger embryos (i.e. any stage from the 2-cell embryo up to approximately 7.5 days of development; Stevens, 1968, 1970; Solter *et al.*, 1970).

Teratomas and teratocarcinomas can also arise spontaneously from germ cells, but the process of tumorigenesis is rather different in males and females. In the ovary, tumor formation occurs when oocytes undergo spontaneous parthenogenesis *in situ* and develop as normal embryos for a brief period; these parthenogenetic embryos subsequently become disorganized and form a tumor in the ovary in the same way as embryos experimentally transplanted to an extra-uterine site. While most of the tumors that form in this way are benign teratomas and consist of only differentiated embryonic tissues, some are malignant teratocarcinomas which contain undifferentiated pluripotent stem cells in addition to the differentiated cell types. This process of spontaneous parthenogenesis, embryonic development and subsequent ovarian tumor formation occurs with high frequency in the LT strain of mouse. Approximately 50% of all adult LT females develop ovarian tumors (Stevens and Varnum, 1974).

Tumors can also form spontaneously in male mice. In contrast to the spontaneous ovarian and embryo-derived tumors, testicular tumors apparently do not arise as disorganized embryos. Instead they are initiated when primordial germ cells begin abnormal proliferation during the development of the fetal testis (reviewed by Stevens, 1975). However, aside from this difference in ontogeny, no consistent biochemical or morphological differences have as yet been detected between the stem cells of spontaneous testicular tumors and the stem cells of either ovarian or embryo-derived teratocarcinomas.

Embryonic cells that have become malignant teratocarcinoma stem cells can readily be "reverted" to normal embryonic behavior by being placed in an embryonic environment. Using a micromanipulator, teratocarcinoma cells can be injected into a host blastocyst and the manipulated embryos can then be transferred to pseudopregnant foster mothers where they develop to term. When the animals that are born are analyzed, cells derived from the injected teratocarcinoma cells and from the host blastocyst can be distinuished by genetic differences between them.

Brinster (1974, 1975) was the first to demonstrate that teratocarcinoma cells could in fact participate in the development of a normal adult mouse. The full potential of embryonal carcinoma cells for normal embryonic behavior *in utero* was, however, only appreciated when Mintz and her colleagues carried out experiments using a variety of genetic markers to distinguish derivatives of the injected teratocarcinoma stem cells from those of the host embryo (Mintz and Illmensee, 1975; Illmensee and Mintz, 1976). Using cells taken directly from transplantable tumors they clearly demonstrated that the tumor stem cells could form virtually every tissue of a normal tumor-free animal. However, their most exciting observation was that in at least one animal, the embryonal carcinoma cells were able to form functional germ cells, and to therefore pass their genes to offspring. Thus, teratocarcinoma stem cells derived from a transplated normal embryo and subsequently carried as a malignant tumor for as long as eight years were apparently able to respond to normal developmental signals when placed in the appropriate environment.

B. *Properties of Established Teratocarcinoma Cell Lines.*

Whatever the origin of the tumor, as long as there is a proliferating population of stem cells it is possible to isolate and establish embryonal carcinoma cell lines from it. Although it is generally true that at their inception teratocarcinoma stem cells are multipotent, they do not necessarily remain so. Aside from differentiation, which ultimately leads to a terminally differentiated phenotype, changes can occur in the stem cells *per se,* either during passage of the tumor *in vivo,* or after establishment of cell lines in *vitro.* Thus, some of the embryonal carcinoma cell lines that have been isolated are multipotent and can differentiate into a wide variety of differentiated cell types, while others have restricted differentiative capacities of various sorts.

The most extreme type of restriction is known as "nullipotency." Such cells are apparently incapable of differentiation. For example, the nullipotent cell lines described by Martin and Evans (1975a,b) were isolated from an undifferentiated "nullipotent" tumor and as yet have not been found to differentiate under any conditions. A less extreme form of restriction occurs when embryonal carcinoma cells become capable of "limited" differentiation. So-called "neuroteratocarcinoma" stem cells which can only give rise to neural cell types are one of the most common examples of such restricted stem cells. Another, well-known example of such a limited stem cell line is the F9 cell line first described by Bernstine, *et al.*, (1973). Although originally reported to be nullipotent, Sherman and Miller (1978) have shown that F9 cells are capable of forming a small amount of endoderm and the cells are now known as "pseudonullipotent" or "quasinullipotent."

It would be difficult to discuss individually all of the many different embryonal carcinoma cell lines that are currently available. Regardless of their differentiative capacity they have a common, highly distinctive morphology which can be taken as characteristic. The cells generally adhere stongly to one another and grow as poorly attached colonies or "epithelioid nests"; in the phase contrast microscope the cell-cell boundaries are very indistinct. The individual cells are rounded or slightly bipolar, with relatively little cytoplasm and a nucleus with one or two prominent nucleoli (Fig. 2).

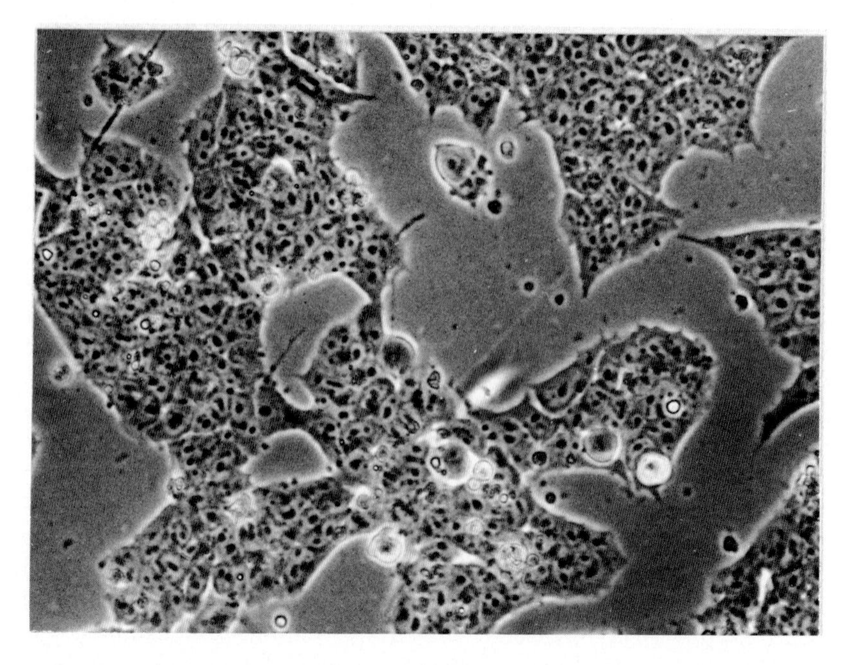

Fig. 2. Phase contrast photomicrograph of an embryonal carcinoma cell culture (approx. 250X).

From the data that are available, it is possible to make the generalization that the most important factor in the differentiation of teratocarcinoma stem cell lines is cell aggregation: differentiation of embryonal carcinoma cells occurs when the cells are cultured at a high local density. In some cases this is accomplished by allowing single cells or small clumps of cells to attach to a tissue culture surface and to grow until they become large, tightly rounded colonies (for example, see McBurney, 1976). With other established lines, for example PCC3, the cells may be cultured to high density as a confluent monolayer (Nicolas *et al.*, 1975, 1976). Such PCC3 cultures become very dense and then apparently undergo a period of partial lysis, after which numerous differentiated cell types appear. For other embryonal carcinoma cell lines differentiation can be obtained by allowing the cells to aggregate and form clumps in bacteriological petri dishes, to which they do not adhere (see below; Sherman, 1975; Sherman and Miller, 1978).

The similarity in the pattern of differentiation of embryonal carcinoma cells and early embryonic cells is demonstrated in studies of certain clonal embryonal carcinoma cell lines (Martin and Evans, 1975a,b; Martin *et al.*, 1977). Such cells are maintained in the undifferentiated state by frequent subculture to a confluent fibroblastic feeder cell layer. To initiate differentiation the cells are seeded in the absence of a feeder layer and the clumps that form are not disaggregated. Instead they are detached and cultured in suspension (by plating in bacteriological petri dishes to which they do not adhere). This allows the clumps to become very rounded, which triggers the differentiative process known as "embryoid body" formation (Pierce and Dixon, 1959; Stevens, 1959). The first stage in the process, similar to that of the first stage in the development of the mouse inner cell mass, is the differentiation of an outer layer of endoderm (Fig. 3). When kept in suspension these two-layered structures continue to differentiate into complex "cystic" embryoid bodies. Their development parallels to a remarkable extent the development of the inner cell mass and its derivatives during the early post-implantation phase of embryonic development (Fig. 4): in normal embryos, and also teratocarcinoma-derived embryoid bodies, a cavity known as the proamniotic cavity forms. It subsequently expands and there is a change in cellular organization and morphology, leading to the formation of a columnar epithelial layer, the embryonic ectoderm of the 6.5 day embryo. At latter times a mesodermal layer forms between the outer endoderm and inner ectodermal layers in the embryo and also in some of the embryoid bodies. The similarities and differences between teratocarcinoma embryoid body development and normal post-implantation development are discussed in detail by Martin (1978).

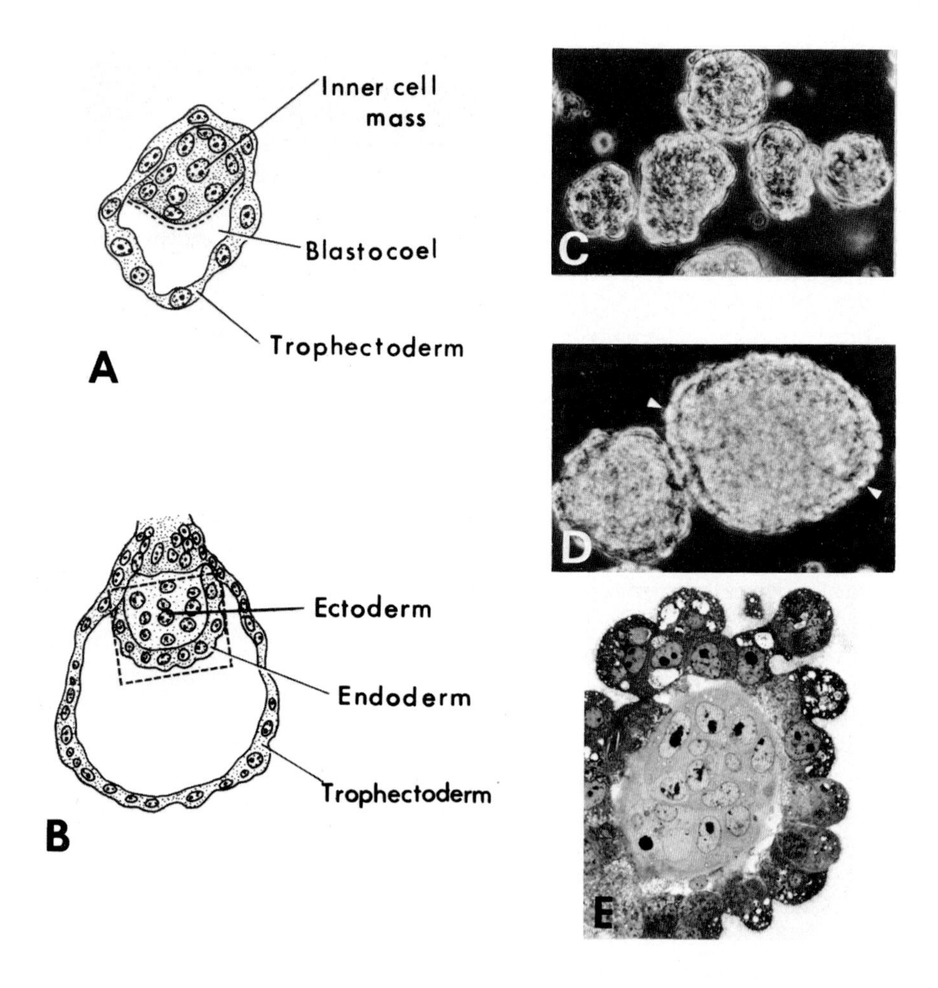

Fig. 3. The relationship of teratocarcinoma embryoid bodies to the inner cell mass of the mouse blastocyst. (A) Diagram of an "early" mouse blastocyst (approximately 64 cells, 3.5 days *post coitum).* The dotted line indicates where the primitive endoderm will form during the next 24 hours of development. (B) Diagram of a "late" or "expanded" mouse blastocyst at around the time of implantation (approx. 4.5 days *p.c.*). The primitive endoderm has formed on the blastocoelic surface of the inner cell mass. The two-layered inner cell mass indicated by the dotted line is the fetal portion of the embryo, to which teratocarcinoma embryoid bodies are analogous. (C) Phase contrast photomicrograph of a clump of teratocarcinoma cells before endoderm formation (approx. 200X). (D) A teratocarcinoma simple embryoid body. Arrows point to places where the outer endoderm cell layer is particularly clear (approx. 200X). (E) Section of a small embryoid body stained with hematoxylin and eosin. The core of embryonal carcinoma is surrounded by a layer of endoderm (approx. 350X).

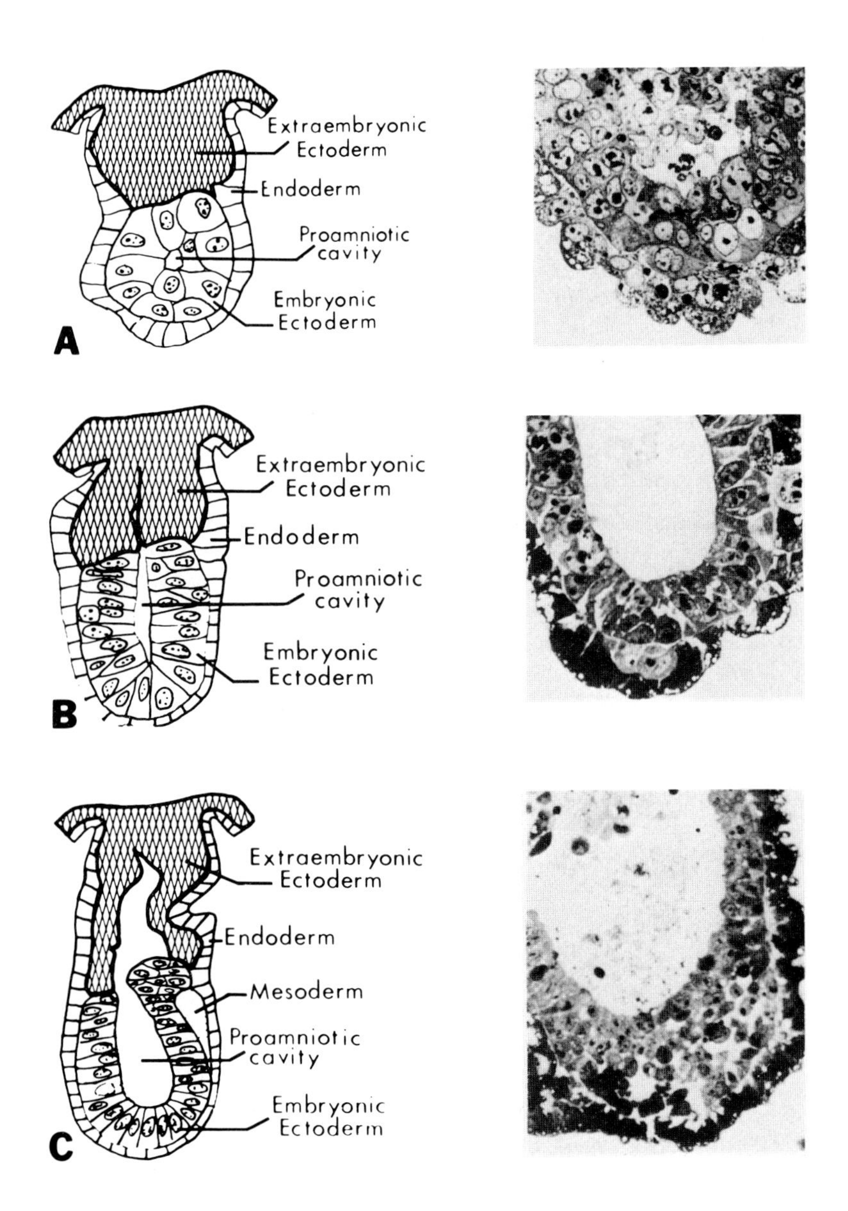

Fig. 4. Similarity between cystic embryoid body development and normal early postimplantation development. (A), (B), (C) are schematic representations of normal embryos at approximately 5, 6, and 7 days of development, respectively. To the right of each are sections of morphologically similar stages in the development of cystic embryoid bodies formed by embryonal carcinoma cells *in vitro* (Mag. approx. 250X for (A) and (B); 200X for (C). Reproduced by permission from Martin (1978).

II. CELL SURFACE MOLECULES STUDIED BY IMMUNOLOGICAL TECHNIQUES

The limited amount of material has made it relatively difficult to study the expression of embryonic cell surface molecules or to determine their role in mammalian development. The similarity between teratocarcinoma stem cells and early embryonic cells, coupled with the ready availability of teratocarcinoma cells lead one to the obvious conclusion that these cells should provide one means of studying the mammalian embryonic cell surface.

Since immunological techniques provide powerful tools for studying the cell surface, it is not surprising that most of the studies of the expression of cell surface molecules of teratocarcinoma cells have been carried out using various immunological probes. Two basic approaches have been taken. The first is to use teratocarcinoma cells as immunogens to raise sera which detect cell surface molecules common to both embryonic and teratocarcinoma cells. The long range goal of such studies is to identify, characterize and ultimately understand the function of the molecules detected by these sera. A second approach is to use sera specific for defined cell surface molecules (such as H-2 and fibronectin) and to ask whether these molecules are expressed by teratocarcinoma cells and early embryonic cells, and if so, what role they play in development. Studies of the latter sort are discussed by Wartiovaara in this volume (see also Jacob, 1977; Gachelin, 1978).

A. *Molecules Detected by Conventional Sera Raised Against Teratocarcinoma Cells.*

Sera raised against teratocarcinoma stem cells can be used to detect antigens on the surface of embryonic cells. The first evidence for this was obtained by Edidin *et al.* (1971). They found that a xenogeneic serum raised in rabbits against mouse teratocarcinoma cells reacted with early mouse embryos. However, even after adsorption, the serum described in that report was not strictly specific for teratocarcinoma cells and early embryos. In 1973 Artzt *et al.* raised a more specific *syngeneic* serum against the F9 line of embryonal carcinoma cells. Their results indicated that the antigen(s) common to teratocarcinoma stem cells and early embryos are not generally expressed by adult tissues, except perhaps for those in immunologically privileged sites such as the testis. Thus, the anti-F9 serum was shown to react with embryonal carcinoma cells, male germ cells and cleavage stage mouse embryos, but was reported to have no reactivity against various differentiated cell types including differentiated derivatives of teratocarcinoma stem cells.

Since these early reports, several different conventional sera have been raised against teratocarcinoma stem cells (reviewed by Gachelin, 1978; Solter and Knowles, 1979). Descriptions of the temporal and topographical expression of the antigens detected by these sera are obviously a prerequisite for understanding any functional significance these antigens might have. Therefore, many of the studies on these different sera have been aimed at defining which particular embryonic cells at various stages of early development express cell surface antigens common to embryonal carcinoma cells. Still other studies have been aimed at determining whether these antigens are coded for by genes in the T/t complex (Artzt *et al.*, 1974; Kemler *et al.*, 1976). Most of the data obtained in these studies serve to support the original conclusion that there are certain cell surface antigens common to embryonic cells, teratocarcinoma stem cells, and germ cells, and also to emphasize the complexity of the problem of studying the cell surface.

There is some evidence that the antigen(s) recognized by at least one of these sera may play an important, stage-specific role in early development. Kemler *et al.* (1977) found that monovalent Fab fragments of rabbit anti–F9 antibodies can reversibly inhibit compaction and subsequent blastulation of cleavage stage embryos and morulae, although this antibody apparently has little effect on cell division. Inner cell masses isolated from early blastocysts are also affected by such Fab fragments and become decompacted a few hours after addition of anti–F9 Fab. In contrast, ICMs isolated from expanded blastocysts do not (Jacob, 1977). Similar effects are observed in cultures of teratocarcinoma stem cells. When monovalent anti–F9 is added to embryonal carcinoma cells shortly after plating, the cells round up and do not adhere to each other as tightly as in the absence of the anti–F9 fragments. Although cell division does not appear to be affected, subsequent differentiation of stem cells is apparently inhibited in antibody-treated cultures (Jacob, 1977).

B. *Molecules Detected by Monoclonal Antibodies*

In recent years the use of antibodies as tools for studying the cell surface has been revolutionized by the development of techniques for producing monoclonal antibodies. It is now appreciated that conventional sera, even when they have been adsorbed and defined as operationally specific, are complex mixtures of antibodies of different specificities, affinities and classes. Since many of the problems that arise with conventional sera can be circumvented by the use of monoclonal antibodies, the emphasis in studies of antigens common to teratocarcinoma cells and embryos has now shifted to the use of monoclonal antibodies.

Interestingly, the first monoclonal antibody used to study antigen expression by teratocarcinoma cells was not raised against embryonal carcinoma cells. Instead, Stern *et al.* (1978), using a monoclonal antibody directed against a glycosphingolipid, the Fossman antigen, serendipitously found that it detects molecules containing similar determinants on teratocarcinoma cells. A variety of other tumor lines did not express this antigen, but testicular cells and mouse embryos were positive when tested with the monoclonal antibody. Studies of antigen expression by developing mouse embryos (Willison and Stern, 1978) indicated that this antigen does not appear on the embryonic cell surface until the start of blastulation. Surprisingly, not all the trophectodermal cells of an individual blastocyst were positive for this antigen. In contrast, all of the cells of the ICM and its first differentiated derivative (endoderm) apparently express this antigen. By the late blastocyst stage, however, it ceases to be expressed by the trophectoderm.

A related temporal and topographical distribution of antigen expression by embryonic cells was found using a monoclonal antibody raised against F9 embryonal carcinoma cells (Solter and Knowles, 1978). Preliminary evidence suggested that the stage-specific embryonic antigen (SSEA-1) detected by this monoclonal antibody, like the Forssman specificity detected by Stern and his collaborators, might be a glycolipid. However, further experiments indicated that SSEA-1 and the Forssman-like antigen are not identical, although it appears that both may be present on the same cells during embryonic development.

Several other monoclonal antibodies raised against teratocarcinoma cells are now available (Goodfellow *et al.*, 1979; Kemler *et al.*, personal communication). The antigens they detect are apparently different from those described above. All of these reagents should prove useful in future studies, in which the goal will be to obtain a panel of monoclonal antibodies that detect different antigens on the cell surface of embryonic cells and to use these reagents to elucidate the role such cell surface molecules may play in embryonic development.

III. GLYCOCONJUGATES EXPRESSED ON THE SURFACE OF TERATOCARCINOMA CELLS

Methods other than immunological techniques have been used to study the cell surface molecules of teratocarcinoma cells. Such studies have focussed on the expression of glycoconjugates on the surface of embryonal carcinoma cells. One reason for this is that, as discussed above, at least some of the antigens recognized by antibodies that bind to embryonal carcinoma cells appear to be glycoconjugates. Also, the availability of various plant lectins which recognize and bind to specific carbohydrate moieties makes it relatively easy to study the expression of sugar-containing molecules on the cell surface.

In order to determine whether different glycoconjugates might be expressed by undifferentiated stem cells and their differentiated derivatives, Gachelin *et al.* (1976) studied the binding to teratocarcinoma cells of five purified plant lectins, each recognizing a different sugar specificity. Two embryonal carcinoma cell lines and a differentiated teratocarcinoma-derived endodermal cell line were used in this study. Considerable variation in the amount of lectin bound per unit of cell surface area was observed among the three different cell lines tested. The most interesting result was that fucose-binding lectin (from *Lotus tetragonolobus)* did not react with the endodermal cell line whereas it did react with both undifferentiated embryonal carcinoma cell lines. These results suggest that differentiation of stem cells to an endodermal cell type is accompanied by the loss of terminal L-fucose residues on cell surface glycoconjugates. This hypothesis could be tested directly by measuring the binding of the fucose-specific lectin to embryonal carcinoma cells in comparison with its binding to the primary endoderm as it is formed during differentiation of the stem cells *in vitro*.

In a subsequent study, embryonal carcinoma cells and their differentiated derivatives were exposed to ^{3}H-fucose and the fucosyl-glycopeptides synthesized by the cells at different stages of differentiation were analyzed (Muramatsu *et al.*, 1978). The data indicated that embryonal carcinoma cells synthesize a class of large fucosyl-glycopeptides which is not synthesized by their various differentiated derivatives. Particularly interesting was the observation that cleavage stage mouse embryonic cells synthesize fucosyl-glycopeptides of similar molecular weight to those found in embryonal carcinoma cells, whereas a variety of normal and transformed adult cells do not contain fucosyl- glycopeptides in this size range. Evidence was presented that these large fucosyl-glycopeptides were mainly derived from the cell surface of the embryonal carcinoma cells. Recent evidence suggests that they are related to the fucose-containing cell surface molecules previously detected by lectin binding on the surface of undifferentiated cells (Muramatsu *et al.*, 1979), as well as to cell surface molecules recognized by peanut agglutinin and anti-F9 antibodies.

Studies using peanut agglutinin (PNA), which recognizes terminal D-galactosyl residues showed specific binding of this lectin to the surface of embyonal carcinoma cells (Reisner *et al.*, 1977). In contrast, most differentiated mouse cells do not bind this lectin unless they have been treated with neuraminidase to expose galactosyl residues. Interestingly, the few mouse cell types that do bind this lectin without prior neuraminidase treatment are immature cells or cells located in immunologically privileged environments. This pattern is somewhat similar to that found for the antigen(s) detected by anti-F9 sera (F-9 antigen). Examination of the possible relationship between the expression of PNA binding sites and the F9 antigen indicated that certain embryonal carcinoma cell lines were heterogeneous with respect to their expression of these two cell surface molecules: 50% of cells of the PCC3/A/1 embryonal carcinoma cell line showed binding of PNA and approximately the same proportion bound anti-F9 serum. Assays for F9 antigen expression on cells separated into PNA-positive and PNA-negative subpopulations showed a great enrichment of F9-positive cells in the former and F9-negative cells in the latter group.

To determine whether there are changes in the expression of PNA binding sites during differentiation, the PCC3/A/1 cells were allowed to differentiate over a period of one month and the number of PNA-positive cells was measured. As with the expression of F9 antigen (Nicolas *et al.*, 1975), the number of PNA-positive cells decreases from approximately 50% in undifferentiated cultures to 1% after differentiation. Although the possibility of selective survival and differentiation of only the PNA-negative cells has not been ruled out, the results suggest that differentiation of the stem cells may be accompanied by changes in the expression of D-galactosyl residues on the cell surface.

Taken together, these results indicate that significant changes in the carbohydrate moieties of cell surface glycoconjugates occur during differentiation of teratocarcinoma stem cells. Although as yet we know nothing about the function of the molecules detected in these studies, it is tempting to speculate that such molecules play a critical role in cell interactions during embryogenesis.

IV. CELL SURFACE MOLECULES IMPLICATED IN INTERCELLULAR ADHESION

It is apparent from the preceding discussion that most of the work on cell surface molecules of teratocarcinoma cells has been aimed primarily at identifying and characterizing them using immunological and biochemical techniques. The ultimate goal of such studies is the more difficult task of determining what function these particular molecules might have in the differentiative process. An alternative approach is to begin with a particular function, and try to identify the cell surface molecules responsible. From what is already known about the differentiation of teratocarcinoma stem cells it is clear that cell-cell recognition and adhesion are important factors in differentiation. An important question, then, is what cell surface molecules mediate these processes for teratocarcinoma and early embryonic cells?

Since cell-cell recognition and adhesion in many systems appears to be mediated by the interaction of a cell surface glycoconjugate, termed a "receptor," with specific carbohydrate-binding molecules (such a lectins; see reviews by Barondes and also by Dazzo, in this volume) a logical working hypothesis is that adhesion of teratocarcinoma cells may involve a similar mechanism.

Oppenheimer *et al.* (1969) were the first to suggest that complex carbohydrates may play a role in the intercellular adhesion of teratocarcinoma cells. In a subsequent study Oppenheimer and Humphreys (1971) demonstrated the presence of a trypsin-labile component in the peritoneal fluid of mice carrying the ascitic (embryoid body) form of a teratocarcinoma. This factor was shown to promote the reaggregation of cells from dissociated embryoid bodies. These results, taken in conjunction with earlier evidence for the involvement of carbohydrate-containing molecules in teratocarcinoma cell adhesion (Oppenheimer *et al.*, 1969), led the authors to speculate that the aggregation factor from the ascitic fluid might be a glycoprotein synthesized by the teratocarcinoma cells and necessary for their aggregation. Further studies (Oppenheimer, 1975) demonstrated the sensitivity of Teratoma Adhesion Factor (TAF) to treatment with β-galactosidase, which is consistent with its being a glycoprotein.

Oppenheimer (1975) also presented evidence that TAF-mediated reaggregation occurred as a consequence of the interaction of TAF with a carbohydrate-recognition component on the surface of the teratocarcinoma cells. Thus, preincubation of cells with 0.025M D-galactose specifically inhibited initial TAF-mediated cell adhesion. The conclusion from this study was that galactose-containing moieties of the TAF mediate cell adhesion by interacting with a cell surface galactose-binding molecule.

These results provide support for the hypothesis that teratocarcinoma cell adhesion is mediated by a carbohydrate-recognition system. However, one difficulty in interpreting these data is that there is no direct evidence that TAF, isolated from the peritoneal fluid of teratocarcinoma-bearing mice, is actually synthesized by teratocarcinoma cells. Furthermore, the results indicating a galactose specificity were obtained with teratocarcinoma-derived cells which, unlike the tumor line of origin (OTT 6050), no longer formed embryoid bodies; in contrast, these cells grew as a single cell suspension in the peritoneal cavity (Oppenheimer, 1975). Such a growth pattern is not typical of most embryonal carcinoma cells, which normally proliferate in aggregates.

With the establishment and characterization of various teratocarcinoma stem cell lines, some of which mimic embryonic behavior *in vitro*, it became possible to re-assess the role of carbohydrates in teratocarcinoma stem cell adhesion. In our study (Grabel *et al.*, 1979) two types of clonal teratocarcinoma stem cell lines were used: the pluripotent PSA1 cell line which differentiates *in vitro* via the formation of embryoid bodies, as described above; and a Nulli cell line that does not differentiate (Martin and Evans, 1975a,b; Martin *et al.*, 1977). Both of these cell lines grow as aggregates *in vitro* and also reaggregate rapidly without the addition of any exogenous factor, following dissociation by pipetting in calcium, magnesium-free phosphate buffered saline.

A. *Detection and Characterization of an Endogenous Lectin-like Molecule on the Surface of Teratocarcinoma Stem Cells*

In order to detect a cell surface carbohydrate-binding component on the surface of these teratocarcinoma stem cells we applied a simple visual assay. This assay is based on the fact that erythrocytes display on their surfaces a variety of carbohydrate structures, termed receptors, which vary with species and treatment of the red blood cells. If the carbohydrate configuration displayed by a particular type of erythrocyte is recognized by a carbohydrate-binding molecule on the surface of the cells to be assayed, the erythrocytes bind to the cells and "rosettes" are thus formed (Fig. 5a). When we tested the ability of teratocarcinoma cells to form rosettes with trypsinized glutaraldehyde-fixed rabbit erythrocytes (GTRs), 50–70% of the teratocarcinoma cells formed such rosettes with three or more GTRs bound to their surface.

If rosette formation is indeed a consequence of recognition between a binding component on the surface of the teratocarcinoma stem cells and glycoconjugate receptors on the GTR surface, addition of saccharides which structurally resemble the erythrocyte receptor should inhibit rosette formation. Our data indicate that certain mannose-rich glycoconjugates such as yeast invertase, horseradish peroxidase (HRP), and several yeast mannans effectively inhibit rosette formation by both pluripotent and nullipotent cells (Fig. 5b, Table I). In contrast, simple sugars such as D-mannose or D-galactose have no effect on rosette formation.

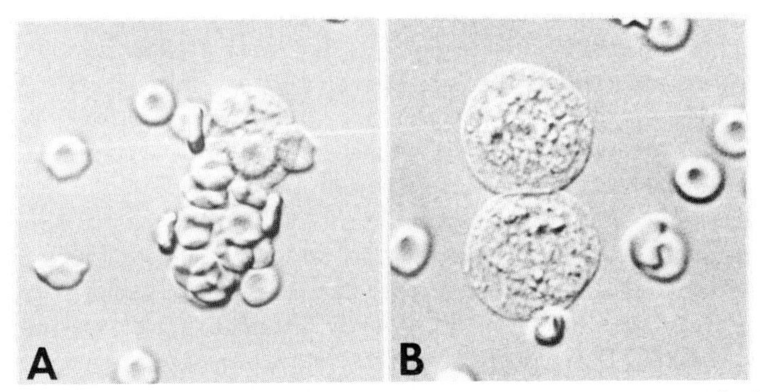

Fig. 5. Inhibition by invertase of erythrocyte binding to teratocarcinoma stem cells. (Hoffman modulation contrast optics, approx. 1750X) (A) Rosettes formed in control assays. (B) Rosette assay of teratocarcinoma stem cells pretreated with 1 mg/ml invertase. Reproduced by permission from Grabel *et al.* (1979).

TABLE I

Inhibition of Nulli Cell-GTR Rosette Formation

Inhibitor	% Inhibition of Rosette Formation
D-Glucose	0
D-Galactose	0
N-acetyl D-galactosamine	4
Lactose	3
D-Mannose	0
N-acetyl D-glucosamine	0
α-methyl D-mannoside	7
Fetuin	0
Asialofetuin	0
Transferrin	0
Horseradish peroxidase	65 56 [a]
Invertase	93 85 [b]
Periodate-oxidized Invertase	26
Mannan, wild type	47
Mannan, mnn 1	34
Mannan, mnn 2	72
Mannan, mnn 4	58

Cells were preincubated with inhibitor for 30 minutes at O°C prior to rosette formation, as described by Grabel, Rosen and Martin (1979). Simple saccharides were added at 0.1M, except for α-methyl D-mannoside which was added at 0.05M. All glycoproteins were added at 1 mg/ml. All results are averages obtained from at least two independent experiments.

a. HRP boiled for 10 minutes.

b Following incubation with invertase, Nulli cells were washed three times by centrifugation prior to the addition of erythrocytes.

That the carbohydrate moieties of the effective glycoproteins were responsible for rosette inhibition was demonstrated in two ways. First, periodate treatment of invertase, which oxidizes certain carbohydrates, markedly reduced its activity as an inhibitor of rosettes. Second, glycopeptide derivatives of invertase, obtained by exhaustive pronase digestion, also inhibited rosette formation.

Several experiments demonstrate that the rosette-mediating component is localized on the teratocarcinoma stem cell surface. When stem cells pretreated with invertase are extensively washed to remove any unbound inhibitor, inhibition by invertase is not significantly reduced (Table I). In contrast, pretreatment of erythrocytes with inhibitor does not prevent rosette formation with untreated stem cells. Brief exposure of the teratocarcinoma stem cells to trypsin significantly reduces their ability to form rosettes. Furthermore, in preliminary experiments HRP binding to the surface of teratocarcinoma stem cells was shown by histochemical staining. These results indicate that rosette formation occurs when a trypsin-labile component on the surface of the teratocarcinoma stem cells binds to a receptor on the GTRs which has structural similarities to certain mannose-rich glycoconjugates.

To determine if the teratocarcinoma carbohydrate-binding component which mediates rosette formation is involved in cell adhesion two types of experiment were carried out. In the first series we found that invertase inhibits the reassociation of mechanically dissociated stem cells. Although the inhibition is partial, remaining cell contacts are looser and more easily disrupted. Such reaggregation experiments must, however, be interpreted with caution since the dissociation procedure may alter the cell surface and lead to artifactual aggregation (Maslow, 1976). More conclusive evidence, not subject to this qualification, comes from a second series of experiments.

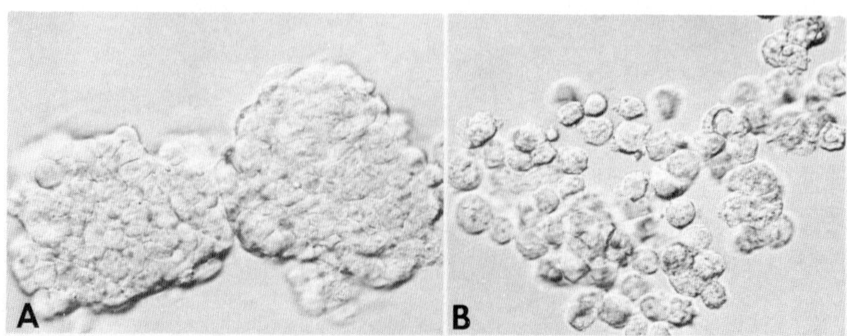

Fig. 6. Effect of invertase on intercellular adhesion. Aggregates of teratocarcinoma stem cells formed during growth *in vitro* were agitated with a vortex, mixed and fixed with glutaraldehyde (approx. 700X), (A) Untreated aggregates; (B) Aggregates treated with invertase. Reproduced by permission from Grabel *et al.* (1979).

If Nulli or PSA1 cells are plated under certain conditions they will grow as aggregates attached to the substratum (Martin and Evans, 1975; Martin *et al.*, 1977). These are released into suspension by gentle pipetting. In both cases, the clumps formed are extremely tight with indistinct cell boundaries (Fig. 6a). Incubation with invertase or glycopeptides derived from it substantially loosens the cell clumps (Fig. 6b). The effect can be quantified by counting the number of cells released from the aggregates after vortex mixing. PSA1 or Nulli aggregates formed during growth of the cells in culture and treated with invertase or its glycopeptide derivatives liberate substantially more cells after agitation than do controls (Fig. 7).

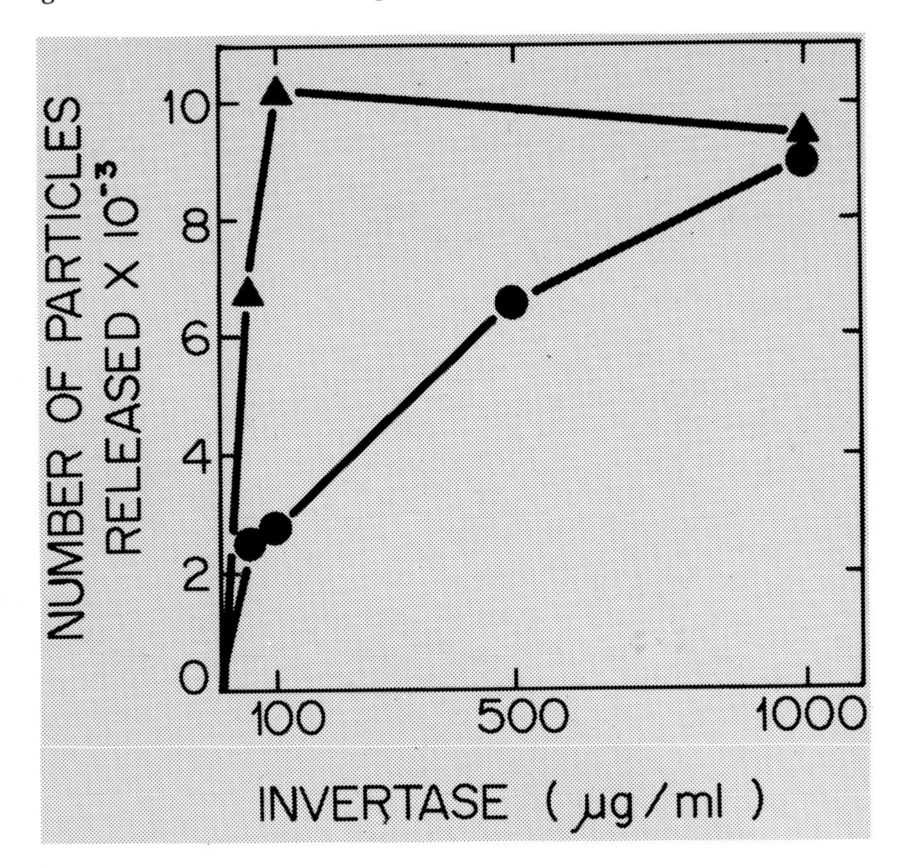

Fig. 7. Dissociation of preformed aggregates by treatment with invertase. Aggregates of Nulli cells (●) or pluripotent PSA1 cells (▲) were treated with invertase and then vortexed. The figure shows the difference in total number of particles released from aggregates exposed to invertase as compared to control aggregates. Reproduced by permission from Grabel *et al.* (1979).

Based upon the results of this study, our working hypothesis is that cell-cell adhesion of teratocarcinoma stem cells is mediated by the interaction of a carbohydrate-binding molecule (possibly a lectin) with its complementary oligomannosyl receptor on the surface of an adjacent cell (Fig. 8). The postulated receptor may be among the mannose-rich glycoproteins known to be present on the surface of mammalian cells (Sarkar and Menge, 1977).

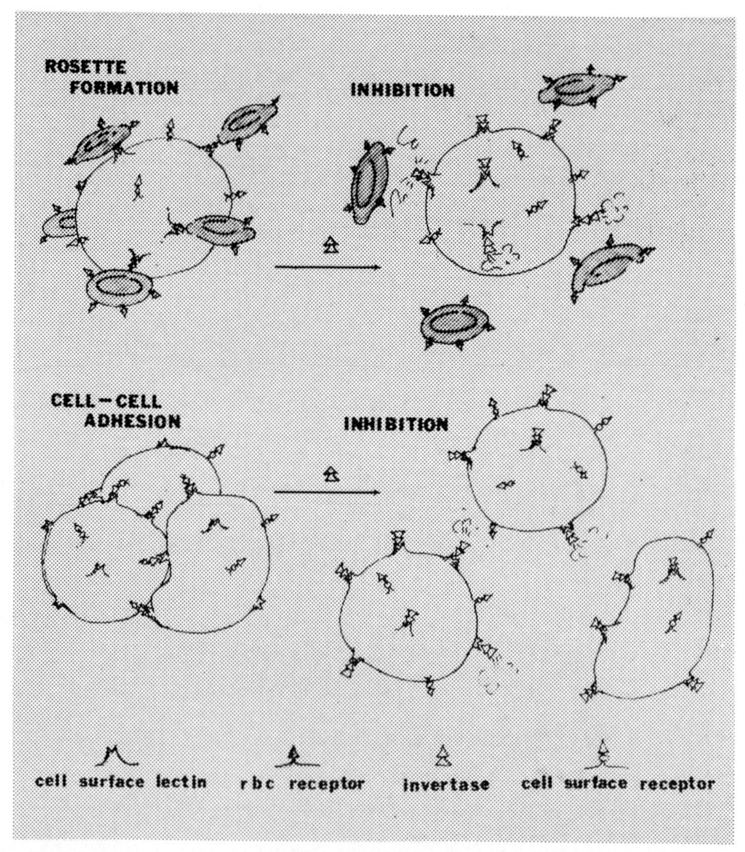

Fig. 8. Schematic representation of how the endogenous lectin-like molecule on the surface of teratocarcinoma stem cells might mediate intercellular adhesion.

The upper panel shows the visual assay for the cell surface lectin-like molecule. The adhesion of erythrocytes (rbc) to the teratocarcinoma stem cells is mediated by a carbohydrate-binding molecule which recognizes oligomannosyl residues on the surface of the erythrocytes. Adhesion of the erythrocytes is inhibited by pretreatment of the stem cells with invertase because it binds to the cell surface carbohydrate-recognition molecule.

The lower panel shows that normal intercellular adhesion occurs when the stem cell surface lectin-like molecule interacts with cell surface glycoconjugate receptors on adjacent cells. If invertase is added it mimics the endogenous cell surface receptor and binds to the lectin-like molecule. Cell-cell contacts are consequently loosened.

B. *Changes with Differentiation*

Since aggregation is thought to be a trigger for the differentiation of teratocarcinoma stem cells (see above) surface components important in stem cell adhesion may also be involved in mediating differentiative transitions. It remains to be determined whether or not the molecules described here are specific to undifferentiated cells and whether they change as the cells differentiate. Preliminary studies have been carried out with two types of endodermal cells. We have found that PYS cells, an established teratocarcinoma-derived cell line thought to resemble parietal yolk sac endoderm (Lehman *et al.*, 1974) can form rosettes with GTRs. In contrast to the results with undifferentiated stem cells described above, invertase is a relatively poor inhibitor of PYS cell rosette formation. On the other hand, blood group substance obtained from hog mucin, a heterogeneous mixture which contains a multiplicity of sugar determinants (Pereira and Kabat, 1979), is a poor inhibitor of stem cell rosette formation, but is superior to invertase as an inhibitor of endoderm cell rosettes. Similar results were obtained with primary endoderm cells isolated from PSA1 embryoid bodies formed *in vitro*. These results suggest that differentiation of stem cells may be accompanied by a change in carbohydrate-binding specificity of cell surface components.

Isolation and characterization of the teratocarcinoma cell surface molecules described here will enable us to further examine their stage-specificity and to determine whether these or similar molecules are expressed by early embryonic cells.

V. SUMMARY

There is an increasing mass of data describing molecules on the embryonic cell surface, some of which are apparently expressed in a stage-specific manner. We know relatively little, however, about which among them play a critical role in mammalian development. In the preceding discussion we have tried to make clear how teratocarcinoma cells can be used to study the molecules that are expressed on the early embryonic cell surface. Two different approaches have been taken to this problem, both capitalizing on the ease with which large numbers of teratocarcinoma stem cells can be obtained, and their remarkable similarity to certain normal embryonic cells.

One approach involves immunological and biochemical identification of cell surface molecules common to teratocarcinoma cells and embryonic cells. The probes that are developed in such studies can then be used to try to determine what function the detected molecules might serve. An example is the series of experiments in which Fab fragments of an anti-F9 embryonal carcinoma cell serum were found to have a reversible effect on embryonic and teratocarcinoma stem cell adhesion (Kemler *et al.*, 1977; Jacob, 1977). The second approach to studying the cell surface is to consider a particular function and to try to identify molecules that serve that function. One example is the study in which we searched for and identified a cell surface carbohydrate-binding molecule which may mediate intercellular adhesion of teratocarcinoma stem cells.

While the studies that have thus far been carried out might be considered preliminary, enough has been accomplished to indicate that teratocarcinoma cells will be a valuable tool for attaining an understanding of the way in which cell surface molecules are involved in early mammalian development.

ACKNOWLEDGEMENTS

We are grateful to Mr. D. Akers for his expert help in preparing the photographic material.

G.R.M. is supported by a Faculty Research Award from the American Cancer Society. L.B.G. is the recipient of a postdoctoral fellowship from the NIH. S.D.R. is supported by a Research Career Development Award from the N.I.H.

REFERENCES

Artzt, K., Dubois, P., Bennett, D., Condamine, H., Babinet, C. and Jacob, F. (1973). *Proc. Nat. Acad. Sci. U.S.A.* **70**, 2988-2992.

Artzt, K., Bennett, D. and Jacob, F. (1974). *Proc. Nat. Acad. Sci. U.S.A.* **71**, 811-814.

Barondes, S. (1979). This volume.

Bernstine, E.G., Hooper, M.L., Grandchamp, S. and Ephrussi, B. (1973). *Proc. Nat. Acad. Sci. U.S.A.* **70**, 3899-3903.

Brinster, R.L. (1974). *J. Exp. Med.* **140**, 1049-1056.

Brinster, R.L. (1975). *In* "Teratomas and Differentiation" (M. Sherman and D. Solter, eds.), pp. 51-58. Academic Press, New York.

Damjanov, I. and Solter, D. (1974). *Current Top. in Pathol.* **59**, 69-130.

Damjanov, I., Solter, D., and Skreb, N. (1971). *Wilhelm Roux' Arch.* **167**, 288-290.

Dazzo, F. (1979). This volume.

Edidin, M., Patthey, H.L., McGuire, E.J. and Sheffield, W.D. (1971). *In* "Embryonic and Fetal Antigens in Cancer" (N.G. Anderson and J.H. Coggin, Jr., eds.), pp. 239-248. Oak Ridge, Tennessee, Oak Ridge National Laboratory.

Gachelin, G. (1978). *Biochim. Biophys. Acta* **516**, 27-60.
Gachelin, G., Buc-Caron, M.H., Lis, H. and Sharon, N. (1976). *Biochim. Biophys. Acta* **436**, 825-832.
Goodfellow, P.N., Levinson, J.R., Williams, V.E. and McDevitt, H.O. (1979). *Proc. Nat. Acad. Sci. U.S.A.* **76**, 377-380.
Grabel, L.B., Rosen, S.D. and Martin, G.R. (1979). *Cell* **17**, 477-484.
Graham, C.F. (1977). *In* "Concepts in Mammalian Embryogenesis" (M. Sherman, ed.), pp. 315-394. MIT Press, Cambridge, Mass.
Hogan, B.L.M. (1977). *In* "Biochemistry of Cell Differentiation" (J. Paul, ed.), Vol. 15, pp. 333-376. University Park Press, Baltimore.
Illmensee, K. and Mintz, B. (1976). *Proc. Nat. Acad. Sci., U.S.A.* **73**, 549-553.
Jacob, F. (1977). *Immunol. Rev.* **33**, 3-32.
Kemler, R., Babinet, C., Condamine, H., Gachelin, G., Guenet, J.L. and Jacob, F. (1976). *Proc. Nat. Acad. Sci. U.S.A.* **73**, 4080-4084.
Kemler, R., Babinet, C., Eisen, H. and Jacob, F. (1977). *Proc. Nat. Acad. Sci. U.S.A.* **74**, 4449-4452.
Kleinsmith, L.J. and Pierce, G.B. (1964). *Cancer Res.* **24**, 1544-1552.
Lehman, J.M., Speers, W.C., Swartzendruber, D.E. and Pierce, G.B. (1974). *J. Cell Physiol.* **84**, 13-28.
Martin, G.R. (1975). *Cell* **5**, 229-243.
Martin, G.R. (1978). *In* "Development in Mammals" (M. Johnson, ed.), Vol. 3, pp, 225-265. Elsevier/North-Holland Biomedical Press, Amsterdam.
Martin, G.R. and Evans, M.R. (1975a). *Proc. Nat. Acad. Sci. U.S.A.* **72**, 1441-1445.
Martin, G.R. and Evans, M.J. (1975b). *In* "Teratomas and Differentiation" (M. Sherman and D. Solter, eds.), pp. 169-187. Academic Press, New York.
Martin, G.R., Wiley, L.M. and Damjanov, I. (1977). *Develop. Biol.* **61**, 69-83.
Maslow, P. (1976). *In* "The Cell Surface in Animal Embryogenesis and Development" (G. Poste and G. Nicholson, eds.). pp. 697-745. Elsevier/North-Holland Press, Amsterdam.
McBurney, M.W. (1976). *J. Cell. Physiol.* **89**, 441-455.
Mintz, B. and Illmensee, K. (1975). *Proc. Nat. Acad. Sci. U.S.A.* **72**, 3585-3589.
Muramatsu, T., Gachelin, G., Nicholas, J.F., Condamine, H., Jakob, H. and Jacob, F. (1978). *Proc. Nat. Acad. Sci. U.S.A.* **75**, 2315-2319.
Muramatsu, T., Gachelin, G., Damonneville, M., Delarbre, C. and Jacob, F. (1979). *Cell* **18**, 183-191.
Nicolas, J.-F., Dubois, P., Jakob, H., Gaillard, J. and Jacob, F. (1975). *Ann. Microbiol. (Inst. Pasteur)* **126A**, 3-22.
Nicolas, J., Avner, P., Gaillard, J., Guenet, J., Jakob, H. and Jacob, F. (1976). *Cancer Res.* **36**, 4224-4231.
Oppenheimer, S. (1975). *Exp. Cell Res.* **92**, 122-126.
Oppenheimer, S. and Hunphreys, T. (1971). *Nature* **232**, 125-127.
Oppenheimer, S., Edidin, M., Orr, C. and Roseman, S. (1969). *Proc. Nat. Acad. Sci. U.S.A.* **63**, 1395-1401.
Pereira, M.G.A., and Kabat, E.A. (1979). *J. Cell. Biol.* **82**, 185-194.
Pierce, G.B. Jr. (1967). *Current Top. Develop. Biol.* **2**, 223-246.
Pierce, G.B. and Dixon, F.J. (1959). *Cancer* **12**, 573-583.
Reisner, Y., Gachelin, G., Dubois, P., Nicholas, J.F., Sharon, N. and Jacob, F. (1977). *Develop. Biol.* **61**, 20-27.
Sarkar, J. and Menge, A. (1977). *J. Supramol. Struct.* **6**, 617-632.
Sherman, M.I. (1975). *In* "Teratomas and Differentiation" (M. Sherman and D. Solter, eds.), pp. 189-205. Academic Press, New York.

Sherman, M.I. and Miller, R.A. (1978). *Develop. Biol.* **63**, 27-24.

Solter, D. and Damjanov, I. (1978). *In* "Methods in Cancer Research," Vol. XVIII, pp. 277-332. Academic Press, New York.

Solter, D. and Knowles, B.B. (1978). *Proc. Nat. Acad. Sci. U.S.A.* **75**, 5565-5569.

Solter, D. and Knowles, B.B. (1979). *Current Top. Develop. Biol.* **13**, 139-166.

Solter, D., Skreb, N. and Damjanov, I. (1970). *Nature* **227**, 503-504.

Stern, P.L., Willison, K.R., Lennox, E., Galfre, G., Milstein, C., Secher, D. and Zeigler, A. (1978). *Cell* **14**, 775-783.

Stevens, L.C. (1959). *J. Nat. Cancer Inst.* **23**, 1249-1295.

Stevens, L.C. (1967). *Adv. Morphogen.* **6**, 1-31.

Stevens, L.C. (1968). *J. Embryol. Exp. Morphol.* **20**, 329-341.

Stevens, L.C. (1970). *Develop. Biol.* **21**, 364-382.

Stevens, L.C. (1975). *In* "Teratomas and Differentiation" (M. Sherman and D. Solter, eds.), pp. 17-32. Academic Press, New York.

Stevens, L.C. and Varnum, D.S. (1974). *Develop. Biol.* **37**, 369-380.

Wartiovaara, J. (1979). This volume.

Willison, K.R. and Stern, P.L. (1978). *Cell* **14**, 785-793.

Endogenous Cell-Surface Lectins: Evidence that They are Cell Adhesion Molecules

Samuel H. Barondes

Department of Psychiatry
University of California, San Diego
La Jolla, California 92093

I. INTRODUCTION

A major function of the cell surface is to mediate specific cellular associations. One way to achieve this is by the mutual binding of specific cell surface molecules of interacting cells, thereby holding the cells together. The purpose of this presentation is to consider some of the problems in identification of such molecules. I will then consider the evidence that endogenous lectins, which appear on the surface of developing cellular slime molds and chick tissues, may play a role in cellular association.

II. SOME GENERAL CONSIDERATIONS ABOUT CELL ADHESION MOLECULES

For the purpose of this paper, molecules that bind cells to other cells or to extracellular materials will be called cell adhesion molecules (CAMs).

ISBN 0-12-612984-3

Such molecules might be "non-specific", and bind any pair of cells together. Alternatively they might bind to specific receptors that are only expressed on certain cells. By mediating binding of these specific cells, the CAMs could also be functioning as "cell recognition molecules", i.e. molecules that determine which cells associate. In this regard it is important to point out that cell recognition molecules need not be CAMs. Rather they might be cell surface molecules confined to certain cells, which when combined with appropriate complementary molecules signal that cellular contact should be established by some other molecular interaction. The latter interacting molecules would be CAMs but need not be cell recognition molecules.

Because of their versatility it is generally assumed that many CAMs are proteins. Cell adhesion is thought to result from interactions between a cell surface protein and some type of complementary cell surface receptor which would also be designated as a CAM. The complementary receptor could be either a protein or another type of molecule. Oligosaccharides which are abundant on the cell surface as components of glycoproteins and glycolipids, are favorite candidates.

Some common conceptions of the molecular basis of cell-cell adhesion are illustrated in Fig. 1. In each case there are two cell surfaces shown bound together by interaction of one or more molecules arising from each cell surface. In cases A-C (Fig. 1) the interacting cell surfaces are indistinguishable but in case D they are distinct. Note also that the interacting molecules might be identical or different. They both could be integral membrane proteins or glycoproteins which are firmly rooted within the lipid bilayer (Fig. 1A, 1B, 1D) or peripheral membrane proteins like the complementary bivalent ligand shown in Fig. C.

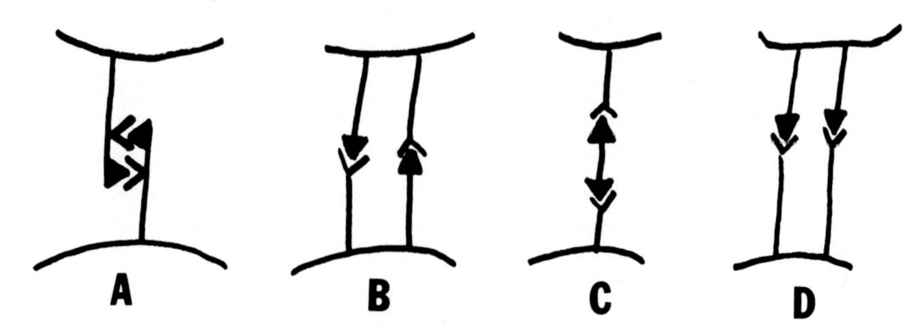

Fig. 1. Models of cell adhesion molecules binding the surfaces of adjacent cells. (A) Single cell adhesion molecule containing two different and complementary sites. (B) Two distinct complementary cell adhesion molecules both bound in the lipid bilayer of the cell surface. (C) Two distinct complementary cell adhesion molecules, one of which is outside the lipid bilayer. (D) As in (B) but each interacting cell contains only one type of cell adhesion molecule.

The adhesive interactions shown in Fig. 1 might bind many different types of cells together since they might all possess sufficient copies of the interacting molecules to bind to each other. If this were the case, the CAMs might serve no cellular recognition function. However, if there were several complementary pairs of cell adhesion molecules displayed to a greater or lesser extent on different cells, a basis for highly complex specific cellular associations could be established. Even a single complementary pair present on all cells, but to a variable extent, could provide the basis for complex patterns of selective cellular association. For example, simply varying the number of CAMs on the cell surface or their time of appearance in development could provide some selectivity. Furthermore, the degree to which the specific molecules are ordered or clustered in the membrane could be of importance. Although it is possible that CAMs play no role in the specific cellular associations generally referred to as cellular recognition, it is easy to see how they could be employed for this purpose.

The adhesive interactions shown in Fig. 1 might be continuously active in maintaining or reestablishing contact between cells that are continuously associating and dissociating. In this view cellular adhesion is in a dynamic equilibrium requiring continuation of those reactions which first held the interacting cells together. However, some cells may be maintaind in a stable association by an additional molecular mechanism. In such cases the cell adhesion molecules that played a role in initial cellular association might disappear once the stable secondary association was established. This possibility is important to keep in mind in developing systems, which might only transiently manifest CAMs. There is indeed evidence, as discussed later, that cellular adhesion reactions appear only at specific stages in development; and that cell-cell interactions may undergo changes with continued contact (Umbreit and Roseman, 1975).

III. CRITERIA FOR IDENTIFICATION OF CELL ADHESION MOLECULES

The first task in attempting to find CAMs is to establish criteria for their identification. Unfortunately the phenomenon under consideration, binding of a cell to another cell or to an extracellular material, involves many more variables than the association of two molecules in solution. For molecular analysis it may be convenient to consider that cells are billiard balls coated with varying amounts of adhesive materials of varying degrees of mutual affinity. But cells are obviously more complex. Some characteristics that distinguish cells from such a model include: 1) irregular and changing shape; 2) motility, which may be influenced by positive and negative chemotactic systems; and 3) continuous reorganization of membrane structure in response to various cues. Failure to acknowledge these properties in evaluating actual experiments on CAMs may lead to serious misinterpretation.

Some commonly cited characteristics of (or criteria for) CAMs are listed in Table I. First, a CAM must be shown to be present on the cell surface. Not all copies of the CAM need be on the cell surface but some must be. In addition, a CAM must be present on the cell surface when the cell develops adhesiveness. If it appears on the surface after adhesiveness develops this argues against its role in cell adhesion; and if its appearance correlates with the development of adhesiveness this is taken as supportive evidence. There should also be something on the cell surface to which the CAM binds which we will refer to as a complementary receptor. As indicated the receptor would also be a CAM; and as shown in Fig. 1A, might be identical with the CAM with which it is also complementary. Identification of the receptor might be made by demonstrating that CAM binds in some specific manner to the cell surface. Specificity is generally inferred from high affinity of binding and saturability of binding, the latter indicating a limited number of specific binding sites. If the CAM is bivalent, binding might be detected by the agglutination of the cells by the CAM, along the lines shown in Fig. 1C.

The other two criteria for a CAM are based on measurements of cell adhesion either by *in vitro* assays or by some more natural method of directly observing cell adhesion. Such adhesion should be blocked by univalent antibodies directed against the CAM, if they are added in sufficient amount and are of sufficient affinity to block the function of the CAM. Addition of haptens which block the active site of a CAM (e.g. sugars which bind the site of a CAM which is complementary to a cell surface heterosaccharide) should also block cell adhesion. Finally a mutant cell line (isolated and maintained in culture) that shows impaired cell adhesion should have absent or abnormal CAM.

Each of these expected characteristics of a CAM is compatible with all the models shown in Fig. 1 as well as more elaborate ones. An additional point expressed in Table I is that, in the experimental application of each one of these criteria, any result found could be a false positive or a false negative. For example, let us consider the most obvious criterion, the fact that a CAM must be present on the cell surface. This could be shown in many ways. If an antibody specific for this molecule is available, evidence that the antibody binds to the cell surface establishes the antigen's cell surface location. Another technique is to incorporate radioactive iodine into cell surface proteins by a reaction which depends on an enzyme, lactoperoxidase. Since the enzyme cannot penetrate the cell membrane, only cell surface proteins would be accessible to iodination. After the iodination reaction is terminated the iodinated proteins could be identified by one of a number of techniques — such as polyacrylamide gel electrophoresis. If a specific protein is iodinated this is taken as evidence that it is present on the cell surface.

Establishing that a particular molecule is present on the cell surface with these techniques is not definitive evidence that it plays a role in cell adhesion. In this context, such a result could be called a false positive. On the other hand the apparent absence of a molecule from the cell surface cannot be taken as incontrovertible evidence that it is not a CAM. There are a number of ways in which this could come about. For example, the molecule might be relatively scarce compared with other cell surface molecules and could be iodinated so sparsely by the lactoperoxidase technique that it might not be detected on gel electrophoresis. The molecule in question could also be hidden from such reagents, or others — such as antibodies raised against the molecule in a soluble form. It might be capable of interacting as a CAM in this cryptic state — or it might become exposed after a preliminary cellular interaction of another type. The point here is that failure to find a molecule on the cell surface could be a false negative in that it does not preclude its being a CAM.

TABLE I

Expected Characteristics of (Criteria for) a Cell Adhesion Molecule (CAM); and False Positives and Negatives

Expected Characteristics of CAM	Reasons For	
	False Positive	False Negative
located on cell surface	coincidental	overlooked; inaccessible
appearance of CAM correlates with development of adhesiveness	coincidental	CAM present before adhesiveness but other factor is limiting
complementary receptor on cell surface (if CAM is isolated as polyvalent molecule it will agglutinate test cells)	binding has other function or is non-specific	CAM binds poorly when solubilized; receptors saturated
binding univalent antibodies or haptens to CAM blocks adhesion	non-specific effect	affinity of univalent antibodies or haptens for CAM relatively low
mutant with impaired cell adhesion has defective CAM	indirect effect	defect in CAM not detected by the measurements used

False positives and false negatives are also possible with each of the other criteria listed in Table I. For example, a correlation between the time during differentiation of synthesis of a putative CAM and the time of development of cellular adhesiveness could be coincidental. On the other hand, it is possible that the putative CAM is present long before the cell converts from a nonadhesive to an adhesive form. One way this could come about is if the complementary substance with which it binds is the limiting factor — either its synthesis or its appearance on the cell surface. Another possibility is that some general property of the cell membrane is the limiting factor. Therefore failure to demonstrate a correlation between the appearance of adhesiveness and the appearance of the putative CAM could be a false negative.

There could also be false positives and false negatives with another relatively straightforward criterion — the presence of complementary cell surface receptors for the putative CAM. Demonstrating such receptors could be a false positive in that the molecule-receptor complex might be doing something other than binding cells together. For example, hormones bind to cells triggering specific reactions — and yet hormones are not CAMs. Even if the putative CAM is polyvalent and binding leads to agglutination of the cells, this could be a false positive. For example, basic proteins normally associated with nucleic acids can agglutinate cells by binding to acidic residues on the cell surface; but it is unlikely that these are CAMs. Failure to demonstrate binding of a putative CAM to a cell or of agglutination of cells by such molecules could also be misinterpreted. The molecule might bind very poorly once solubilized since it might be in a somewhat denatured form. Another possibility is that receptors present on the surface of the cell are already saturated with CAMs so that adding more does not lead to measurable binding.

The requirement that reagents that bind CAMs should block the adhesion reaction is also subject to misinterpretations. The reagents, be they univalent antibodies or specific ligands, could produce a false positive by nonspecifically affecting cellular interactions. Or there might be a false negative since the affinity of the antibodies or ligands for a CAM might be so much lower than that of the CAM for its complementary cell surface receptor that there is little competition and no measurable block in adhesion. Finally, even studies of mutants which show abnormal cell adhesion are subject to misinterpretation. A false positive could be produced if a mutation affected a critical early step in differentiation which in turn blocked many subsequent steps in development. These later steps could include synthesis of the putative CAM as well as another molecule which is the actual direct mediator of cell adhesion. Even a point mutation in the putative CAM found in cells that showed impaired adhesion could be a false positive in that the effect could be indirect. By this I mean that the affected molecule might not directly bind cells together but might influence some other process which itself directly mediated the adhesion. On the other hand a false negative is possible in that a mutant cell with impaired cell adhesion could appear to have normal putative CAMs since their abnormality might not be detectable by the measurement technique being used. Only determination of the complete structural sequence of the potential CAM might turn up the defect.

How then are we to conclusively identify CAMs? Since satisfying any single criterion is not convincing, multiple experimental approaches are required. Even then interpretations must be made with caution. They should be heavily dependent on estimates of the validity of the cell adhesion assays employed for many of these studies.

IV. CRITERIA FOR AN ASSAY OF SPECIFIC CELL ADHESION

An important prerequisite for studying CAMs is to establish some quantitative measurement of cell adhesion which is valid and "meaningful" — i.e. measures something identical with or closely related to the cell adhesion which occurs *in vivo*. Establishing valid assays is probably the major stumbling block in this field. The nature of these assays and criteria for evaluating them will now be considered.

In many cell adhesion assays a comparison is made of the rate of binding of cells to each other in gyrated suspension (For example see Orr and Roseman, 1969). In some cases single cells prepared by dissociation of tissues are gyrated together and the rate of formation of clumps of cells is observed. Dissociation itself poses a major problem in that it might remove or destroy CAMs. Often the cells must be incubated under proper conditions to recover from dissociation before the assay is done. Clump formation may then be measured either by microscopic inspection or by the use of an electronic particle counter. The windows of the counter can be set so that observations are made of the disappearance of particles with the dimensions of single cells and also of the formation of particles of sizes corresponding to clumps of various numbers of cells. The rate of clump formation is taken as a measure of the adhesiveness of the cells. Using this assay the relative adhesiveness of different cell populations can be estimated. In addition the role of potential CAMs can be evaluated with this assay. In some cases the effect of these CAMs on cell adhesion can be studied directly. In others effects of reagents that react with the CAMs can be evaluated. For example, univalent antibodies raised against purified CAMs would be expected to block cell adhesion. In addition, the assay might be useful in searching for CAMS. For example, if crude antisera raised against whole cells block the measured cell adhesion, the relevant cell surface antigens, which might be CAMs, could be identified by showing that they adsorbed the blocking activity of the antiserum.

Another commonly used assay measures the binding of a population of single cells to an immobilized layer of cells (Walther *et al.,* 1973.) The single cells, called probe cells, are labelled with radioactivity or in some other manner whereas the layer of cells immobilized on a solid support (e.g. a tissue culture dish) are unlabelled. The rate of binding of labelled cells to the layer is a measure of adhesion of the probe cells to the layered cells. Actual measurement may be by counting of radioactivity or by autoradiography. If the probe cells are labelled with a dye, probe cell binding can be directly observed by microscopy. Relative binding of different types of cells to self or other can be determined by varying the cell type in probe and layer. Applications of this assay are similar to the assay measuring clump formation in gyrated suspension.

Such assays have the merit of convenience and reproducibility. However, it is difficult to know the relationship between the parameters they measure and cell adhesion *in vivo.* How does binding in a gyrated suspension relate to binding *in vivo*? How can one decide if the affinity of a cell for a layer of cells bound to a plastic dish is a measure of its adhesiveness *in vivo*? The fact that adhesion can be measured in these circumstances provides us with little confidence that it is related to the behavior of the cells in the organism.

Some criteria which may be used in attempting to determine if a cell adhesion assay is valid include: tissue-specificity, developmental stage-specificity and the effect of specific reagents. Each criterion is the expectation of a correlation between adhesion as measured in the assay and some biological function of the cell. For example, if liver cells stick better to liver cells than to brain cells and if the latter stick better to brain cells, this indicates cell type specificity or tissue-specificity which validates the assay. If selectivity is demonstrated with a wide range of cell types, confidence is increased.

Developmental stage-specificity is another criterion that has some value. If cells at a specific embryonic age stick better to like cells than to those of a different age, this is taken as evidence validating the assay. However, it remains possible that the adhesion measured in this assay is due to a property of the cell surface which appears with differentiation but which normally has nothing to do with cell-cell association; and the important cell adhesion property might not be assayed under the experimental conditions.

The criterion of blockade by specific reagents can also be used to validate a cell adhesion assay. For example, consider the finding that antibodies (or their univalent derivatives) raised against a tissue-specific, or a developmental stage-specific cell surface component blocks the cell-cell adhesion observed in a given assay. This suggests that this specific component is a CAM and that the assay is valid since it is dependent on the action of this component. However, the finding does not necessarily validate the assay since binding antibodies to cell surface molecules not normally involved in cell adhesion might also be inhibitory. As already emphasized interpretations based on these assays must be made with caution.

V. STUDIES WITH CELLULAR SLIME MOLDS

Given the complexities of the problem of cell adhesion, it is particularly important to choose a favorable system for its analysis. The cellular slime molds are becoming a popular choice for this work. In the presence of bacteria which they feed upon, these simple eukaryotic cells exist as unicellular amoebae which show no tendency to associate. However, within 9 - 12 hours of exhaustion of the food supply the amoebae became adhesive and aggregate. In the ensuing 12 hours this aggregate differentiates into a fruiting body consisting of about 20% stalk cells and 80% spore cells.

Several properties of the cellular slime molds make them advantageous for studying cell adhesion: 1) cells can be isolated in a non-adhesive form and at various stages of development of cell-cell adhesion; 2) a large number of cells at an identical stage of development can be raised so that considerable material is available for biochemical studies; 3) the culture conditions are simple and resemble the natural environment of these organisms so that slime mold cells can be harvested in a state in which they show normal cell association; 4) there are a number of species of cellular slime molds that have been shown to display species-specific cellular association raising the possibility of studying species-specific cell-cell adhesion; 5) conditions have been worked out for *in vitro* assays of cell adhesion which measure a form of adhesion which appears "meaningful" since it appears with differentiation (Gerisch 1968; Rosen *et al.*, 1973) and is influenced by species-specific factors (McDonough *et al.*, 1979).

Gerisch and his colleagues (see Müller and Gerisch 1978) were the first to use cellular slime molds to attempt to determine the molecular basis of cell adhesion. Their experimental strategy was as follows: a) raise antibodies to cohesive slime mold cells; b) make univalent antibody fragments (Fab fragments) so that when the antibodies are added to the cells they will not agglutinate them (as could be the case with divalent immunoglobulin) but might inhibit cell adhesion by blocking appropriate cell surface molecules; c) identify the cell surface components to which the Fab binds. Müller and Gerisch (1979) indeed found that Fab fragments blocked cell adhesion and that, when they fractionated the membrane proteins of adhesive slime mold cells, one glycoprotein adsorbed all this blocking activity. The purified glycoprotein, called "contact site A" could be a CAM — fulfilling the criteria that it is developmentally regulated, present on the cell surface and that Fab directed against it blocks cell adhesion measured by a quantitative assay.

A major purpose of this presentation is to consider another candidate for a CAM in cellular slime molds, endogenous lectins. Extracts of aggregating cellular slime molds contain such proteins that can be isolated in soluble form and assayed as agglutinins of erythrocytes (Rosen *et al.*, 1973). The agglutinins, called lectins are divalent or polyvalent carbohydrate binding proteins which presumably bind erythrocytes together by combining with saccharides on their cell surfaces. This is supported by the finding that agglutination can be blocked by specific simple sugars.

Extracts of vegetative cellular slime molds contain little or no detectable lectin activity, suggesting a role for the lectin in development. In the case of *D. discoideum*, the species most carefully studied, two discrete lectins are synthesized with differentiation (Frazier *et al.*, 1975). Each has been purified to homogeneity and is a homotetramer with subunit molecular weights of either 24,000 or 26,000. Other species also apparently contain more than one lectin (Barondes and Haywood, 1979). The lectins from all the species that have been studied are similar proteins but have discriminable physicochemical properties. The relative inhibitory potency of a number of simple saccharides on their hemagglutination activity is similar but not identical (Barondes and Haywood, 1979).

The developmentally regulated lectins in slime molds are present on the cell surface. This has been shown in a number of ways. For example, antibodies raised against the purified proteins bind to the surface of partially differentiated cells as shown by immunofluorescent and immunoferritin techniques (Chang *et al.*, 1975, 1977). When the antibodies were applied to fixed permeable cells, most of the binding was intracellular (Chang *et al.*, 1977). Although precise quantitation of the amount of lectin on the cell surface is difficult we found (unpublished) that about 2% of the total cellular lectin in *D. purpureum* (in the range of 2×10^5 molecules/cell) can be stripped from the surface of intact cells with appropriate sugars. After such stripping about 8×10^5 molecules remain on the cell surface as shown by binding specific antibodies, so that total cell surface lectin represents about 10% of total cellular lectin.

About 2 x 10^5 molecules of lectin per cell was found associated with a plasma membrane fraction from *D. discoideum* (Siu *et al.*, 1978).

Evidence for cell surface receptors for slime mold lectins has been provided in a number of ways. First fixed slime mold cells bound added exogenous slime mold lectin with affinities as high as 10^9 M^{-1} (Reitherman *et al.*, 1975). Such binding of slime mold lectins to slime mold cell surfaces was blocked by appropriate sugars. Added lectin agglutinated fixed slime mold cells which had lost the endogenous cell surface lectin activity during fixation (Reitherman *et al.*, 1975). Fixed partially differentiated slime mold cells were agglutinated at lower concentrations of slime mold lectin than fixed vegetative cells, but this was not the case with several plant lectins (Reitherman *et al.*, 1975). Lectins from one species of slime mold bound avidly to the surface of other species; but a small degree of selective lectin affinity has been observed (Reitherman *et al.*, 1975). Binding of ferritin labelled lectin to the surface of fixed slime mold cells has also been shown (Chang *et al.*, 1977). Furthermore, living slime molds which already contain considerable endogenous lectin on their cell surface can bind a comparable amount of added slime mold lectin (unpublished). This appears to be specific binding since it is blocked by specific hapten sugars.

The finding that slime molds contain cell surface lectins and receptors at a stage when they display cell-cell adhesion raised the possibility that the lectins and receptors are cell adhesion molecules. Were this the case then univalent antibody fragments which bind to the lectin or haptens which bind to the active site might be expected to inhibit cell-cell adhesion. These expectations have been met only in one specific set of experimental circumstances. In studies with *Polysphondylium pallidum* we found that developmentally regulated cell-cell adhesion of the type shown above can be demonstrated under a number of conditions including the presence of hypertonic medium or metabolic inhibitors. Under these conditions cell-cell adhesion was blocked by two materials that bind the lectin — univalent antibodies or asialofetuin (Rosen *et al.*, 1977). The latter is a modified glycoprotein which attached to the lectin's carbohydrate-binding site. However, these reagents had little or no effect when normal medium was used (Rosen *et al.*, 1977). In similar experiments with *D. discoideum* and *D. purpureum* we have been unable to find any conditions where cell adhesion is measurable and univalent immunoglobulin fragments have a specific inhibitory effect. However, these results are difficult to interpret since most of the cell surface lectin does not bind the univalent antibodies (unpublished).

In contrast with these ambiguous results, strong support for a role of cell surface slime mold lectins in cell adhesion has been provided by studies with a mutant isolated by Ray *et al.* (1979) which shows a block of differentiation at the stage when the cells form aggregates. This mutant of *D. discoideum* has an abnormality in the major lectin from *D. discoideum*, discoidin-I, which comprises 90 — 95% of the total cellular lectin. The abnormal lectin is synthesized as the cells differentiate and is detected by a radioimmunoassay. However its carbohydrate binding site is abnormal since it fails to bind to Sepharose or to agglutinate appropriate test erythrocytes. An interesting property of this mutant is that, when mixed with normal slime mold cells, the mutant cells are incorporated into a normal fruiting body. Presumably the normal cells contain sufficient discoidin-I on their surface to lead to aggregation with the mutant cells. Although evidence of this type might be a false positive, as considered above, it is a substantial addition to the case that the lectins in slime molds are CAMs.

VI. DEVELOPMENTALLY REGULATED LECTINS IN EMBRYONIC CHICK TISSUES

Because of these findings in slime molds we sought developmentally regulated lectins in a higher organism, the embryonic chick. This organism was chosen for investigation since its development has been studied extensively, and since it is readily available and inexpensive. We initially concentrated on developing muscle because it is a relatively simple and abundant tissue. We found that embryonic chick pectoral muscle contains two lectins both of which show changes in activity with development. The first to be identified, which we designate lectin-1, is present at very low levels in pectoral muscle from 8 day old chick embryos, rises to a maximum in 16 day embryo muscle and then declines to very low levels in the adult (Nowak *et al.*, 1976). At its maximum it constitutes about 0.1% of the protein in muscle extracts. It has been purified to homogeneity by affinity chromatography and is a dimer with subunit molecular weight of 15,000 (Nowak *et al.*, 1977; Den and Malinzak, 1977). Antibodies raised to it have been used to demonstrate that it is predominately intracellular although some is detectable on the surface of cultured myoblasts. Existence of an unsaturated receptor on the cell surface is suggested by the finding (unpublished) that addition of exogenous lectin to cultured cells renders them more agglutinable by specific antibody raised against the lectin.

Thus far, there has been little direct study of the possible role of this lectin in cell adhesion. It has been reported that thiodigalactoside, which is a potent inhibitor of the hemagglutination activity of lectin-1, blocks fusion of myoblasts from an established rat cell line (Gartner and Podleski, 1975). However, thiodigalactoside did not affect the fusion of chick myoblasts in primary cultures (Den *et al.*,1976).

Studies with other tissues suggest that lectin-1 may have been adapted for more than one function. Lectin with indistinguishable physicochemical, immunological and carbohydrate-binding properties has been purified from extracts of embryonic chick brain and liver (Kobiler *et al.*, 1978) and is also found in some adult chicken tissues (Beyer *et al.*, 1979). Although lectin-1 may be identical in all these tissues there are notable differences in its developmental regulation (Kobiler and Barondes, 1977) and site of localization. Whereas the lectin in chick brain appears to show the same type of developmental regulation as in chick muscle, falling to low levels in the adult, levels of lectin-1 are very high in adult chicken liver, intestine and pancreas. Furthermore, there are differences in the distribution of the lectin in the various tissues. Lectin-1 is predominantly intracellular in embryonic neurons (Gremo *et al.*, 1978), myoblasts (Nowak *et al.*, 1977) and the goblet cells of the intestine (Beyer *et al.*, 1979). As indicated previously, it is also detectable on the cell surface of cultured myoblasts (Nowak *et al.*, 1977) and has been detected on the surface of some neurons from the optic tectum (Gremo *et al.*, 1978). In contrast, in the pancreas it is localized to the extracellular space between the lobules (Beyer *et al.*, 1979). This localization raises the question that lectin-1 may play some role in cell-matrix or matrix-matrix interactions in the pancreas.

Much less is known about the second lectin (lectin-2) from embryonic skeletal muscle. It too changes in activity with muscle development (Mir-Lechaire and Barondes, 1978), and is present in other tissues (Kobiler and Barondes, 1979). It differs strikingly from lectin-1 in that it agglutinates different test erythrocytes and is best inhibited by a specific group of glycosaminoglycans including dermatan sulfate, heparin and heparin sulfate (Kobiler and Barondes, 1979). It also differs in many ways from fibronectin (Ceri *et al.*, 1979). Unfortunately this lectin has resisted extensive attemps at purification. The major difficulty is that the lectin activity is associated with a macromolecular complex with apparent molecular weight >10^6 daltons. Attempts at dissociation have thus far been either unsuccessful or associated with marked loss of lectin activity. The active substance does not therefore appear to be similar to the other lectins we have described thus far. Although detailed studies are limited because of failure of the purification efforts, one clue to its function is that it is secreted into the medium of differentiating muscle cultures (Ceri *et al.*, 1979). This finding and the fact that it interacts with glycosaminoglycans, especially those found in the substrate attached material of cultured muscle (Ceri *et al.*, 1979), suggests an extracellular function, perhaps a role in adhesion of cells to extracellular matrix.

VII. CONCLUSION

There is considerable evidence, that the lectins in slime molds play a role in cell-cell adhesion. Four out of our five criteria are met and failure to consistently meet the fifth criterion could be a false negative. However, despite this extensive support the case is not conclusively proved. Evidence that lectins play a role in cell adhesion in chick tissue is even weaker and largely indirect.

The difficulty of conclusively evaluating the hypothesis that endogenous lectins play a role in cell adhesion was not anticipated. The systems studied are quite favorable and the finding that the lectins are abundant, easily purified and good antigens has provided excellent reagents (pure lectin, high titer antibodies) for evaluating lectin function. The fact that the problem has not yet been resolved underscores the challenging nature of the analysis of cell adhesion.

ACKNOWLEDGEMENTS

This work was supported by grants from the USPHS and the McKnight Foundation.

REFERENCES

Barondes, S.H. and Haywood, P.L. (1979). *Biochim. et Biophys. Acta* **550**, 297-308.
Beyer, E.C., Tokuyasu, K. and Barondes, S.H. (1979). *J. Cell Biol.***82**, 565-571.
Ceri, H., Shadle, P., Kobiler, D. and Barondes, S.H. (1979). *J. Supramol. Struct.*, in press.
Chang, C.-M., Reitherman, R.W., Rosen, S.D. and Barondes,S.H. (1975). *Exp. Cell Res.* **95**, 136-159.
Chang, C.-M, Rosen, S.D. and Barondes, S.H. (1977). *Exp. Cell Res.* **104**, 101-109.
Den, H. and Malinzak, D.A. (1977). *J. Biol. Chem.* **252**, 5444-5448.
Den,H., Malinzak, D.A. and Rosenberg, A. (1976). *Biochem. Biophys. Res. Commun.* **69**, 621-627.
Frazier, W.A., Rosen, S.D., Reitherman, R.W. and Barondes, S.H. (1975). *J. Biol. Chem.* **250**, 7714-7721
Gartner, T.K. and Podleski, T.R. (1975). *Biochem. Biophys. Res. Commun.* **67**, 972-978.
Gerisch, G. (1968). *Current Topics in Develop. Biol.* (A.A. Moscona and A. Monroy, eds.) Vol. 3, pp. 159-197. Academic Press, New York.
Gremo, F., Kobiler, D. and Barondes, S.H. (1978). *J. Cell Biol.* **78**, 491-499.
Kobiler, D. and Barondes, S.H. (1977). *Develop. Biol.* **60**, 326-330.
Kobiler, D. and Barondes, S.H. (1979). *F.E.B.S. Lett.* **101**, 257-261.
Kobiler, D., Beyer, E.C. and Barondes, S.H. (1978). *Develop. Biol.* **64**, 265-272
McDonough, J.P., Springer, W.R. and Barondes, S.H. (1979)., *Exp. Cell. Res.*, in press.
Mir-Lechaire, F. and Barondes, S.H. (1978). *Nature* **272**, 256-258.
Müller, K. and Gerisch, G. (1978). *Nature* **274**, 445-449

Nowak, T.P., Haywood, P.L. and Barondes, S.H. (1976). *Biochem. Biophys. Res. Commun.* **68**, 650-657
Nowak, T.P., Kobiler, D., Roel, L.E. and Barondes, S.H. (1977). *J. Biol. Chem.* **252**, 6026-6030.
Orr, C.W. and Roseman, S. (1969). *J. Membr. Biol.* **1**, 110-116
Raper, K.B. and Thom. C. (1941). *Amer. J. Bot.* **28**, 69-78.
Ray, J., Shinnick, T. and Lerner, R.A. (1979). *Nature* **279**, 215-221
Reitherman, R.W., Rosen, S.D., Frazier, W.A. and Barondes, S.H. (1975). *Proc. Nat. Acad. Sci. U.S.A.* **72**, 3541-3545.
Rosen, S.D., Kafka, J.A., Simpson, D.L. and Barondes, S.H. (1973). *Proc. Nat. Acad. Sci. U.S.A.* **70**, 2554-2557
Siu, C.H., Loomis, W.F. and Lerner, R.A. (1978). *In* "The Molecular Basis of Cell-Cell Interaction", (R.A. Lerner and D. Bergsma, eds.) pp. 439-458, Alan R. Liss, New York.
Rosen, S.D., Chang, C.-M. and Barondes, S.H. (1979). *Develop. Biol.* **61**, 202-213.
Umbreit, J. and Roseman, S. (1975). *J. Biol. Chem.* **250**, 9360-9368.
Walther, B.T., Ohman, R. and Roseman, S. (1973). *Proc. Nat. Acad. Sci. U.S.A.* **70**, 1569-1573.

Subject Index

SUBJECT INDEX

F

G

H

I

SUBJECT INDEX

T

U

V

X